DIVERSE APPLICATIONS OF NANOTECHNOLOGY IN THE BIOLOGICAL SCIENCES

An Essential Tool in Agri-Business and Health Care Systems

DIVERSE APPLICATIONS OF NANOTECHNOLOGY IN THE BIOLOGICAL SCIENCES

An Essential Tool in Agri-Business and Health Care Systems

Edited by

Khalid Rehman Hakeem, PhD
Majid Kamli, PhD
Jamal S. M. Sabir, PhD
Hesham F. Alharby, PhD

First edition published 2022

Apple Academic Press Inc.
1265 Goldenrod Circle, NE,
Palm Bay, FL 32905 USA

4164 Lakeshore Road, Burlington,
ON, L7L 1A4 Canada

CRC Press
6000 Broken Sound Parkway NW,
Suite 300, Boca Raton, FL 33487-2742 USA

4 Park Square, Milton Park,
Abingdon, Oxon, OX14 4RN UK

© 2022 by Apple Academic Press, Inc.

Apple Academic Press exclusively co-publishes with CRC Press, an imprint of Taylor & Francis Group, LLC

Reasonable efforts have been made to publish reliable data and information, but the authors, editors, and publisher cannot assume responsibility for the validity of all materials or the consequences of their use. The authors, editors, and publishers have attempted to trace the copyright holders of all material reproduced in this publication and apologize to copyright holders if permission to publish in this form has not been obtained. If any copyright material has not been acknowledged, please write and let us know so we may rectify in any future reprint.

Except as permitted under U.S. Copyright Law, no part of this book may be reprinted, reproduced, transmitted, or utilized in any form by any electronic, mechanical, or other means, now known or hereafter invented, including photocopying, microfilming, and recording, or in any information storage or retrieval system, without written permission from the publishers.

For permission to photocopy or use material electronically from this work, access www.copyright.com or contact the Copyright Clearance Center, Inc. (CCC), 222 Rosewood Drive, Danvers, MA 01923, 978-750-8400. For works that are not available on CCC please contact mpkbookspermissions@tandf.co.uk

Trademark notice: Product or corporate names may be trademarks or registered trademarks and are used only for identification and explanation without intent to infringe.

Library and Archives Canada Cataloguing in Publication

Title: Diverse applications of nanotechnology in the biological sciences : An Essential Tool in Agri-Business and Health Care
  Systems / edited by Khalid Rehman Hakeem, PhD, Majid Kamli, PhD, Jamal S. M. Sabir, PhD, Hesham F. Alharby, PhD.
Names: Hakeem, Khalid Rehman, editor. | Kamli, Majid, editor. | Sabir, Jamal S. M., editor. | Alharby, Hesham F., editor.
Description: First edition. | Includes bibliographical references and index.
Identifiers: Canadiana (print) 20220154686 | Canadiana (ebook) 20220154821 | ISBN 9781774638408 (hardcover) |
  ISBN 9781774638415 (softcover) | ISBN 9781003277255 (ebook)
Subjects: LCSH: Nanobiotechnology. | LCSH: Nanomedicine. | LCSH: Agriculture.
Classification: LCC TP248.25.N35 D58 2022 | DDC 660.6—dc23

Library of Congress Cataloging-in-Publication Data

....

CIP data on file with US Library of Congress

....

ISBN: 978-1-77463-840-8 (hbk)
ISBN: 978-1-77463-841-5 (pbk)
ISBN: 978-1-00327-725-5 (ebk)

ABOUT THE EDITORS

Prof (Dr.) Khalid Rehman Hakeem

Khalid Rehman Hakeem, PhD, is Professor at King Abdulaziz University, Jeddah, Saudi Arabia. After completing his doctorate (Botany; specialization in Plant Ecophysiology and Molecular Biology) from Jamia Hamdard, New Delhi, India, in 2011, he worked as a lecturer at the University of Kashmir, Srinagar, for a short period. Later, he joined Universiti Putra Malaysia, Selangor, Malaysia, and worked there as Post Doctorate Fellow in 2012 and Fellow Researcher (Associate Prof.) from 2013 to 2016. Dr. Hakeem has more than 10 years of teaching and research experience in plant ecophysiology, biotechnology and molecular biology, medicinal plant research, plant-microbe-soil interactions as well as in environmental studies. He is the recipient of several fellowships at both national and international levels; also, he has served as the visiting scientist at Jinan University, Guangzhou, China. Currently, he is involved with a number of international research projects with different government organizations.

So far, Dr. Hakeem has authored and edited more than 70 books with international publishers, including Springer Nature, Academic Press (Elsevier), and CRC Press. He also has to his credit more than 155 research publications in peer-reviewed international journals and 65 book chapters in edited volumes with international publishers.

At present, Dr. Hakeem serves as an editorial board member and reviewer of several high-impact international scientific journals from Elsevier, Springer Nature, Taylor and Francis, Cambridge, and John Wiley. He is included in the advisory board of Cambridge Scholars Publishing, UK. He is also a fellow of Plantae group of the American Society of Plant Biologists, member of the World Academy of Sciences, member of the International Society for Development and Sustainability, Japan, and member of Asian Federation of Biotechnology, Korea. Dr. Hakeem has been listed in Marquis Who's Who in the World, since 2014–2019. Currently, Dr. Hakeem is engaged in studying the plant processes at ecophysiological as well as molecular levels.

Dr. Majid Kamli

Majid Kamli, PhD, is Associate Professor in the Department of Biological Sciences, Faculty of Science at King Abdulaziz University, Jeddah, Saudi Arabia. Dr. Kamli has considerable experience in teaching and research in molecular biology. He has published more than 30 research articles in well-reputed journals and participated in various scientific conferences and seminars. Dr. Kamli's current work focuses on the application of biogenic-metallic nanoparticles in multidrug-resistant in Candida species. His PhD (Biotechnology) was earned at Jamia Millia Islamia, New Delhi, India, where his research topic was "Molecular Characterization of Genetic Determinant arsB, an Arsenic Resistant Gene, of Arsenic Resistant Bacteria." He has previously held research associate positions at the Pohang University of Science and Technology (POSTECH) Republic of Korea; at Weill Cornell Medical College, Cornell University, Doha, Qatar; and a postdoctoral fellow position at the School of Biotechnology, Yeungnam University, Republic of Korea. His postdoctoral research was focused on the functional study of genes highly upregulated during myogenic satellite cell differentiation, TGF-β signaling in breast cancer, and study of human genetic variations in patients suffering from different neurodegenerative diseases, especially HSP (hereditary spastic paraplegia) and movement disorders. Before joining King Abdulaziz University, Dr. Kamli worked as Research Professor at the College of Medicine, Wonkwang University, Republic of Korea.

Prof. Jamal Sabir

Jamal S. M. Sabir, PhD, is currently a Professor in the Department of Biological Sciences and Director at the Centre of Excellence in Bionanoscience Research at Faculty of Science at King Abdulaziz University, Jeddah, Saudi Arabia. Dr. Sabir received his PhD (Genetics and Biotechnology) from Sussex University, UK. Dr. Sabir has extensive teaching and research experience. He is a member of several scientific societies and has held various prominent administrative positions.

Dr. Sabir has worked on various projects, including international collaborative joint research projects, where he worked as a principal investigator. The works included research on genomics, bionanoscience, biomonitoring, nanotechnology, MERS coronavirus, and H5N8 virus. He has also supervised MSc and PhD projects. Dr. Sabir has extensively published research articles in high impact journals in his field, and has received several awards for his research publications from King Abdulaziz University, Jeddah. He has collaborated and contributed to more than 150 research articles and five books. He is part of groups that have ten patents and discoveries in biotechnology. Dr. Sabir is committed to his research, and his research interests include infectious diseases and genomics.

Dr. Hesham F. Alharby

Hesham F. Alharby, PhD, is Associate Professor in the Department of Biological Sciences at King Abdulaziz University (KAU), Jeddah, Saudi Arabia. He earned his PhD from the School of Biological Sciences at the University of Western Australia, Perth, Australia. Dr. Alharby's work focuses on plant biology, mainly in eco-physiology and molecular biology. He has published more than 40 papers in peer-reviewed international journals and has attended several international conferences. He was a head of laboratories at Teachers College, Jeddah, Saudi Arabia in 2005. He is a Head of the Plant Section in the Department of Biological Sciences at KAU. He is also working as the Deputy Head, Center of Excellence in Bionanoscience Research.

DEDICATION

This book is dedicated to the

Center of Excellence in
Bionanoscience Research
King Abdulaziz University
Jeddah, Saudi Arabia

CONTENTS

Contributors .... *xiii*

Abbreviations .... *xvii*

Preface .... *xxi*

1. **Nanotechnology Applications in Nanomedicine: Prospects and Challenges** 1

   Arpita Dey, Smhrutisikha Biswal, and Somaiah Sundarapandian

2. **Application of Nanotechnology in Drug Delivery** 39

   Muzafar Ahmad Rather, Showkeen Muzamil Bashir, Showkat Ul Nabi, Salahi Uddin Ahmad, Jiyu Zhang, and Minakshi Prasad

3. **Application of Nanoparticles in Biomedical Imaging** 73

   Afodun Adam Moyosore

4. **Nanoparticles in Medical Imaging: A Perspective Study** 109

   Mujtaba Aamir Bhat, Ishfaq Ahmad Wani, Naseer Ahmad Hajam, Safikur Rahman, and Arif Tasleem Jan

5. **Nanotechnology in Healthcare Management** 139

   Ifrah Manzoor, Muzafar Ahmad Rather, Saima Sajood, Showkeen Muzamil Bashir, Sohail Hassan, Manzoor-u-Rehman, and Rabia Hamid

6. **Overview and Emergence of Nanobiotechnology in Plants** 173

   Saheed Adekunle Akinola, Rasheed Omotayo Adeyemo, and Ismail Abiola Adebayo

7. **Nanofertilizers and Nanopesticides: Application and Impact on Agriculture** 199

   Zeenat Javeed, Umair Riaz, Ghulam Murtaza, Shahzada Munawar Mehdi, Muhammad Idrees, Qamar Uz Zaman, and Waqas Khalid

8. **Greener Methods of Nanoparticle Synthesis** 213

   Mohd Yousuf Rather, Somaiah Sundarapandian, and Mohammed Latif Khan

9. **Nanoscience in Biotechnology** 241

   Charu Gupta, Mir Sajad Rabani, Mahendra K Gupta, Shivani Tripathi, and Anjali Pathak xii

10. Plant Product-Based Nanomedicine for Malignancies: Types and Therapeutic Effects..269

Zuha Imtiyaz, Tabish Mehraj, Andleeb Khan, Mir Tahir Maqbool, Rukhsana Akhter, Mufeed Imtiyaz, Wajhul Qamar, Azher Arafah, and Muneeb U. Rehman

11. Toxicity of Nanoparticles in Plants..303

Syed Ali Zulqadar, Mohammad Ali Kharal, Umair Riaz, Rashid Iqbal, Behzad Murtaza, Ghulam Murtaza, and Muhammad Akram Qazi

12. Plant-Based Nanoparticles and Their Applications..................327

Ishani Chakrabartty

Index..*341*

CONTRIBUTORS

Ismail Abiola Adebayo
Department of Microbiology/Immunology, Faculty of Biomedical Sciences,
Kampala International University, Western-Campus, Ishaka-Bushenyi, Uganda;
E-mail: ibnahmad507@gmail.com; Ismail.abiola@kiu.ac.ug

Rasheed Omotayo Adeyemo
Department of Microbiology/Immunology, Faculty of Biomedical Sciences, Kampala International
University, Western-Campus, Ishaka-Bushenyi, Uganda
Department of Paraclinical Sciences, University of Pretoria, Pretoria, South Africa

Salahi Uddin Ahmad
Lanzhou Institute of Husbandry and Pharmaceutical Sciences, Chinese Academy of Agricultural
Sciences, Lanzhou 730050, China
Faculty of Veterinary, Animal and Biomedical Sciences, Sylhet Agricultural University, Sylhet-3100,
Bangladesh

Rukhsana Akhter
Department of Biochemistry, School of Biological Science, University of Kashmir, Hazratbal,
Srinagar, J&K, India

Saheed Adekunle Akinola
Department of Microbiology/Immunology, Faculty of Biomedical Sciences,
Kampala International University, Western-Campus, Ishaka-Bushenyi, Uganda
Food Security and Safety Niche, Faculty of Natural and Agricultural Sciences,
North-West University, Mmabatho, South Africa

Azher Arafah
Department of Clinical Pharmacy, College of Pharmacy, King Saud University, Riyadh, Saudi Arabia

Showkeen Muzamil Bashir
Division of Veterinary Biochemistry, Faculty of Veterinary Sciences and Animal Husbandry,
SKUAST-Kashmir, Shuhama, Srinagar, Jammu and Kashmir, India;
E-mail: showkeen@skuastkashmir.ac.in/showkeen.muzamil82@gmail.com

Mujtaba Aamir Bhat
School of Biosciences and Biotechnology, Baba Ghulam Shah Badshah University, Rajouri, India

Smhrutisikha Biswal
Department of Physics, School of Physical, Chemical and Applied Sciences, Pondicherry University,
Puducherry, India; E-mail: smspandian65@gmail.com

Ishani Chakrabartty
Department of Science, P. A. First Grade College (affiliated to Mangalore University, Mangalore),
Nadupadav, Mangalore, Karnataka, India

Arpita Dey
Department of Ecology and Environmental Sciences, School of Life Sciences, Pondicherry University,
Puducherry, India

Charu Gupta
School of Studies in Microbiology, Jiwaji University, Gwalior, Madhya Pradesh, India

Mahendra K Gupta
Microbiology Research Lab, School of Studies in Botany, Jiwaji University, Gwalior, Madhya Pradesh, India

Naseer Ahmad Hajam
School of Biosciences and Biotechnology, Baba Ghulam Shah Badshah University, Rajouri, India

Rabia Hamid
Department of Nanotechnology, University of Kashmir, Hazratbal, Srinagarar, Jammu and Kashmir, India; E-mail: rabeyams@gmail.com

Sohail Hassan
Department of Microbiology, University of Veterinary and Animal Sciences, Lahore, Pakistan

Muhammad Idrees
Department of Computer Science and Engineering, University of Engineering and Technology, Lahore, Narowal Campus, Pakistan

Mufeed Imtiyaz
Khwaja Yunis Ali Medical College, Enayetpur Sharif Sirajgonj, Bangladesh

Zuha Imtiyaz
Clinical Drug Development of Herbal Medicine, College of Pharmacy, Taipei Medical University, Taipei, Taiwan

Rashid Iqbal
Department of Agronomy, Faculty of Agriculture and Environmental Sciences, The Islamia University of Bahawalpur, Bahawalpur, Pakistan

Arif Tasleem Jan
School of Biosciences and Biotechnology, Baba Ghulam Shah Badshah University, Rajouri, India; E-mail: atasleem@bgsbu.ac.in

Zeenat Javeed
Soil and Water Testing Laboratory for Research, Bahawalpur, Pakistan

Waqas Khalid
Department of Biomedical Engineering Technology, NFC Institute of Engineering and Technology Multan, Pakistan

Andleeb Khan
Department of Pharmacology and Toxicology, College of Pharmacy, Jazan University, Jazan, Saudi Arabia

Mohammed Latif Khan
Department of Botany, Dr. Harisingh Gour Vishwavidyalaya (A Central University), Sagar, Madhya Pradesh, India

Mohammad Ali Kharal
Department of Agriculture Extension, Khairpur Tamewali, Government of Punjab, Pakistan

Ifrah Manzoor
Division of Veterinary Biochemistry, Faculty of Veterinary Sciences and Animal Husbandry, SKUAST-Kashmir, Shuhama, Srinagar, Jammu and Kashmir, India
Department of Biochemistry, University of Kashmir, Hazratbal, Srinagarar, Jammu and Kashmir, India

Contributors

Mir Tahir Maqbool
National Center for Natural Products Research, School of Pharmacy, University of Mississippi, University, MS, United States

Shahzada Munawar Mehdi
Rapid Soil Fertility Survey and Soil Testing Institute, Lahore, Punjab, Pakistan

Tabish Mehraj
Department of Pharmaceutical Sciences, School of Pharmacy, University of Mississippi, University, MS, United States

Afodun Adam Moyosore
Department of Anatomy, Faculty of Biomedical Sciences, Kampala International University, Kampala, Uganda
Department of Medical Imaging, Ultrasound and Doppler Unit, Crystal Specialist Hospital, Dopemu-Akowonjo, Lagos, Nigeria; E-mail: adam.afodun@kiu.ac.ug; afodunadam@yahoo.com

Behzad Murtaza
Department of Environmental Sciences, COMSAT Vehari Campus, Islamabad, Pakistan

Ghulam Murtaza
Institute of Soil & Environmental Sciences, University of Agriculture, Faisalabad, Pakistan
Assistant Soil Fertility Office, Bahawalpur, Pakistan

Showkat Ul Nabi
Division of Veterinary Clinical Medicine, Ethics and Jurisprudence, F.V.Sc. & A.H, SKUAST-Kashmir, Shuhama-190006, J&K, India

Anjali Pathak
Microbiology Research Lab, School of Studies in Botany, Jiwaji University, Gwalior, Madhya Pradesh, India

Minakshi Prasad
Department of Animal Biotechnology, College of Veterinary Sciences, Lala Lajpat Rai University of Veterinary and Animal Sciences, Hisar 125004, Haryana, India

Wajhul Qamar
Department of Clinical Pharmacology & Toxicology and Research Centre, College of Pharmacy, King Saud University, Riyadh, Saudi Arabia

Muhammad Akram Qazi
Rapid Soil Fertility Survey and Soil Testing Institute, Lahore, Punjab, Pakistan

Mir Sajad Rabani
Microbiology Research Lab, School of Studies in Botany, Jiwaji University, Gwalior, Madhya Pradesh, India

Safikur Rahman
Department of Botany, M.S College, B. R. Ambedkar University, Muzaffarpur, Bihar, India

Muzafar Ahmad Rather
Division of Veterinary Biochemistry, Faculty of Veterinary Sciences and Animal Husbandry, SKUAST-Kashmir, Shuhama, Srinagar, Jammu and Kashmir, India

Mohd Yousuf Rather
Department of Ecology and Environmental Sciences, Pondicherry University, Puducherry, India
Department of Botany, Dr. Harisingh Gour Vishwavidyalaya (A Central University), Sagar, Madhya Pradesh, India

Manzoor-u-Rehman
Division of Veterinary Biochemistry, Faculty of Veterinary Sciences and Animal Husbandry, SKUAST-Kashmir, Shuhama, Srinagar, Jammu and Kashmir, India

Muneeb U Rehman
Department of Clinical Pharmacy, College of Pharmacy, King Saud University, Riyadh, Saudi Arabia; E-mail: muneebjh@gmail.com

Umair Riaz
Soil and Water Testing Laboratory for Research, Bahawalpur, Pakistan; E-mail: umairbwp3@gmail.com, umair.riaz@uaf.edu.pk

Saima Sajood
Department of Biochemistry, University of Kashmir, Hazratbal, Srinagarar, Jammu and Kashmir, India

Somaiah Sundarapandian
Department of Ecology and Environmental Sciences, School of Life Sciences, Pondicherry University, Puducherry, India

Shivani Tripathi
Microbiology Research Lab, School of Studies in Botany, Jiwaji University, Gwalior, Madhya Pradesh, India

Ishfaq Ahmad Wani
School of Biosciences and Biotechnology, Baba Ghulam Shah Badshah University, Rajouri, India

Qamar Uz Zaman
Department of Environmental Sciences, University of Lahore, Lahore Campus, Pakistan

Jiyu Zhang
Lanzhou Institute of Husbandry and Pharmaceutical Sciences, Chinese Academy of Agricultural Sciences, Lanzhou 730050, China

Syed Ali Zulqadar
Soil and Water Testing Laboratory for Research, Bahawalpur, Pakistan

ABBREVIATIONS

AD	Alzheimer's disease
ADDL	amyloid beta-derived diffusible ligand
ADMET	absorption, distribution, metabolism, excretion, and toxicity
AFM	atomic force microscope
AgNPs	silver nanoparticles
AI	artificial intelligence
AI	active ingredients
AIE	aggregate-induced emissions
ASTM	American Society of Testing and Materials
AuNPs	gold nanoparticles
BAuA	Bundesanstalt für Arbeitsschutz und Arbeitsmedizin
BBB	blood–brain barrier
BSI	British Standards Institution
CalTech	California Institute of Technology
CLIO	cross-linked iron oxide
CMC	chemistry, manufacturing, and controls
CmR-CAs	cancer-marking contrast
CNS	central nervous system
CNTs	carbon nanotubes
CRC	colorectal cancer
CRISPR	clustered regularly interspaced short palindromic repeats
CT	computed tomography
CVD	chemical vapor deposition
DLS	dynamic light Scattering
DMF	N,N-dimethylformamide
DOTA	tetraazacyclododecane tetra acetic acid
Dox	doxorubicin
DTPA	diethylenetriaminepentaacetic acid
EPR	enhanced permeability retention
EPR	enhanced permeation and retention
FA	ferulic acid
FDA	Food and Drug Administration
FEM	field electron microscope
FGF	fibroblast growth factor

FIM	field ion microscope
FMT	ferumoxytol
FNPs	fluorescent nanoparticles
FRET	fluorescence resonance energy transfer
FRET	Förster resonance energy transfer
GA	gambogic acid
GBCA	gadolinium-based contrast substance
GCC	guanylyl cyclase C
GLAD	glancing angle deposition
GMP	good manufacturing practice
GMP	genetically modified plants
GNPs	gold nanoparticles
GO	graphene oxide
GQDs	graphene quantum dots
GSHPx	glutathione peroxidase
HA	hyaluronic acid
HcG	human chorionic gonadotropin
HDL	high-density lipoprotein
HDACs	histone deacetylases
HFD	high-fat diet
HUVECs	umbilical-cord vein endothelial cells
IFP	interstitial fluid pressure
ISLNs	intelligent SLNs
ISO	International Organization for Standardization
JNMF	Jawaharlal Nehru Memorial Fund
LbL	layer-by-layer
LMD	laser microdissection
LPH	lipid–polymer hybrid nanoparticles
LSPR	localized surface plasmon resonance
MDA	malondialdehyde
MDR	multidrug-resistant
MEMS	microelectromechanical systems
MINT	molecular imaging in nanotechnology and theranostics
MNPs	magnetic nanoparticles
MNPs	metal nanoparticles
MPI	magnetic particle imaging
MPS	mononuclear phagocyte system
MRI	magnetic resonance imaging
MRs	microrobots

Abbreviations

MS	molecular sentinel
MSNP	mesoporous silica NP
MSTF	methionine serum thymus factor
MT	microtissues
NBCDs	nonbiological complex drugs
NdFeB	neodymium/iron/boron
ND	nanodiamond
NEMS	nanoelectromechanical systems
NHCs	nanoscale hexagonal columns
NIOSH	National Institute of Occupational Safety and Health
NIRF	near-infrared fluorescence
NLC	nanostructured lipid carriers
nm	nanometers
NNI	National Nanotechnology Initiative
NPs	nanoparticles
NRs	nanorods
NSCs	neural stem cells
ORR	objective response rate
OS	overall survival
OTC	over-the-counter
PAI	photoacoustic imaging
PAMAM	polyamidoamine dendrimers
PD	Parkinson'sdisease
PDT	photodynamic
PDGF	platelet-derived growth factor
PEG	polyethylene glycol
PET	positron emission tomography
PFOB	perfluoroalkyl bromide
PHP	purple heart plant
PHSNs	porous hollow silica nanoparticles
PLE	Pongamia pinnata leaf extract
PLGA	poly(lactic-co-glycolic acid)
PNPs	polymeric nanoparticles
PPI	polypropylene imine
PPM	phenotypic personalized medicine
PSF	point spread function
PTSD	posttraumatic stress disorder
PVD	physical vapor deposition
PVP	poly(N-vinylpyrrolidone)

QD	quantum dots
QEMSA	quantum dot electrophoretic mobility shift assay
QPOP	quadratic phenotypic optimization platform
QSAR	quantitative structure–activity relationship
QSPR	quantitative structure–property relationship
RES	reticuloendothelial system
RNA	ribonucleic acid
ROS	reactive oxygen species
SCCP	Scientific Committee on Consumer Products
SEM	scanning electron microscope
SERS	surface-enhanced Raman spectroscopy
SErRSNPs	surface-enhanced (resonance) Raman scattering nanoparticles
siRNA	small inhibitory RNA
SLN/SLC	solid lipid nanoparticles/carriers
SLNs	solid lipid nanoparticles
SNPs	silver nanoparticles
SOD	superoxide dismutase
SPECT	single-photon emission computer tomography
SPF	sun protection factor
SPIONs	superparamagnetic iron oxide nanoparticles
SPM	scanning probe microscopes
SPR	surface plasmon resonance
SSS	sensitive sensor systems
STM	scanning tunneling microscope
SWCNTs	single-walled carbon nanotubes
TB	pulmonary tuberculosis
TMV	tobacco mosaic virus
TNFα	tumor necrosis factor-alpha
UGC	University Grants Commission
VEGF	vascular endothelial growth factor
ZnO	zinc oxide

PREFACE

Nanotechnology is a study of microscopic structures with a scale of 0.1–100 nm. It has played an important role in various scientific fields such as medicine, drug development, plant and agricultural sciences, electronics, and space industries. Nowadays, nanotechnology has gained significant interest in applied medical sciences. Medical nanotechnology or nanomedicine is a medical term for the application of nanotechnology. It is considered a new method for detecting, diagnosing, and treating diseases, and this technology overcomes existing limitations placed by existing diagnostic and therapeutic strategies. Small nanoparticles can cross biological barriers, including skin surface epithelium, intraluminal epithelium, and help improve drug delivery systems. In medical research, the diagnosis of cancer, diabetes, cardiovascular, and pain control has been applied. Several types of nanoparticles are used for prevention and disease treatment in the medical sciences. Nanoparticles such as gold can be used as probes to identify nucleic acid sequences besides being used in clinical cancer and other diseases. Further, nanotechnology developed better imaging and diagnostic methods. Nanomedicine is a new applied science and will play a key role in human care.

Similarly, nanobiotechnology is an essential method in plant and agricultural sciences and is considered the main economic driver. It is assessed that incorporating cutting-edge nanotechnology into agribusiness will drive global monetary growth to approximately US$ 3.4 trillion by 2020, which demonstrates how agri-nanobiotechnology plays a pivotal position in the agricultural sector without adversely affecting the environment and other biosafety regulatory issues. Agri-nanobiotechnology is an innovative green technology for global food security, biodiversity, and climate change solutions.

This book presents the role of nanobiotechnology in medical sciences and plants. It includes the original, latest, and updated information on the role of nanobiotechnology in health care systems and agriculture. The book covers the application of nanotechnology in drug delivery and diagnostics. Further, it emphasizes how nanomedicine can treat different types of cancers and can improve medical imaging for the diagnostics of different kinds of diseases. Moreover, in this book, nanobiotechnology in plants has been well-documented with its application in nanofertilizers and nanopesticides and application of plant-based nanoparticles for agriculture.

This book is written in a very lucid and comprehensible language and supplemented with illustrations, figures, tables, etc. Therefore, it can also be used as a reference book by teachers and students at the graduate and undergraduate levels in colleges and universities.

We are thankful to the contributors for readily accepting our invitation for not only sharing their knowledge and research but for venerably integrating their expertise in dispersed information from diverse fields in composing the chapters and enduring editorial suggestions to finally produce this venture. We also thank Apple Academic Press team for their generous cooperation at every stage of the book production.

—Editors

CHAPTER 1

NANOTECHNOLOGY APPLICATIONS IN NANOMEDICINE: PROSPECTS AND CHALLENGES

ARPITA DEY[1], SMHRUTISIKHA BISWAL[2], and
SOMAIAH SUNDARAPANDIAN[1*]

[1]*Department of Ecology and Environmental Sciences,
School of Life Sciences, Pondicherry University, Puducherry, India*

[2]*Department of Physics, School of Physical, Chemical and Applied
Sciences, Pondicherry University, Puducherry, India*

Corresponding author. E-mail: smspandian65@gmail.com

ABSTRACT

The field of nanomedicine is booming with substantial progress made so far and possibilities of achieving essential milestones in versatile clinical applications, albeit faced with barriers that limit its clinical translation. Nanomedicines have witnessed significant encouragement due to its ability to address diverse disease challenges, from diagnostics to treatment, supported by current research. This scientific discipline is comparatively young and integrates fundamental scientific concepts with newer findings to address challenging medical goals. This chapter focused on the progress of nanomedicine so far and assessed potential opportunities that it holds, discusses the challenges and risks in nanomedicine design and development in bringing nanomedicine formulations in clinical applications from bench to bedside to assess the future scope.

1.1 INTRODUCTION

Modern nanotechnology applications are filled with anticipation to achieve many possibilities and endeavors from scientific communities of different disciplines. It concerns matter at the tiniest scale and understanding their unique properties that can span at the translational scope of diverse fields such as material science, chemistry, electronics, and information technology. Although modern-day nanotechnology was not evolved until the latter decades of the last century, nanomaterials' use dates back centuries ago.

1.1.1 NANOTECHNOLOGY

The U.S. National Nanotechnology Initiative (NNI) described nanotechnology as "The understanding and control of matter at dimensions between approximately 1 and 100 nm, where unique phenomena enable novel nanotechnology applications. Encompassing nanoscale science, engineering, and technology, nanotechnology involves imaging, measuring, modeling, and manipulating matter at this length scale." To go with the popular definitions, nanoscience-based technology is designing functional systems operated at the nanoscale (1–100 nm). It involves manipulating matter at the atomic, molecular, and supramolecular scale. Nanoscience and nanotechnology are interdisciplinary across physics, chemistry, biology, and material science and engineering.

Nanotechnologies, as briefed by Gabriel A. Silva, are *technologies that use engineered materials or devices with a structural or functional arrangement on the nanoscale, implying that the dynamic behavior of the device or material can be designed and controlled at the laboratory.* They can, therefore, be defined by their functional properties and their interaction (Silva, 2010).

The word "Nano" is derived from the Greek word "*νᾶνος*" (Latin nanos), which translates to a dwarf. By scale, a nanometer is one-billionth of a meter (10^{-9} m). At this scale of dimensions, materials exhibit unique physical, chemical, and biological properties much different from the properties of their bulk counterpart, single atoms, and molecules. For example, nanomaterials have a high specific surface-area-to-volume ratio than the higher scale dimensions of the same material with the same mass, affecting their chemical or electrical nature. In addition to that, in such a small scale of measurement, matter starts exhibiting quantum confinement characteristics affecting their optical, magnetic, and electrical behavior. These important nanomaterials' essential properties can be fine-tuned to produce superior and novel structures, design systems, and fabricate devices for better performance.

In nature, matter does exist in nanosized forms. Magnetite (Fe_3O_4) nanoparticles of specific morphology are synthesized by the bacteria, *Magnetospirillum magnetotacticum* which uses the magnetism caused by the particles to find a favorable direction for its growth. Bacteria like the *Lactobacillus* sp. can take up metal ions and reduce them inside the cell to synthesize nanoparticles.

The oldest of colored glasses is the fourth-century AD Lycurgus Cup made by the Romans contained nanosized particles of gold (Au) and silver (Ag) in a very small amount, as well as late medieval church windows with luminous colored glass, was also due to application of Au and Ag nano-sized particles. Later on, in the Islamic world ceramic glazes contained Ag or copper (Cu) and/or other metal nanoparticles, and later in Europe in the 9th–17th centuries, Ottoman techniques in the 13th–18th centuries used carbon nanotubes (CNTs) and nanoscale wires of cementite creating a moiré pattern in "Damascus" saber blades, and 16th-century renaissance pottery in Italy used nanomaterials without their knowledge about nanomaterials (Poole and Owens, 2003; Pradell et al., 2007; Reibold et al., 2006). In 1856, Faraday prepared colloidal gold and called the prepared colloidal gold particles as the "divided state of gold." There was the use of nanoporous ceramic filters to separate viruses in the 19th century. In the year 1990, Max Plank and Albert Einstein deducted theoretical evidence of the tiny particles that showed a new kind of behavior (Krukemeyer et al., 2015). Subsequently, in the year 1902, the ultramicroscope designed by Richard Zsigmondy and Henry Siedentopf helped identify structures smaller than 4 nm in Ruby glasses (Mappes et al., 2012). In 1912 with the help of immersion ultramicroscopy developed by Zsigmondy, colloidal solutions could be possible to study. The field electron microscope (FEM) was developed by Erwin Müller (1936) and later on (1951), the field ion microscope (FIM), which helped to see individual atoms and their surface arrangement (Yamamoto et al., 2014).

The possibility of the emergence of such a science field was first conceptualized and addressed by Richard Feynman in a talk "There's Plenty of Room at the Bottom" at the American Physical Society Meeting at the California Institute of Technology (CalTech) on December 29, 1959. He talked about the "bottom-up" approach and put forward the proposal of single molecular and atomic assembly to meet the functional designing requirements. His insight was that physics laws do not necessarily dismiss the probability of arranging a structure through controlling and designing atomic arrangements accurately, which could open a new era of material science to achieve new heights (Chen and Liang, 2018).

The word "nanotechnology" was first used in 1974 by Norio Taniguchi, who defined nanotechnology as "nanotechnology mainly consists of the processing of separation, consolidation, and deformation of materials by one atom or one molecule."

Modern-day nanotechnology began with the invention of the scanning tunneling microscope (STM) in 1981 by Gerd Binnig and Heinrich Rohrer, which led to the development of scanning probe microscopes (SPM) in 1981 and the atomic force microscope (AFM) in 1982. Robert Curl, Harold Kroto, and Richard Smalley found the fullerenes or buckyballs, stable forms of carbon (chemical formula C60/C70), formed by graphite evaporation in an inert atmosphere (1985). Iijima observed CNTs, a fullerene family member, by TEM (Iijima, 1991). The discovery of fluorescent carbon dots (C-dots) (<10 nm) in 2004 by Xu and his coworkers and their low toxicity and good biocompatibility advanced the field of bioimaging, biosensor, and drug-delivery applications (Bayda et al., 2020). Andre Geim and Konstantin Novoselov discovered graphene, a two-dimensional allotrope of C, in the same year (Novoselov et al., 2004). All these materials paved the way for sophisticated nanotechnology applications in different fields of science.

1.1.2 NANOMATERIALS

Nanomaterials don't have a universal definition. Most of the attempts to define nanomaterials have focused on the size (from 1 to 100 nm), and a few describe a nanomaterial by the percentage (~50% by number) of the particles in the 1- to 100-nm range. Such quantification is erroneous and arbitrary. The use of terms that can lead to flexibility and freedom in interpreting nanomaterials' definitions is too ambiguous. Many particles that show nano-structure-specific behavior don't fulfill the size criteria as per conventional understanding; therefore, limiting a report by size is erroneous. But from a legal perspective, it is essential to have a powerful and accurate framework for the use of nanomaterials outside research. So, it is recommended not to have any standard definitions of nanomaterials instead of focusing on their characteristics, functions, and toxicity profile governed by their physico-chemical and morphological properties (Nature Nanotechnology, 2019).

Approaches of nanomaterial synthesis: The two primary approaches of nanomaterial synthesis are (1) top-down or breakdown techniques and (2) bottom-up or buildup approaches. Top-down or breakdown approaches for materials or devices are processed from bulk material and typically involve solid-phase methods. Examples are dry and wet mechanical grinding, mechanochemical milling and ultrasonic wave methods, mechanical alloying, etc. In bottom-up approaches, materials are synthesized by atom-by-atom and by specific molecules in the nanoscale range (1–100 nm) through chemical and physical methods utilizing self-assembly positional assembly atoms and molecules (Iqbal et al., 2012; Silva, 2010). Bottom-up approaches can be broadly categorized into solid-phase and liquid-phase methods. Examples of solid-phase methods are chemical, for example, chemical vapor deposition (CVD) and physical, for example, physical vapor deposition (PVD). Liquid-phase methods include sedimentation, chemical reduction, indirect reduction, spray pyrolysis, solvothermal, etc. Another bottom-up approach of synthesis that has been encouraged and showed promising efficiency in nanomaterial synthesis is biosynthesis. Biosynthesis of nanomaterials involves plant, microbe, fungi, etc. extracts containing phytochemicals to synthesize nanoparticles by redox reactions. The biosynthesis method complies with green chemistry principles, as it avoids or minimizes the use of hazardous reagents and by-products, ensuring energy efficiency and is economical.

The synthesis of nanomaterials essentially requires that the use of the device and synthesis processes must control size, shape, size distribution, crystal structure, and composition. They should also ensure purity, control of aggregation, stabilization of physical properties, highly reproducible, and easy to scale up, higher yield, and economical. Major challenges in the synthesis of nanomaterials lie therein and control the atoms in bottom-up approaches for self-assembly.

1.1.2.1 NANOMATERIALS USED IN NANOMEDICINES

While new nanomaterials are being continuously explored for theranostics and nanomedicine, some common nanostructures that are used in nanomedicines are:

Polyfunctionalized carbon materials: Carbon materials like single-wall [SWCNTs (0.4–2 nm)] and multiwall CNTs [MWCTs (2–100 nm)], Helical microtubules of carbon, fullerene (mainly C_{60}), graphene oxide (GO), and nanodiamond (ND) are used in tumor targeting and bioimaging. The major limitations of their use are their long-term toxicity and lack of clinical trials.

Metal nanostructures: Metal nanostructures, especially transition metal-based NPs such as gold nanoparticles (AuNPs), serve as carriers of therapeutic cargos, having diagnostic and medicinal properties of their own. For example, the plasmon behavior of AuNPs is applied in photothermal therapy and imaging techniques as scaffolds for cell surface sensing. Passive tumor targeting and active targeting of gold-based-nanomedicine have reached anticancer clinical trials. Nanocrystalline silver is widely established as antimicrobial agents. Histone deacetylases (HDACs) inhibitors like SAHA and sodium butyrate incorporated with gold nanoparticles showed good therapeutic potential in cancer cells.

Superparamagnetic iron oxide nanoparticles (spions) have found applications in magnetic resonance imaging (MRI) and hyperthermia. They have entered into the therapeutic application and are being applied in cancer treatments.

Quantum dots (QD): QDs are often used for fluorescent imaging in nanomedicine.

Silica nanoparticles are mesoporous nanoparticles (from 1.5 to 3 nm size) with a larger active surface area with honeycomb structures, capable of multiple functionalizations, enhanced targeting specificity, and higher drug-carrying capacity, low toxicity, and higher biocompatibility.

Organic–inorganic hybrid nanoparticles are hybrid of both organic and inorganic nanocarriers with the properties of enhanced selectivity and pH-sensitive release of drugs. Examples include lipid bilayer mesoporous silica nanoparticles as cargo for carrying Zoledronic acid drug in vivo (Desai et al., 2017).

Graphene: Functionalized graphene and its oxide have promising opportunities for anti-inflammatory and hydrophobic anticancer drugs delivery, with high drug loading capacity. In biosensing, graphene-based Förster resonance energy transfer (FRET) biosensors show great promises. Photoluminescence of GO-based nanomaterials can be utilized in biomedical imaging and in photothermal in vivo cancer treatment. These systems can also be applied in stem cell treatment using the good electrical coupling between graphene and neurons.

Liposomes, including lipids, proteins, albumin, vesicles, are among the most successful drug carriers. Liposomes are spherical vesicles with single or multiple phospholipid bilayers. They have excellent biocompatibility, biodegradability, and, therefore, low cytotoxicity, pharmacokinetics and help in pH and temperature-sensitive release of drugs. The ability to chemically modify liposomes by the attachment of biomolecules improves their targeted delivery and carrying specificity. These properties have advanced a field of liposomology. For example, the incorporating liposomes with biocompatible polymers have been applied in cancer therapy for specific release of chemo-therapeutic drugs (Torchilin, 2005).

Macromolecules such as polymers, copolymers, antibodies, and proteins are extensively used as drug nanocarriers. The challenges of polymer nanoparticles lie in drug loading, release, and miscibility. Polymeric nanoparticles (PNPs) are more stable, with homogeneity in particle size, greater circulation times, and controlled drugs while releasing drugs with better drug loading capacities than liposomes. Dendrimers have symmetric branches around a linear polymer core (De Souza et al., 2020). Dendrimers with functional branch termini can encapsulate drugs and traverse biological barriers. Modifying these hyperbranched dendrimers physicochemically and biologically gives an edge to develop them as ideal drug-delivery vehicles, high solubility, reduced systemic toxicity, and selectivity of drug–dendrimer conjugates help to target solid tumors.

Solid lipid nanoparticles (SLNs) have the advantages of high physical stability, in vitro and in vivo, controlled carrier release, site-specific targeting, but are limited by low drug loading. An unstable lipid matrix, Intelligent SLNs (ISLNs), uses external impulses to initiate drug release and have strong clinical potential. However, rapid cellular clearance, low encapsulation efficacy, and inefficient drug release are the major drawbacks.

Polymeric micelles are self-assembling colloidal particles (10- to 100-nm size range) having a core–shell structure that is hydrophobic and hydrophilic, respectively. Their higher retention time can be utilized to carry therapeutic agents into target cells with a higher specification, in addition to that, the stability and pharmacokinetics of their structure for controlled release (Gothwal et al., 2016; Zhu and Liao, 2015).

Lipid–polymer hybrid nanoparticles (LPH) are hybrid nanostructures. They have a hydrophobic polymeric core surrounded by a hydrophilic polymeric shell and a lipid monolayer interface in between. This highly stable hybrid shows better drug loading, controlled drug delivery, and targeting cells, but limited by synthetic polymer toxicity in addition to their instability in the high ionic environment (Hadinoto et al., 2013; Mukherjee et al., 2019).

1.1.3 NANOBIOTECHNOLOGY

One of nanotechnology applications' most-talked fields is nanobiotech-nology, which is comparably new and requires equal contribution from physicists, chemists, and biologists. Nanobiotechnology can be defined as nanotechnology applications in understanding and developing biological fields, which may involve working with natural starting materials or biological design principles or wider biological applications, including medicines. As defined by Krukemeyer, et al., *"Nanobiotechnology is concerned with molecular intra- and intercellular processes and is of crucial importance for nanotechnology applications in medicine. This manifests itself in the diverse interplay between medically relevant nanotechnologies and possible nanobiotechnology applications in human medicine"* (Krukemeyer et al., 2015).

1.1.4 NANOMEDICINES AND NANOTECHNOLOGY

Nanomedicine denotes the application of nanotechnology in medicine (the U.S. NNI). Patrick Boisseau and Bertrand Loubaton prefer to say it as "nanotechnology-enabled medicine" (Boisseau and Loubaton, 2011). The European Technology Platform on Nanomedicine describes nanomedicine as *"the application of nanotechnology to health which exploits the improved and often novel physical, chemical, and biological properties of materials at the nanometric scale for the prevention, early and reliable diagnosis and treatment of diseases"* (EC Publication Office). The European Science Foundation defines nanomedicine as *"Nanomedicine uses nano-sized tools for the diagnosis, prevention, and treatment of disease and to gain an increased understanding of the complex underlying pathophysiology of the disease. The ultimate goal is to improve quality of life."* (Duncan et al., 2005).

In the last few decades, nanotechnology research and medicine applications have gained significant momentum of interest among scientific communities, mostly due to the anticipations and facts that it may bring breakthroughs in medical treatment, diagnosis, and prevention in the coming years. Nanomedicine has emerged as a fast-progressing discipline in biosciences research (Moghimi et al., 2005; Wagner et al., 2006; Wu et al., 2020). Optimistic progress has been made in developing diagnostic devices, contrasting and fluorescent imaging agents, theranostics, and target-specific drug delivery. Development of sophisticated nanoelectronics with miniaturization in design holds great promise for improved biomedical applications, particularly in implanted electronics inside the human body for simulation and monitoring of vital signals, nanobioelectronic system that triggers enzymatic activities, the release of drugs from nanomembranes, nanogenerators for self-sustained implants systems, as well as in biophysical studies.

The unique properties and behaviors of three-dimensional nano-structures hold great promises for multifaceted biological applications (Wu et al., 2020). However, nanomedicine's successful translation from the preclinical trials to the therapeutic applications in patients is still challenging.

Nanomedicine can deliver drugs at the target sites with higher accuracy and efficiency, overcoming biological barriers, ineffective delivery of hydrophobic drugs and biologics (Howard, 2016; Moghimi et al., 2001) but at the same time, the multiple components of 3D nanostructures are complex in behavior too and these products, to be successfully translated to enter the clinics, require careful design and engineering, reproducible scale-up production, accurate characterization of physicochemical properties, consistency of the products in terms of stable physiochemical properties, biological interaction, and pharmacokinetics. The safety, legal, and ethical issues of administering nanomedicine must be ensured before entering clinical trials (Wu et al., 2020).

1.1.4.1 DEVELOPMENT OF NANOMEDICINE: HISTORICAL PERSPECTIVE

Nanomedicine has emerged as a prime scientific endeavor in the current century. Still, references of its use were found in ancient documentation and hundreds of years ago, although the numbers of such literature are few. The use of gold metal in medicine is not new. The first use of nanomedicine in the form of colloidal gold can be flipped back to ancient times. In 1890, the German bacteriologist Robert Koch discovered that gold could act against bacterial infections. Much earlier than that, fine particles of gold were being used in medical preparations in the Indian medical system called "Ayurveda," which dates back to 500 and 1000 BCE. In ancient Egypt, over 5000 years ago, mention of gold used in dentistry was found. The alchemists in Alexandria used gold in preparation of colloidal elixir for health. The great alchemist Paracelsus, who is considered the founder of modern medicine, used fine particles of metals and minerals, including gold, for treatments (Pradeep, 2007). Metchnikov and Ehrlich (Nobel Prize for Medicine in 1908) worked on phagocytosis respective cell-specific diagnostic therapy and considered the modern pioneers of nanomedicine (Cooper, 2008). With the inventions of advanced microscopy in the 1960s, cell structures and cell constituents' discovery became possible.

In 1960, the conception of nanosurgery and nanodrug delivery devices was first proposed by Richard Feynman, and that was preset to imagine that nanotechnologies may bring breakthroughs in the field of biology and medical sciences soon (Feynman, 1959). Nanomedicine's interdisciplinary science is being researched for medicines, pharmacology, and medical technology and has established itself through scientific optimism and success only since the 1990s. After that, significant advances in nanomedicines were in the liposome and DNA-based drugs, polymer, and antibody–drug conjugates, polymer nanocapsules in the 1970s, protein–drug conjugates block-copolymer micelles, anti-arthritis nanoparticles in the 1980s, and silver nanoparticles for antimicrobial activities in the 1990s and polymer–protein conjugates early in 21st century.

Nadrian Seeman described the first concept of DNA nanotechnology in 1982. The uses of DNA and other biopolymers are found in sensing and diagnostic. In 2006, Paul Rothemund designed the "scaffolded DNA origami," in his one-pot reaction, which helped improve the functional complexity and morphology of self-assembled DNA nanostructures (Bayda et al., 2020). Doxil/Caelyx (liposomal doxorubicin) was the first nanomedicine approved by FDA in 1995.

1.2 PROMISES AND PROSPECTS OF NANOMEDICINES: RECENT ADVANCES

The last few decades have seen a sharp rise in the interest and optimism in research for the applications of nanomaterials in nanomedicine such as in the diagnosis, drug delivery, regenerative medicine, antibacterial activities, biomarker detection such as in nanobiochips, nanoelectrodes, or nanobiosensors, and molecular imaging.

The intrinsic properties and the unique chemistry in a variety of reactions of these nanomaterials make them ideal for application in combination therapy, including cell target, controlled drug release, photothermal and photodynamic therapies, and bioimaging (Jiang et al., 2018; Lucherelli et al., 2020; Lu et al., 2018; Ménard-Moyon et al., 2015). Tumor-targeting therapeutics with several functional molecules, surface modification of targeting moieties such as proteins and peptides, antibodies and cytotoxic drugs, and nanoparticles have been developed. Nanomaterials are specifically designed to carry therapeutic molecules and modulate essential biological processes (Bayda et al., 2020).

The key concerns that arise while developing nanomaterials for the drug-delivery system are their biocompatibility, biodegradability, the drug dose, their biodistribution, and interactions with the biological environment after administration (Fadeel et al., 2018; Lucherelli et al., 2020).

The possibilities of nanotechnology in medicine expand in diverse medical science and technology applications in diagnostic, therapeutic, regenerative medication, low-cost-quick diagnosis for genetic disorders, microbial infections, and diagnosis before the manifestation of symptoms, neurodegenerative diseases, cancer, organ implantation, and stimulation of neuronal activities.

1.2.1 TARGETED DRUG DELIVERY

The highly potential advantages of nano-based drug carriers for drug delivery that are being widely anticipated are in developing personalized drugs and therapy, to facilitate hydrophobic drugs for intravenous administration, improving the chemical stability of a drug, in bioconjugation of targeting moieties with the drugs, combining diagnostics and therapy for improved theranostics, controlled drug release, and utilization of the enhanced permeability and retention (EPR) effect in some tumors (McGoron, 2020). The EPR effect has been illustrated in Figure 1.1. Two main routes for tumor targeting are passive targeting involving the EPR phenomena and the active targeting that involves covalent attachment of drugs using linkers and a receptor.

In addition to pharmacokinetic and therapeutic properties, a drug must be targeted and delivered to specific molecules for significant activity. The bioavailability of a drug in vivo is influenced by the size of the drug molecules and their solubility parameters. The aim of the modern drug-delivery system must ensure their efficiency for in situ disease cells, appropriate dosage and should not affect normal cells. Nanotechnology in targeted drug delivery has been developed in the forms of drug nanoparticles, polymeric nanosystem, and inorganic biodegradable platform as the carrier of drugs and nanocarrier molecules' surface functionalization. Nano drug-delivery system helped in better cellular uptake compared to the conventional medicines as well as reduced the toxic side effects, thereby has the potential in easing chemotherapeutics in cancer treatment.

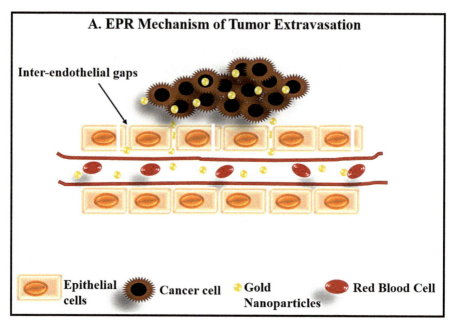

FIGURE 1.1 Passive entry of nanoparticles into tumor cells by interendothelial gaps and in the absence of effective lymphatic drainage system through EPR mechanism.

Sources: Maeda et al. (2001); Matsumura and Maeda (1986); Pandit, et al. (2020).

1.2.2 NANOBOTS IN DRUG DELIVERY

To avoid the harmful side effects of chemotherapeutic drugs on healthy cells, there are constant efforts by scientists to deliver drugs to specific sections. One such experiment has been with microrobots (MRs) of different shapes. These MRs carrying the medications with the help of an applied magnetic field reach targeted microtissues (MT) and attach there so that they are not removed via body fluids circulation.

A drug-loaded micro-robotic needle that accurately targets and remains attached to cancerous tissue in lab experiments without the continuous application of a magnetic field may enhance precision in drug delivery. A recent designing and fabrication of a functional MR like a corkscrew with a microneedle using laser lithography at its end to eliminate the use of an energy-consuming magnetic field for an extended time was succeeded (Lee et al., 2020). The MR was layered with nickel and titanium oxide postlithography to manipulate it magnetically and ensure its biocompatibility. After fixation of the MR in the target tissue and cells, the needles are flushed out at high speed. The use of computational techniques helped in precise automatic targeting and fixation only in seven seconds. The microneedles were incorporated with an anticancer drug and tested against human colorectal cancer cells in a microchamber, which achieved effective targeting and cell damage.

1.2.3 BIOCOMPATIBLE MAGNETIC MICRO- AND NANODEVICES

Single cell-specific therapeutics and diagnosis with the help of nanoparticles hold great potential in drug or gene delivery system development. Developing hard-magnetic nanomagnets can bring important developments in medicine and in the fabrication of smaller devices to minimize the effects of invasive surgeries. Targeted delivery and cell transfection by magnetically propelled micromotors are possible with responsive nanoscale actuation and transport to the site. Major challenges in using magnetic nanoparticles (MNPs) in biomedicine are their high toxicity, difficulty in fabrication, low chemical stability, and weak magnetic moments. Widely commercialized, neodymium iron boron (NdFeB) supermagnets are challenging to fabricate or applied at microscopic scales.

Iron platinum (FePt) alloy is an attractive system in micro-and nanodevices and for next-generation active gene and drug-delivery probes, controlled by a magnetic field. It was used to fabricate fully biocompatible drill-shaped nanopropellers of a bacterium size for gene delivery. Highly magnetic nanostructures (50% stronger than the strongest known micromagnets [NdFeB], chemically stable, and biocompatible based on the FePt-L10 alloy) were designed for magnetic targeting. The FePt nanopropellers were synthesized by specialized high-vacuum nanofabrication "Glancing Angle Deposition" (GLAD) method (Kadiri et al., 2020). GLAD can simultaneously fabricate billions of nanorobots with high precision in time and easy scalability. The biocompatible, nontoxic nanopropellers, coated with DNA coding for a green fluorescent protein, were used for the precise transportation of DNA inside lung carcinoma cells with subsequent enhanced green fluorescence expression.

The Fe–Pt system has the potential to be developed further for the fabrication of micro-robotics, nanodevices, antibiotics delivery, multimodal therapeutics, diagnostics, and delivery, and to overcome the problem of antimicrobial resistance.

1.2.4 LOCALIZED TREATMENT

Targeting drug carriers to the specific cells is of utmost importance to increase their efficacy. Nanoengineered systems can target targeted drug delivery to the disease cells but potentially get cleared by the lymphatic system that limits their activity.

"Magnetically responsive" small drug-delivery vehicles, by incorporating superparamagnetic iron oxide nanoparticles (SPIONs) into layer-by-layer (LbL) microcapsules, presented a new approach to localized drug delivery through the use of permanent and alternating electromagnetic fields for drug movement and release (Read et al., 2020). Microcapsules with SPIONs were rapidly engulfed by immune cells. Still, the magnetic field inhibited the activity of phagocytosing cells, thus retaining drugs in the target sites, prolonging the release of active drugs. The absence of cellular reactive oxygen species (ROS) generation after their introduction into cells and intact cell viability suggested their safety, feasibility, and biocompatibility for improved local drug delivery.

1.2.5 NANOTECHNOLOGY IN CANCER THERAPY

Application of nanoparticles has improved the delivery of cancer therapeutics and, in chemotherapeutics as carriers of drugs, has mostly increased patient tolerance by reducing systemic toxicity.

Fenton reaction–based catalysis, radiation, and photodynamic therapy and nanomaterials help generate ROS inside tumors to cause apoptosis, which subsequently leads to irreversible tumor cell death.

One of the major drawbacks of cancer therapy drugs currently is their hydrophobicity, limiting their therapeutic efficiency and creating unwanted toxicity. Nanomaterials' ability to complex with hydrophobic drugs helps to overcome these barriers in improving drug activity (Lucherelli et al., 2020; Yang et al., 2013).

Recently, in a study, graphene's chemical multifunctionalization, known for its good biocompatibility and biodegradability, combined with other therapies, greatly enhanced targeting cancer therapy. The course is a step toward combined chemo and phototherapy based on graphene against cancer and holds great promise for therapeutic applications of multifunctional materials like graphene (Lucherelli et al., 2020).

Magnetic nanomedicine is being studied and developed as a drug carrier for chemotherapy, in hyperthermia and synergistic chemotherapeutics.

Magnetic fluid hyperthermia is entering clinical treatment as their capacity to heat generation helps in treating malignancy. In a recent report by Chen et al. (2020), a magnetic hydrogel complex was tested in the lab for effective synergistic magnetic hyperthermia and chemotherapy. The three drugs doxorubicin, ferumoxytol (FMT), and medical chitosan were approved in the clinic. The complex works by converting to a hydrogel at hyperthermia temperature in an alternating magnetic field and shows temperature-dependent drug-release behavior. It showed it enhanced synergistic efficiency of 32.4% cell-apoptosis on colon carcinoma cells in vitro, compared to the individual drugs. The study holds a promise for the clinical development of magnetic nanomedicine.

Accurate orientation in drug delivery to the target tumor region, long active hours of the drugs, hyperthermia efficiency, and biosafety are the major criteria for approval of magnetic nanomedicine. In the study, FMT acted as a magnetic heat agent, and the hydrogel with chemotherapeutic drug infers dual-mode treatment (Chen et al., 2020). The nanodrug complex converts to hydrogel inside the tumor environment. Thus, the magnetic hyperthermia and chemotherapy can be applied for the localized tumor treatment, avoiding the normal cells and tissues.

Maeda and colleagues first demonstrated the EPR principle in 1986 (Matsumura and Maeda, 1986). It involves the buildup of molecules and particles inside tumors, because of the hyperpermeability in the vasculature and poor lymphatic drainage inside tumors. Until recently, the EPR concept was not mired in controversy as recent findings demonstrated that the EPR effect is not the dominant mechanism because of the heterogeneity in tumors and nanoparticles response (Hansen et al., 2015).

The heterogeneity of the vasculature and complex factors in the tumor microenvironment have challenged the role of the EPR effect in nanoparticle–tumor interaction. Nanoparticles, as previously thought, do not enter tumor vasculature via interendothelial gaps but enter the cells through transcytosis. New evidence suggested that the mechanism of tumor extravasation by nanoparticles could be through the active process of endothelial transcytosis (De Lázaro and Mooney, 2020). Transcytosis was demonstrated as a busy entry route for molecules through vesicles to cross different biological barriers. It also acts as a trans-endothelial nutritional pathway (Nel et al., 2017). The transcytosis process has been illustrated graphically in Figure 1.2. A total of 97% of nanoparticles are transported through the process of transcytosis, and the lower number of gaps on the endothelial lining prevents the accumulation of nanoparticles in the tumor blood vessels (Sindhwani et al., 2020).

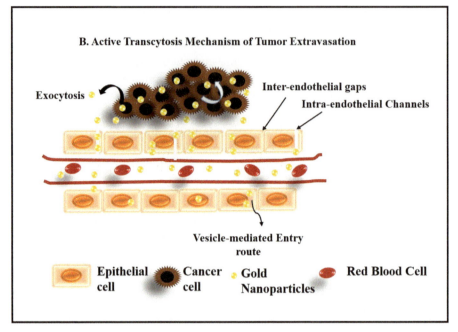

FIGURE 1.2 Active transcytosis process of nanoparticles entry into tumor cells through vascular endothelium and intraendothelial channels.

Sources: De Lázaro and Mooney (2020), Sindhwani et al. (2020).

The findings brought a fundamental paradigm shift in understanding cancer nanomedicine and encourage researchers to develop techniques on the basis of these understanding to enhance delivery efficiency, for example, by utilizing tumor-penetrating peptides (Liu et al., 2017) and by designing polymer conjugates. These, with the help of a cationization process, facilitate active transcytosis through adsorption (Zhou et al., 2019). Still, there is limited understanding of tumor extravasation mechanisms, the specific interactions of nanoparticle, and the use of preclinical cancer models. Nanoparticle parameters such as composition, structure, surface chemistry, and surface charge control their interactions in vivo barriers and the mechanism of transcytosis.

1.2.5.1 CANCER IMMUNOTHERAPY

Modulation of host anticancer immunity by using nanomaterials opened novel cancer therapeutics. Cancer immunotherapy utilizes the body's innate immune system against cancer cells. It enhances anticancer immunity by selectively regulating critical signaling pathways within immune cells and inducing significant antitumor effects and improving cancer therapeutics (Jiang et al., 2017). The optimal antitumor immune responses are linked with reducing the intrinsic immune suppressive signals inside the tumor. For example, inorganic nanocrystals [magnesium silicide (Mg_2Si), for example] can control local oxygen concentration inside a tumor and scavenge excess immune-suppressive extracellular ions, potentially reprogram the tumor microenvironment for better therapeutic results (Zhang et al., 2017).

The challenges in cancer immunotherapy are faced in their scalability, reproducibility in production of immune nanomedicine, which is also in compliance with "chemistry, manufacturing, and controls (CMC)," and "good manufacturing practice (GMP)" (Shi et al., 2017). Also, most of the preclinical trials of immune nanomedicine are carried out in small rodents, limited by the erroneous representation of human physiology, thereby obstructing clinical trials. Continuous exploration of the immune system's interaction with nanomaterials will help design effective and safe optimal immune nanomedicines (Liu et al., 2018).

1.2.6 GENE EDITING

Since the discovery of Clustered Regularly Interspaced Short Palindromic Repeats (CRISPR) in 2012, gene editing has gained extensive research momentum. The CRISPR-Cas9 (CRISPR-associated protein 9) is a recently developed gene-editing tool, a technology inspired by bacteria for new treatment of genetic diseases or disorders. CRISPR is composed of a scissor-like protein called Cas9, and a guide RNA molecule called sgRNA. The sgRNA guides the Cas9 protein to reach the target gene in the nucleus to edit the mistakes with the host cells' repair system's help. But to deliver the gene-clipping tool CRISPR-Cas9 in the cytosol and then to the nucleus across the cell membrane directly and effectively overcoming the cell's defense system is a significant challenge (Mout et al., 2017). In a recent report, CRISPR/Cas9-ribonucleoprotein (Cas9-RNP)-based genome editing was able to specifically target gene and avoid integrational mutagenesis (Mout et al., 2017). The study found that Cas9–sgRNA complex coengineered with the cationic arginine gold nanoparticles (ArgNPs) showed high efficiency (~90%) toward direct cytoplasmic and nuclear delivery besides approximately 30% gene editing efficiency. The Cas9 protein was also designed with an atomic sequence to release the inside nucleus by tweaking the Cas9 protein, and the delivery process was real-time monitored using advanced microscopy.

The current method of ex vivo CRISPR gene editing in hematopoietic stem and progenitor cells employs an electroporation method, sometimes with a virus transduction method, which can help in better gene editing at genetic loci. Still, it causes cellular toxicity, too, therefore needs improvement. It was also found that the engraftment kinetics of nanoformulation-treated primary cells were better than the nontreated cells in mice (Shahbazi et al., 2019).

1.2.7 CELL IMAGING

Superresolution fluorescence microscopy overcame the limitation of the resolution issue at the submicron or nanometer scale. Still, it was impeded by aberrations while imaging inside whole cell or tissue, such as in cancer and brain cells. A new imaging technology, named in situ point spread function (PSF) retrieval [INSPR] to measure wavefront distortions induced by a cell or a tissue, from the signals generated by single molecules attached to the cellular structures of interest, as a light source, has helped in the visualization of nanostructures deep inside cells and tissues (Xu et al., 2020). The three-dimensional superresolution imaging technology retrieves the wavefront distortion from the induced distortions and the 3D responses from the recorded single-molecule dataset containing emission patterns of molecules at arbitrary locations. It can pinpoint biomolecules' positions with high precision and accuracy at the nanometer scale and imaging of intra- and extracellular architectures deep inside whole cells and tissues. This imaging technology may provide better help in neurodegenerative diseases such as Alzheimer's by understanding complex smaller cellular components and their interactions and understanding brain impairments, autism, and other brain diseases.

In a recent publication, NaTb $(WO_4)_2$ nanoparticles were successfully explored for fluorescence and cell imaging. These NaTbW green fluorescent nanoparticles showed excellent biocompatibility against *Staphylococcus aureus*, *Escherichia coli*, and mammalian cancer HeLa cells, in vitro. The high biocompatibility and the intracellular green fluorescence make them suitable for functioning as active bioprobes (Munirathnappa et al., 2020).

The use of bimodal imaging of biohybrids combined with bacterial autofluorescence and magnetic resonance contrast imaging for tumor visualization and precise tumor positioning is also being explored (Huo et al., 2020).

1.2.7.1 *IN VIVO IMAGING TO UNDERSTAND CANCER MECHANICS*

The interaction between cells and their microenvironment determines cellular functionalities. Understanding this is essential in regenerative medicine, cancer therapeutics, and other diseases, including heart disease, inflammation, and understanding physiology such as cell migration and organism development.

A nanoparticle-based in vivo imaging technique that quantified the nanomechanical properties of individual living cells in breast carcinoma and normal breast epithelial cells in vivo in an animal model using microrheology with high resolution was developed recently (Wu et al., 2020). The technique helps different nanoparticles be embedded into multiple cell types, such as tumor cells, to reveal their structure, physical properties, and cells' physical change as they form a tumor. In that study, nanoparticles were used to compare the mechanical properties between cells in vitro, both standards 2D and 3D, and in living animals to understand cancer cell mechanics, what happened to the cancer cells when they started growing as tumors. The measured pliability of the normal cells was found to be steady over time. Still, the growing cancer cells mechanically stiffened with time, and this fundamental behavioral finding gives an insight into cancer metastasis and lethality.

The intravital particle tracking method can reveal a new understanding of cellular mechanics in vivo and the role of microenvironment in cellular functions and tell us more about cellular models' validity. The integration of sophisticated imaging and particle tracking technologies has promising applications in nanomedicine in disease diagnosis and treatment, and nanoparticles might monitor the cells and their disease biology.

1.2.8 *GAS NANOMEDICINE*

Gas nanomedicine is an emerging and promising cutting-edge technology, which uses therapeutic gas molecules such as H_2, NO, O_2, SO_2, CO_2, CO, and H_2S in inflammatory diseases, and especially in cancer treatment to selectively killing cancer cells without damaging normal cells, thus can mitigate a major drawback of traditional cancer therapies. But these gases and associated prodrugs are lacking in performance for active accumulation inside tumor cells and controlled gas release, thus limiting therapeutic efficacy and cause side effects. The bioavailability and biosafety of these therapeutic gases must be ensured for the development of precision and advanced gas delivery nanomedicines. Currently, multifunctional nanoplatforms are being researched for nanomedicines for improvement in cancer treatment. The strategies for advanced gas-releasing nanomedicines for cancer treatment are categorized based on four major aspects, which are (1) stimuli-responsive strategies for controlled gas release, (2) catalytic strategies for controlled gas release, (3) tumor-targeted gas delivery strategies, and (4) multimodel combination strategies based on gas therapy.

Although many "from bench to bedside" clinical trials in gas nanomedicine have been approved and executed intensively, there are many gaps that need to be addressed between gas therapy and nanomedicines. These include poor gas release controllability, high tissue penetration stimuli-responsive gas release, prodrug-free catalytic generation of gases, and highly toxic decomposition products. The areas of development in gas nanomedicine, as in stimuli-responsive gas-releasing nanomedicines, advanced stimuli sources (specific enzymes, microwaves, NIR-II, electricity, and magnet fields), have not been researched so far, functional nanocarriers can also be developed for advanced gas-releasing nanomedicines. In addition to the multistep targeting techniques to improve nanomedicines' targeted efficiency, the direct nuclei-targeted delivery of these gases can also be experimented to enhance their anticancer efficacies.

The H_2 gas therapy is a highly encouraged method because of its high biosafety and wide therapeutic applications, but the development and application of H_2-releasing nanomedicines for cancer therapy have not entered clinical trials. In the recent pandemic of novel Covid-19 pneumonic infection, hydrogen gas therapy is being used in improving oxygen inhalation, relieve hypoxia, and scavenge inflammation in patients.

1.2.9 NANOTECHNOLOGY IN OBESITY CONTROL

In a recent study, researchers blocked a gene's activity in immune cells, called macrophages, the key inflammatory cells to control obesity as obesity is associated with chronic low-grade inflammation. A nanoparticle-based siRNA delivery suppressed the macrophage Asxl2 expression to prevent high-fat diet (HFD)-induced obesity (Zou et al., 2020). The study concluded that targeting a single myeloid lineage gene could limit obesity by regulating energy expenditure.

1.2.10 REGENERATIVE MEDICINE

In a recent study, researchers demonstrated that "nanostimulators"–nanoparticles seeded with tumor necrosis factor-alpha (TNFα) tethered directly to stem cells to release blood vessel growth factors and regulate inflammation in damaged muscle to aid in stem cells' regenerative powers to address muscle ischemia locally. The nanoparticles bind to a receptor on the stem cells' surface, allowing cells to stimulate cellular secretory activity in situ and release the beneficial factors for a longer time, thus providing localized, targeted, and extended delivery of TNFα. This provides a significant benefit in regenerative medicines. The in situ stimulation of stem cells can replace the expensive and arduous preconditioning process and improve clinical applications. Further work must be directed toward optimizing the factors for stem cell preparation and to study their long-term effects (Leong et al., 2020).

1.2.11 MICROBIAL-NANOHYBRIDS

"Microbiotic nanomedicine" is defined as "*the application of nanomaterials-hybridized microorganisms in the theranostics of a variety of pathological diseases*" (Huo et al., 2020).

One step toward addressing the challenges of ineffective targeting, insufficient drug accumulations at pathological sites, inaccurate dosages, and biosafety that is gaining research interest is "*nanomaterials-microorganism integrated microbiotic nanomedicine* (microbial-nanohybrids)" for promising nanomedical therapeutics especially for targeted drug delivery by utilizing tropism mechanism of microbes (Huo et al., 2020). But, the satisfactory therapeutic performance and use of microbes for the treatments of cancer and smallpox, for example, come with the drawback of causing infections. Therefore, the interest in microbiotic therapy was weakened eventually (Huo et al., 2020). To overcome the problem, the integration of microbes and nanomaterials into a microbiotic-nanohybrid therapeutics may help in enhancing movement of nanomedicine to target sites and reduce infections and biotoxicities caused by microbes.

In the complex in vivo tumor microenvironment, microbial nanohybrids may help in integrating the varied therapeutic potential of the nanomaterials and the biological functions of the microbes. To ensure their nontoxicity and biocompatibility, attenuation and deactivation of microbes and selection of biocompatible nanomaterials are necessary before the hybridization. As the nanohybrids have to cross the vascular system to reach the targeted pathological sites, they often disintegrate in the highly complex physiological environment disrupting the therapy and diagnosis. Therefore, chemical conjugation of nanohybrids and biological reconstructions is necessary for their stability before their introduction as therapeutics. Homogeneity in the distribution of nanoparticles in and around the microbe with the system's synergetic functionalities together determines the tumor targeting, reaching the tumor microbiome and providing high therapeutic prospects for tumor treatments. The critical development of the tumor xenografts and vasculature restricts most of the chemotherapeutic drugs and nanomedicines from entering the deep interior of the tumor and can induce metastasis and the passive delivery of carrier and drugs turns a failure. For this, self-propelled Janus nanoparticles were developed for stimuli-responsive locomotion, but these are limited by the particles' nondirectional motion. Biologically active microbes can navigate deep inside the tumors at stimulated external chemical and magnetic signals. Additionally, few microbes provide better locomotion inside tumor cells and tissues with the help of their flagellum.

Also, tumor immunotherapy is developing for cancer treatment, where the immune system is stimulated to release TNF and other cytokines for the destruction of the tumor cells. Microbes in the synergetic therapy with nanohybrids help in regulating the tumor immune microenvironment.

Also, the property of microbes' natural tropism and inclusion of MNP into nanohybrids helps in developing efficient magnetic targeting and improved drug delivery for diverse theranostics. Hyperthermia can cause necrosis of the tumor cells, but with a relatively inadequate accumulation of nanomaterials, it will lead to limited performance. The synergetic combination of photothermal therapy and chemotherapy using microbial nanohybrids has been tested to overcome these limitations. In one such report, the Pd@Au nanoparticles were deposited onto Fe_3O_4NPs-coated spirulina biotemplates, subsequently encapsulated by doxorubicin chemotherapy drugs. These showed strong reactivity under the influence of magnetic fields with satisfactory drug release and were successful in the cellular level (Wang et al., 2019).

1.2.12 *INFLAMMATORY DISEASES*

Uncontrolled inflammation is associated with numerous pathologies. This kind of acute inflammation is commonly linked with shifts in redox balance, causing oxidative stress in tissues and cells (Prauchner, 2017). Uninterrupted positive feedback loops between pro-inflammatory signals and oxidative stress contribute to the uncontrolled hyper inflammation, as suggested by the last shreds of evidence (Biswas, 2016; Lugrin et al., 2014).

The adenosine and multidrug therapy approach have the potential to treat inflammation, and based on that, a novel nanoparticle formulation had been developed for targeted delivery of adenosine and tocopherol to the acute inflammation sites. Conjugating therapeutic drugs to squalene, an endogenous lipid, can improve blood-circulation time, accurate targeting of inflammation cells, and higher biocompatibility (Dormont et al., 2020).

To encounter the crosstalk between oxidative stress and inflammatory responses, multidrug nanoparticles by conjugating squalene to adenosine, an endogenous immunomodulator, followed by encapsulation of α-tocopherol, a natural antioxidant into the design was developed, which showed effective drug loading and nontoxicity (Dormont et al., 2020).

The stable multidrug nanoparticles' bioconjugation system of adenosine and squalene with encapsulation of tocopherol nanoformulation was successful in better encapsulation. It targeted delivery at inflammation sites addressing the vascular endothelial barrier dysfunction, conferring significant survival possibilities in animal models. The approach of selective delivery of immunomodulators and antioxidants conjugates could help treat acute inflammation with reduced biotoxicity (Dormont et al., 2020).

1.2.13 NANOTECHNOLOGY IN NEURODEGENERATIVE DISEASE: TREATING ALZHEIMER'S B-AMYLOID PLAQUES

Alzheimer's disease (AD) builds up plaques of self-assembled β-amyloid (Aβ) peptides in the brain, causing loss of neural connectivity and cell death. The main aim of treatment focuses on the prevention of plaque formation. The β-amyloid peptides are formed from the breakdown of an amyloid precursor protein, and these discarded peptides are eliminated in a healthy brain. Still, in AD, these dangerous peptides are self-assembled and aggregated, forming plaques.

The Aβ nanodepleters device design that has ultra-large mesoporous silica nanostructures and anti-Aβ single-chain variable fragments was reported (Jung et al., 2020). It was aimed at reaching and eliminating Aβ monomers aggregates and thereby mitigates Aβ-induced neurotoxicity in vitro. The surface of the nanodepleters was coated with a protein antibody binding to the Aβ peptides. The spherical and porous surface structure increases the available surface area around the eighth time the loading efficiency of Aβ-targeting agent, scFvs, blocking Aβ monomers' aggregation in a concentration-dependent manner, hence, reduced Aβ-induced neurotoxicity. It acts as a "cage" that traps and eliminates the peptides from the brain. Suppression of Aβ plaque formation in vivo was found up to 30%, and so the study suggested that Aβ nanodepleters can be developed as a potential anti-amyloidosis material.

The approach of antibody engineering and nanotechnology can also be employed to remove protein aggregation in other neurodegenerative diseases, like Huntington's disease and Parkinson's disease. They can also be improved for use in fluorescent imaging and MRI.

1.2.14 NANOINFORMATICS

Nanoinformatics, the application of computational techniques to nano-medicine, is still to be developed as an area of research. Powerful machine learning algorithms, quantitative structure–activity relationship (QSAR), quantitative structure–property relationship (QSPR), data mining, network analysis, and ADMET (absorption, distribution, metabolism, excretion, and toxicity) and predictive analysis tools are being applied for understanding and predicting cellular behavior and in developing efficient nanosystem to overcome the challenges in nanomedicines. Nanoinformatics has the potential for nanoparticle design, addressing in vitro barriers, facilitating chemotherapy, in modeling of the tumor cells, and identifying drug-resistant tumors. Applying hyperthermia in targeted drug delivery and gene therapy is developing as the state-of-the-art nanoinformatic techniques for cancer treatment while reducing side effects associated with conventional chemotherapy treatment.

1.2.15 BIOELECTRONICS

The bioelectronic medicines work by modulating electrical signaling within peripheral organs, with precise control of physiological functions, and it can overcome the drawbacks of implantable devices in interfacing with vascularized organs. The bioelectronic strategy was implemented in developing a transgene-free magnetothermal stimulation approach for precise control of hormone secretion from the adrenal gland by a method of hysteretic heating of MNPs remotely. The method is remotely applied in alternating magnetic fields to trigger endogenously expressed heat-sensitive transient receptor potential vanilloid family member 1 (TRPV1) channels and Ca^{2+} influx into the cells in rodents (Rosenfeld et al., 2020). The study demonstrated that MNPs could stay inside the tissue for six months and help in chronic stimulation of the adrenal gland without causing cellular dysfunction and tissue damage. Alterations in normal levels of stress hormones such as adrenaline and cortisol are correlated with mental health disorders such as posttraumatic stress disorder (PTSD) and major depression, and the use of MNPs to stimulate adrenal glands to control the release of hormones approach may facilitate to treat hormone-linked stress disorders by modulating peripheral organ function. This kind of magnetothermal deep organ stimulation can advance the understanding of organ function and development of bioelectronic medicines.

1.2.16 COMBINATORIAL NANOMEDICINE

Combination therapy has been useful in various treatments by bringing marked improvements in objective response rate (ORR), overall survival (OS) for oncology, infectious diseases, autoimmune deficiency, and regenerative medicine. Barriers are seen with conventional drug development, such as exorbitant costs, substantial time to development, uncertainty in efficacy and safety, also significantly impair nanomedicine translation. These also include the inability to optimize drug doses both in single-drug therapies and in combination therapy. Combination therapy is the standard clinical practice when patients require multiple treatments simultaneously. In single-drug regimens, variability in genetics, environment, and physiology is usually not correctly considered by the standard treatment and dosing protocol, and it is difficult to individualize drugs. For comorbidity treatments, personalizing multidrugs is almost impossible and would require numerous trials for the desired effect. Clinical optimization of combination therapies is a crucial parameter for successful translation of preclinical to clinical stage. By focusing on determining optimized combinations and drug–dose ratios, it can be a possible design to develop novel treatment beyond conventional combination therapy.

1.2.17 ARTIFICIAL INTELLIGENCE (AI) IN NANOTHERAPEUTICS

In the nanomedicine field, multifunctional approaches have evolved gradually, integrating therapeutics and diagnostics together and designing multiple nanomedicine-functionalized therapies. It has greatly improved treatment outcomes through targeted and multiagent drug synergism, but at fixed doses, similarly like conventional combination therapeutic strategies (Ho et al., 2019). The drug synergy is variable with time and dose and differs from patient to patient, and that throws a serious challenge. Also, conventional techniques that utilize omics-based information to design drugs for patients lack in a higher degree of actionability in the clinic. In attempts to overcome these hurdles, an AI interface, along with nanomedicine, evolved to determine optimized drug and dose parameters in combinatorial nanotherapeutics to realize its full potential. Drug synergy is important for effective treatment results, but synergism and optimization meet different endpoints. To close this gap, the AI can help in reconciling the drug and dose parameter space that the combination therapy opens up into an accurately optimized, actionable treatment response in the clinic and optimizes the formula of both nanotechnology-modified and unmodified drug combinations. In a recent study, a mechanism-independent, deep learning algorithm–based AI platform called quadratic phenotypic optimization platform (QPOP) was developed to optimized design of a novel combination therapy against multiple myeloma that demonstrated significant accuracy, efficacy, and safety, both in vitro and in vivo and this may act as an actionable platform for patient-specific responses (Ho et al., 2019).

Phenotypic personalized medicine: Advanced technologies like phenotypic personalized medicine (PPM) bolsters combinatorial nanomedicine development, excluding the need to optimize drug–dose ratios for the standard of drug development and testing (Silva et al., 2016). PPM, a top-down technology, was developed to select drugs and optimize drug ratios at the same time. It is independent of algorithms, predictive modeling, and the disease mechanism information suggests the best treatment, supported by experimental data (Silva et al., 2016).

By calibrating individual response to therapy, PPM deduced parabolic maps that enabled optimization of patient-specific drug dose (Silva et al., 2016). It uses an unbiased optimization process to identify and validate unexplored drug combinations unlike combination therapy using presumed synergistic dose, which provides an important step in an optimal treatment outcome and also in realizing new therapeutic approach by understanding key signaling pathways, such as in identifying optimal drug combinations in targeting glucose metabolism in lung cancer treatment (Ho et al., 2016).

AI as in next-generation medicine may prove its potential to develop an advanced road map in nanotherapeutics, to have better and affordable treatment access for patients.

1.3 CHALLENGES THAT LIE AHEAD

Despite the advancement in the frontiers of nanomedicines, nanotherapeutics are limited by significant challenges. One of the critical drawbacks of nanomedicines' clinical translation is overcoming physiological barriers like blood–brain barrier (BBB), as the endothelial cells of the capillary wall restrict the movement of drugs in the brain placental barrier and heart barrier. Moreover, the crosstalk between nanomedicines and the natural immune system of the human body is much complex. Inconsistency in distributing drugs to the specific sites and in specified time reduces targeting efficiency, undesired and insufficient distribution and accumulation of drugs, therapeutic inefficacy, and induced biotoxicity. In the last 30 years, nanomedicines have been developed at an incredibly high rate. However, the number of drugs approved for clinical use in cancer therapy by the Food and Drug Administration (FDA) and European Medicines Agency has been only 10. Only 14% of the nanomedicines have demonstrated efficacy in phase III clinical trials (He et al., 2019). One of the great challenges faced by nanomedicines is their poor accumulation in tumors and ineffective pharmacokinetics. Also, the high attrition rate has questioned the prospects of nanomedicines to enter the next-generation medicine. Therefore, it is important to appraise the nanomedicines and therapeutics and identify the barriers to addressing clinical translation challenges.

1.3.1 CROSSING BIOLOGICAL BARRIERS

Before reaching the target cells or tissues, drugs have to cross multiple biological membrane barriers and ensure long-time stability. Most of the nanodrug formulations fail in that prospect. After the oral delivery of smaller drugs and intravenous administration (IV) of larger molecular drugs of peptides, proteins, and polynucleotides, the drugs in circulation have to cross different biological barriers to reach the target disease sites. For example, oral nanoformulations need to cross intestinal epithelium proving high stability in the gastrointestinal tract and high systemic bioavailability. In the central nervous system (CNS), the BBB constrains the diffusion of 99% of macro- or hydrophilic molecules into the cerebrospinal fluid. It is an excellent barrier in delivering drugs in the brain (Kabanov and Gendelman, 2007; Wang and Wu, 2017; Wu and Moghimi, 2016).

1.3.2 SPECIFYING TARGET

High heterogeneous nature of tumor vasculature distribution, differential permeability, poor perfusion due to high cancer cell density at larger places in tumor and dense tumor stroma, high interstitial fluid pressure (IFP), poor lymphatic drainage in tumors, all these factors together cause reduced extravasation and trans-vascular transport of drugs (Jang et al., 2003; Minchinton and Tannock, 2006; Maeda et al., 2001; Jain, 1987).

Owing to the great porousness of tumor vasculature and poor lymphatic drainage, the EPR effect increased the passive accumulation of nanoparticles inside the tumor and improved tumor drug delivery. The issue with the first-generation nanomedicine, based on passive targeting, is to effectively control the pharmacokinetics and biodistribution of nanoparticles and moderating its physicochemical functions (Golden et al., 1998; Wu et al., 2020). To improve the drug delivery, active biological transport pathways can help to achieve efficacy in targeted drug delivery, minimal or no side effects, and reduction in systemic drug contacts. Active targeting nanoparticles incorporate in their design some targeting moieties, thus reaching target sites with enhanced efficacy, dismiss toxic interaction with healthy cells, and facilitating better cellular uptake of drugs (Lammers et al., 2008; Wu et al., 2017).

The active targeting of drugs depends on the nanocarrier agents and the affinity of the target and its specific ligand, the density of targeting ligands per drug nanocarrier to ensure the drugs reach the specific targets and an optimal internalization happens (Xu et al., 2015).

For example, few receptors find overexpression in malignant cells and ligands and/or monoclonal antibodies incorporated on the surface of nanomaterials specifically target these receptors to increase the drug agents' cellular internalization endocytosis.

Nanomedicines can also work on targeting tumor endothelial cells to enable a tumor-penetrating function to polymers, liposomes, and other nanoparticles (Danhier et al., 2009; Gupta et al., 2005; Hölig et al., 2004; Schiffelers et al., 2003; Skotland et al., 2015). The EPR effect has aided the design of nanomedicines for solid tumors. However, the transition from bench to patient's bedside after three decades of research to develop cancer nanomedicines has not been encouraging. A few (<10) drugs are currently in phase III or IV clinical trials and only 10 have received regulatory approval.

1.3.3 CHARACTERIZATION OF NANOFORMULATIONS DESIGNED FOR NANOMEDICINES

Nanomedicines are 3D complex with different components. Each serves different functions; therefore, it is difficult to characterize, quantify, and predict the interrelationships and interactions and understand its diverse physicochemical and biological behaviors (Doane and Burda, 2012; Wicki et al., 2015). The composition, structure, size, surface properties, porosity, charge, and stability of all these key physicochemical properties influence its biological interactions. It is important to include all these factors individually and their complex web of interactions when formulating nanomedicines. Alterations in any one factor may lead to significant changes in functional properties (Aillon et al., 2009; Dobrovolskaia and McNeil, 2007; Fubini et al., 2010; Kettiger et al., 2013; Nel et al., 2009; Paciotti et al., 2004). In addition to technical ground, challenges in nanodrugs and nanocarriers lie in regulatory aspects.

1.3.4 CLINICAL TRANSLATION OF NANOMEDICINE

Although multiple nanodrug formulations are tested in the lab, substantial innovations are missing in clinical translation. Most of these clinical-stage nanomedicines have improved the pharmacokinetics and eliminating toxicity of chemotherapies, improvement in therapeutic efficiency is not much or absent even for the so far best nanoformulations (De Lázaro and Mooney, 2020).

The drawback of existing preclinical tumor experimental models incorrectly mimicking biological structure and function accurately is one of the causes of failure of clinical translation of cancer nanomedicines. The aim of the clinically approved nanomedicines is to boost drug half-life and achieve passive targeting. Current nanomedicine is seeing a transition to active targeting systems, controlled release in response to environmental cues, and theranostic activity to be developed as the next-generation nanotherapeutics.

Transfer of nanomedicine from bench to bedside requires multidisciplinary contributions, and it must consider clinical, ethical, and societal perceptions. The design and development of nanomedicines and devices must keep in mind the material's toxicological profile. Also, the potential environmental impact must be assessed in the process of manufacturing and subsequent trials and clinical applications. Risk–benefit analysis for both acute and chronic effects of nanomedicines is of utmost necessity. Also proactive risk management before designing and testing the new nanomedicines must be encouraged.

There are challenging paradigms in the current understanding of nano-therapeutics and the behavior of biological molecules in response to these therapeutics. To fill that gap, it would require extensive research support from all the concerned departments.

1.3.5 *REGULATORY AND ETHICAL ISSUES*

The US FDA's decades-old classification method for review and approval purposes of health-care products has been proven challenging in the case of nanomedicine evaluation due to their novel and cross-category characteristics (Paradise, 2019). In addition to that, their unknown risk profiles and biological interaction features bring forth novel ethical and regulatory challenges for clinical trials and translation. In July 2007, FDA's Nano-technology Taskforce had concluded that nanomedical products, including the combination product mechanism, were subject to traditional regulatory approaches, as they couldn't warrant any novel regulatory frameworks.

In the case-by-case approach, the combinations of product therapy for evaluating risks and safety of products have created serious concerns among legal experts, medical practitioners, as well as scientists because these frameworks are not adequate and consistent. In addition to that, the toxicity of nanoparticles and their human health implications through exposures and administration routes, their ability to cross biological barriers, and long-term interaction profile raise their safety concerns too, hence putting forward some ethical constraints.

The core challenges are in the adequacy and efficacy of existing regulatory frameworks for the rapidly emerging complex nanotherapeutics integrated with different biological fields, questions in traditional risk–benefit measures' pre- and proprotocol design and clinical trials, also on the requirements of explicit labeling for public health literacy needs.

1.4 CONCLUDING REMARKS AND FUTURE DIRECTION

Nanotechnology is being extensively applied to fabricate medical devices, biosensing, drug screening, assays, tissue engineering, drug delivery, and imaging drug delivery, release, and efficacy. Over the years, some drug products are available over-the-counter (OTC) such as sunscreens, skincare products, handwash soaps, and dietary supplements. A small number of nano-based drugs have also received approval by the FDA, for example, Doxil and Abraxane. Many drugs are widely being investigated at the preclinical and clinical stages. So far, the patient response in cancer therapy has been modest. There is massive room for improvement in the facilitation of personalized and targeted medicines and theranostics. The future of nanomedicine is aimed at early detection of the molecular level pathological changes, by sophisticated and straightforward imaging methods and minimally invasive treatment, and toward individually tailor-made medicines.

There have been mixed opinions from the scientific community about the prospects of nanomedicines for successful establishment in the human healthcare system. While some are optimistic about the manifold benefits and breakthroughs that nanomedicines can bring in upcoming decades, many are doubtful about the satisfactory progress made so far in nanotechnology-based medical research, and some believe that nanomedicines are yet to make any advanced impact in clinics, not completely undermining the achievement till now. So far, nanomedicines, approved globally, have been fewer in number, and the higher attrition rate has invited uncertainty about their promises. Although nanomedicines' success from the proof-of-concept idea to commercialization seems daunting, industry partnerships and capital funding in nanomedicine research have been encouraging, and some efforts led to regulatory approvals of cancer nanomedicines too. These success stories will inspire the scientific community to continue research and development in nanomedicine for a transition from idea to actualization.

Taken as a whole, the abovementioned discussion says that the possibilities of nanotechnology in nanomedicine and theranostics are many, but they are not without loopholes. There is a need for collective efforts from material scientists, biologists, and clinical scientists to address the fundamental challenges in understanding the mechanisms of nanomedicines' interaction with the biological system to overcome the barriers in clinical translation of nanomedicines and develop improved medical technologies.

ACKNOWLEDGMENT

The author AD would like to thank University Grants Commission (UGC), India and Jawaharlal Nehru Memorial Fund (JNMF), New Delhi, India, for providing financial support during her research. She would also like to acknowledge Professor A. Yogamoorthi, Department of Ecology & Environmental Sciences, Pondicherry University for his inspiration and support.

KEYWORDS

- **nanotherapeutics**
- **target-specific drug delivery**
- **EPR**
- **transcytosis**
- **clinical translation**

REFERENCES

Aillon, K. L.; Xie, Y.; El-Gendy, N.; Berkland, C. J.; Forrest, M. L. Effects of Nanomaterial Physicochemical Properties on in Vivo Toxicity. *Adv. Drug Deliv. Rev.* **2009**, *61* (6), 457–466.

Bayda, S.; Adeel, M.; Tuccinardi, T.; Cordani, M.; Rizzolio, F. The History of Nanoscience and Nanotechnology: From Chemical–Physical Applications to Nanomedicine. *Molecules* **2020**, *25* (1), 112.

Biswas, S. K. Does the Interdependence between Oxidative Stress and Inflammation Explain the Antioxidant Paradox? *Oxid. Med. Cell. Long.* **2016**.

Boisseau, P.; Loubaton, B. Nanomedicine, Nanotechnology in Medicine. *Comptes Rendus Physique* **2011**, *12* (7), 620–636.

Chen, B.; Xing, J.; Li, M.; Liu, Y.; Ji, M. DOX@ Ferumoxytol-Medical Chitosan as Magnetic Hydrogel Therapeutic System for Effective Magnetic Hyperthermia and Chemotherapy in Vitro. *Colloids Surf. B: Biointerf.* **2020**, 110896.

Chen, S.; Liang, X. J. Nanobiotechnology and Nanomedicine: Small Change Brings Big Difference. *Sci. China Life Sci.* **2018**, *61* (4), 371.

Cooper, E. L. From Darwin and Metchnikoff to Burnet and Beyond. In *Trends in Innate Immunity*, Vol. 15; Karger Publishers, 2008; pp 1–11.

Danhier, F.; Vroman, B.; Lecouturier, N.; Crokart, N.; Pourcelle, V.; Freichels, H.; Jérôme, C.; Marchand-Brynaert, J.; Feron, O.; Préat, V. Targeting of Tumor Endothelium by RGD-Grafted PLGA-Nanoparticles Loaded with Paclitaxel. *J. Contr. Releas.* **2009**, *140* (2), 166–173.

De Lázaro, I.; Mooney, D. J. A Nanoparticle's Pathway into Tumours. *Nat. Mater.* **2020**, *19* (5), 486–487.

De Souza, C.; Lindstrom, A. R.; Ma, Z.; Chatterji, B. P. Nanomaterials as Potential Transporters of HDAC Inhibitors. *Med. Drug Disc.* **2020**, 100040.

Desai, D.; Zhang, J.; Sandholm, J.; Lehtimäki, J.; Grönroos, T.; Tuomela, J.; Rosenholm, J. M. Lipid Bilayer-Gated Mesoporous Silica Nanocarriers for Tumor-Targeted Delivery of Zoledronic Acid in Vivo. *Mol. Pharm.* **2017**, 14(9), 3218–3227.

Doane, T. L.; Burda, C. The Unique Role of Nanoparticles in Nanomedicine: Imaging, Drug Delivery and Therapy. *Chem. Soc. Rev.* **2012**, *41* (7), 2885–2911.

Dobrovolskaia, M. A.; McNeil, S. E. Immunological Properties of Engineered Nanomaterials. *Nat. Nanotechnol.* **2007,** *2* (8), 469.

Dormont, F.; Brusini, R.; Cailleau, C.; Reynaud, F.; Peramo, A.; Gendron, A.; Mougin, J.; Gaudin, F.; Varna, M.; Couvreur, P. Squalene-Based Multidrug Nanoparticles for Improved Mitigation of Uncontrolled Inflammation. *Sci. Adv.* **2020,** eaaz5466.

Duncan, R.; Kreyling, W. G.; Biosseau, P.; Cannistraro, S.; Coatrieux, J.; Conde, J. P.; Hennick, W.; Oberleithner, H.; Rivas, J. ESF Scientific Forward Look on Nanomedicine. *Eur. Sci. Foundation Policy Brief.* **2005,** 1–6.

European Technology Platform on Nanomedicine, Nanotechnology for Health, Vision Paper and Basis for a Strategic Research Agenda for Nanomedicine, EC Publication Office, Sept 2005.

Fadeel, B.; Bussy, C.; Merino, S.; Vázquez, E.; Flahaut, E.; Mouchet, F.; Evariste, L.; Gauthier, L.; Koivisto, A. J.; Vogel, U.; Martin, C. Safety Assessment of Graphene-Based Materials: Focus on Human Health and the Environment. *ACS Nano* **2018,** *12* (11), 10582–10620.

Feynman, R. P. Plenty of Room at the Bottom. In APS Annual Meeting, Dec 1959.

Fubini, B.; Ghiazza, M.; Fenoglio, I. Physico-Chemical Features of Engineered Nanoparticles Relevant to Their Toxicity. *Nanotoxicology* **2010,** *4* (4), 347–363.

Golden, P. L.; Huwyler, J.; Pardridge, W. M. Treatment of Large Solid Tumors in Mice with Daunomycin-Loaded Sterically Stabilized Liposomes. *Drug Deliv.* **1998,** *5* (3), 207–212.

Gothwal, A.; Khan, I.; Gupta, U. Polymeric Micelles: Recent Advancements in the Delivery of Anticancer Drugs. *Pharm. Res.* **2016,** *33* (1), 18–39.

Gupta, B.; Levchenko, T. S.; Torchilin, V. P. Intracellular Delivery of Large Molecules and Small Particles by Cell-Penetrating Proteins and Peptides. *Adv. Drug Deliv. Rev.* **2005,** *57* (4), 637–651.

Hadinoto, K.; Sundaresan, A.; Cheow, W. S. Lipid–Polymer Hybrid Nanoparticles as a New Generation Therapeutic Delivery Platform: A Review. *Eur. J. Pharm. Biopharm.* **2013,** *85* (3), 427–443.

Hansen, A. E.; Petersen, A. L.; Henriksen, J. R.; Boerresen, B.; Rasmussen, P.; Elema, D. R.; Rosenschöld, P. M. A.; Kristensen, A. T.; Kjær, A.; Andresen, T. L. Positron Emission Tomography Based Elucidation of the Enhanced Permeability and Retention Effect in Dogs with Cancer Using Copper-64 Liposomes. *ACS Nano* **2015,** *9* (7), 6985–6995.

He, H.; Liu, L.; Morin, E. E.; Liu, M.; Schwendeman, A. Survey of Clinical Translation of Cancer Nanomedicines—Lessons Learned from Successes and Failures. *Acc. Chem. Res.* **2019,** *52* (9), 2445–2461.

Ho, D.; Wang, P.; Kee, T. Artificial Intelligence in Nanomedicine. *Nanoscale Horizons* **2019,** *4* (2), 365–377.

Ho, D.; Zarrinpar, A.; Chow, E. K. H. Diamonds, Digital Health, and Drug Development: Optimizing Combinatorial Nanomedicine, Therapy of Tuberculosis Using a Macrophage Cell Culture Model. *Proc. Natl. Acad. Sci. U.S.A.* **2016,** *113* (15), E2172–E2179.

Hölig, P.; Bach, M.; Völkel, T.; Nahde, T.; Hoffmann, S.; Müller, R.; Kontermann, R. E. Novel RGD Lipopeptides for the Targeting of Liposomes to Integrin-Expressing Endothelial and Melanoma Cells. *Protein Eng. Design Select.* **2004,** *17* (5), 433–441.

Howard, K. A. Nanomedicine: Working Towards Defining the Field. In *Nanomedicine*; Springer: New York, NY, 2016; pp 1–12.

Huo, M.; Wang, L.; Chen, Y.; Shi, J. Nanomaterials/Microorganism-Integrated Microbiotic Nanomedicine. *Nano Today* **2020,** *32,* 100854.

Iijima, S. Helical Microtubules of Graphitic Carbon. *Nature* **1991,** *354* (6348), 56–58.

Iqbal, P.; Preece, J. A.; Mendes, P. M. Nanotechnology: The "Top-Down" and "Bottom-Up" Approaches. In *Supramolecular Chemistry*; John Wiley & Sons, Ltd.: Chichester, UK, 2012.

Jain, R. K. Transport of Molecules across Tumor Vasculature. *Cancer Metast. Rev.* **1987,** *6* (4), 559–593.

Jang, S. H.; Wientjes, M. G.; Lu, D.; Au, J. L. S. Drug Delivery and Transport to Solid Tumors. *Pharm. Res.* **2003,** *20* (9), 1337–1350.

Jiang, W.; Mo, F.; Lin, Y.; Wang, X.; Xu, L.; Fu, F. Tumor Targeting Dual Stimuli Responsive Controllable Release Nanoplatform Based on DNA-Conjugated Reduced Graphene Oxide for Chemo-Photothermal Synergetic Cancer Therapy. *J. Mater. Chem. B* **2018,** *6* (26), 4360–4367.

Jiang, W.; Von Roemeling, C. A.; Chen, Y.; Qie, Y.; Liu, X.; Chen, J.; Kim, B. Y. Designing Nanomedicine for Immuno-Oncology. *Nat. Biomed. Eng.* **2017,** *1* (2), 1–11.

Jung, H.; Chung, Y. J.; Wilton, R.; Lee, C. H.; Lee, B. I.; Lim, J.; Lee, H.; Choi, J. H.; Kang, H.; Lee, B.; Rozhkova, E. A. Silica Nanodepletors: Targeting and Clearing Alzheimer's β-Amyloid Plaques. *Adv. Funct. Mater.* **2020,** *30* (15), 1910475.

Kabanov, A. V.; Gendelman, H. E. Nanomedicine in the Diagnosis and Therapy of Neurodegenerative Disorders. *Progress Polym. Sci.* **2007,** *32* (8–9), 1054–1082.

Kadiri, V. M.; Bussi, C.; Holle, A. W.; Son, K.; Kwon, H.; Schütz, G.; Gutierrez, M. G.; Fischer, P. Biocompatible Magnetic Micro-and Nanodevices: Fabrication of FePt Nanopropellers and Cell Transfection. *Adv. Mater.* **2020,** 2001114.

Kettiger, H.; Schipanski, A.; Wick, P.; Huwyler, J. Engineered Nanomaterial Uptake and Tissue Distribution: From Cell to Organism. *Int. J. Nanomed.* **2013,** *8,* 3255.

Krukemeyer, M. G.; Krenn, V.; Huebner, F.; Wagner, W.; Resch, R. History and Possible Uses of Nanomedicine Based on Nanoparticles and Nanotechnological Progress. *J. Nanomed. Nanotechnol.* **2015,** *6* (336), 2.

Lammers, T. G. G. M.; Hennink, W. E.; Storm, G. Tumour-Targeted Nanomedicines: Principles and Practice. *Br. J. Cancer* **2008,** *99* (3), 392–397.

Lee, S.; Kim, J. Y.; Kim, J.; Hoshiar, A. K.; Park, J.; Lee, S.; Kim, J.; Pané, S.; Nelson, B. J.; Choi, H. A Needle-Type Microrobot for Targeted Drug Delivery by Affixing to a Microtissue. *Adv. Healthcare Mater.* **2020,** *9* (7), 1901697.

Leong, J.; Hong, Y. T.; Wu, Y. F.; Ko, E.; Dvoretskiy, S.; Teo, J. Y.; Kim, B. S.; Kim, K.; Jeon, H.; Boppart, M.; Yang, Y. Y. Surface Tethering of Inflammation-Modulatory Nanostimulators to Stem Cells for Ischemic Muscle Repair. *ACS Nano* **2020,** *14* (5), 5298–5313.

Liu, X.; Lin, P.; Perrett, I.; Lin, J.; Liao, Y. P.; Chang, C. H.; Jiang, J.; Wu, N.; Donahue, T.; Wainberg, Z.; Nel, A. E. Tumor-Penetrating Peptide Enhances Transcytosis of Silicasome-Based Chemotherapy for Pancreatic Cancer. *J. Clin. Invest.* **2017,** *127* (5), 2007–2018.

Liu, Z.; Jiang, W.; Nam, J.; Moon, J. J.; Kim, B. Y. Immunomodulating Nanomedicine for Cancer Therapy. *Nano Lett.* **2018,** *18* (11), 6655–6659.

Lu, Y. J.; Lin, P. Y.; Huang, P. H.; Kuo, C. Y.; Shalumon, K. T.; Chen, M. Y.; Chen, J. P. Magnetic Graphene Oxide for Dual Targeted Delivery of Doxorubicin and Photothermal Therapy. *Nanomaterials* **2018,** *8* (4), 193.

Lucherelli, M. A.; Yu, Y.; Reina, G.; Abellán, G.; Miyako, E.; Bianco, A. Rational Chemical Multifunctionalization of Graphene Interface Enhances Targeting Cancer Therapy. *Angew. Chem. Int. Ed.* **2020.**

Lugrin, J.; Rosenblatt-Velin, N.; Parapanov, R.; Liaudet, L. The Role of Oxidative Stress during Inflammatory Processes. *Biol. Chem.* **2014,** *395* (2), 203–230.

Maeda, H.; Sawa, T.; Konno, T. Mechanism of Tumor-Targeted Delivery of Macromolecular Drugs, Including the EPR Effect in Solid Tumor and Clinical Overview of the Prototype Polymeric Drug SMANCS. *J. Contr. Releas.* **2001,** *74* (1–3), 47–61.

Mappes, T.; Jahr, N.; Csaki, A.; Vogler, N.; Popp, J.; Fritzsche, W. The Invention of Immersion Ultramicroscopy in 1912—the Birth of Nanotechnology? *Angew. Chem. Int. Ed.* **2012,** *51* (45), 11208–11212.

Matsumura, Y.; Maeda, H. A New Concept for Macromolecular Therapeutics in Cancer Chemotherapy: Mechanism of Tumoritropic Accumulation of Proteins and the Antitumor Agent Smancs. *Cancer Res.* **1986,** *46* (12 Part 1), 6387–6392.

McGoron, A. J. Perspectives on the Future of Nanomedicine to Impact Patients: An Analysis of US Federal Funding and Interventional Clinical Trials. *Bioconjugate Chem.* **2020,** *31* (3), 436–447.

Ménard-Moyon, C.; Ali-Boucetta, H.; Fabbro, C.; Chaloin, O.; Kostarelos, K.; Bianco, A. Controlled Chemical Derivatisation of Carbon Nanotubes with Imaging, Targeting, and Therapeutic Capabilities. *Chem.–Eur. J.* **2015,** *21* (42), 14886–14892.

Minchinton, A. I.; Tannock, I. F. Drug Penetration in Solid Tumours. *Nat. Rev. Cancer* **2006,** *6* (8), 583–592.

Moghimi, S. M.; Hunter, A. C.; Murray, J. C. Nanomedicine: Current Status and Future Prospects. *FASEB J.* **2005,** *19* (3), 311–330.

Moghimi, S. M.; Hunter, A. C.; Murray, J. C. Long-Circulating and Target-Specific Nanoparticles: Theory to Practice. *Pharmacol. Rev.* **2001,** *53* (2), 283–318.

Mout, R.; Ray, M.; Yesilbag Tonga, G.; Lee, Y. W.; Tay, T.; Sasaki, K.; Rotello, V. M. Direct Cytosolic Delivery of CRISPR/Cas9-ribonucleoprotein for Efficient Gene Editing. *ACS Nano* **2017,** *11* (3), 2452–2458.

Mukherjee, A.; Waters, A. K.; Kalyan, P.; Achrol, A. S.; Kesari, S.; Yenugonda, V. M. Lipid–Polymer Hybrid Nanoparticles as a Next-Generation Drug Delivery Platform: State of the Art, Emerging Technologies, and Perspectives. *Int. J. Nanomed.* **2019,** *14,* 1937.

Munirathnappa, A. K.; Maurya, S. K.; Kumar, K.; Navada, K. K.; Kulal, A.; Sundaram, N. G. Scheelite like NaTb (WO4) 2 nanoparticles: Green Fluorescence and in Vitro Cell Imaging Applications. *Mater. Sci. Eng.: C* **2020,** *106,* 110182.

Nanomaterials definition matters. *Nat. Nanotechno.* [Internet] **2019,** *14* (3): 193. https://doi.org/10.1038/s41565-019-0412-3

Nel, A.; Ruoslahti, E.; Meng, H. *New Insights into "permeability" as in the Enhanced Permeability and Retention Effect of Cancer Nanotherapeutics.* ACS Publications, 2017.

Nel, A. E.; Mädler, L.; Velegol, D.; Xia, T.; Hoek, E. M.; Somasundaran, P.; Klaessig, F.; Castranova, V.; Thompson, M. Understanding Biophysicochemical Interactions at the Nano–Bio Interface. *Nat. Mater.* **2009,** *8* (7), 543–557.

Novoselov, K. S.; Geim, A. K.; Morozov, S. V.; Jiang, D.; Zhang, Y.; Dubonos, S. V.; Grigorieva, I. V.; Firsov, A. A. Electric Field Effect in Atomically Thin Carbon Films. *Science* **2004,** *306* (5696), 666–669.

Paciotti, G. F.; Myer, L.; Weinreich, D.; Goia, D.; Pavel, N.; McLaughlin, R. E.; Tamarkin, L. Colloidal Gold: A Novel Nanoparticle Vector for Tumor Directed Drug Delivery. *Drug Deliv.* **2004,** *11* (3), 169–183.

Pandit, S.; Dutta, D.; Nie, S. Active Transcytosis and New Opportunities for Cancer Nanomedicine. *Nat. Mater.* **2020,** *19* (5), 478–480.

Paradise, J. Regulating Nanomedicine at the Food and Drug Administration. *AMA J. Ethics* **2019,** *21* (4), 347–355.

Poole, C. P.; Owens, F. J. *Introduction to Nanotechnology*; John Wiley & Sons: New York, NY, 2003.

Pradeep, T. *Nano: The Essentials*. Tata McGraw-Hill Education, 2007.

Pradell, T.; Climent-Font, A.; Molera, J.; Zucchiatti, A.; Ynsa, M. D.; Roura, P.; Crespo, D. Metallic and Nonmetallic Shine in Luster: An Elastic Ion Backscattering Study. *J. Appl. Phys.* **2007**, *101* (10), 103518.

Prauchner, C. A. Oxidative Stress in Sepsis: Pathophysiological Implications Justifying Antioxidant Co-therapy. *Burns* **2017**, *43* (3), 471–485.

Read, J. E.; Luo, D.; Chowdhury, T. T.; Flower, R. J.; Poston, R. N.; Sukhorukov, G. B.; Gould, D. J. Magnetically Responsive Layer-by-Layer Microcapsules Can Be Retained in Cells and under Flow Conditions to Promote Local Drug Release without Triggering ROS Production. *Nanoscale* **2020**, *12* (14), 7735–7748.

Reibold, M.; Paufler, P.; Levin, A. A.; Kochmann, W.; Pätzke, N.; Meyer, D. C. Carbon Nanotubes in an Ancient Damascus Sabre. *Nature* **2006**, *444* (7117), 286–286.

Rosenfeld, D.; Senko, A. W.; Moon, J.; Yick, I.; Varnavides, G.; Greguréc, D.; Koehler, F.; Chiang, P. H.; Christiansen, M. G.; Maeng, L. Y.; Widge, A. S. Transgene-Free Remote Magnetothermal Regulation of Adrenal Hormones. *Sci. Adv.* **2020**, *6* (15), eaaz3734.

Schiffelers, R. M.; Koning, G. A.; ten Hagen, T. L.; Fens, M. H.; Schraa, A. J.; Janssen, A. P.; Kok, R. J.; Molema, G.; Storm, G. Antitumor Efficacy of Tumor Vasculature-Targeted Liposomal Doxorubicin. *J. Contr. Releas.* **2003**, *91* (1–2), 115–122.

Shahbazi, R.; Sghia-Hughes, G.; Reid, J. L.; Kubek, S.; Haworth, K. G.; Humbert, O.; Kiem, H. P.; Adair, J. E. Targeted Homology-Directed Repair in Blood Stem and Progenitor Cells with CRISPR Nanoformulations. *Nat. Mater.* **2019**, 1.

Shi, J.; Kantoff, P. W.; Wooster, R.; Farokhzad, O. C. Cancer Nanomedicine: Progress, Challenges and Opportunities. *Nat. Rev. Cancer* **2017**, *17* (1), 20.

Silva, A.; Lee, B. Y.; Clemens, D. L.; Kee, T.; Ding, X.; Ho, C. M.; Horwitz, M. A. Output-Driven Feedback System Control Platform Optimizes Combinatorial Therapy of Tuberculosis Using a Macrophage Cell Culture Model. *Proc. Natl. Acad. Sci. U.S.A.* **2016**, *113* (15), E2172–E2179.

Silva, G. A. Neuroscience Nanotechnology: Progress, Opportunities and Challenges. In *Nanoscience and Technology: A Collection of Reviews from Nature Journals*, 2010; pp 251–260.

Sindhwani, S.; Syed, A. M.; Ngai, J.; Kingston, B. R.; Maiorino, L.; Rothschild, J.; MacMillan, P.; Zhang, Y.; Rajesh, N. U.; Hoang, T.; Wu, J. L. The Entry of Nanoparticles into Solid Tumours. *Nat. Mater.* **2002**, 1–10.

Skotland, T.; Iversen, T. G.; Torgersen, M. L.; Sandvig, K. Cell-Penetrating Peptides: Possibilities and Challenges for Drug Delivery in Vitro and in Vivo. *Molecules* **2015**, *20* (7), 13313–13323.

Torchilin, V. P. Recent Advances with Liposomes as Pharmaceutical Carriers. *Nat. Rev. Drug Disc.* **2005**, *4* (2), pp. 145–160.

Wagner, V.; Dullaart, A.; Bock, A. K.; Zweck, A. The Emerging Nanomedicine Landscape. *Nat. Biotechnol.* **2006**, *24* (10), 1211–1217.

Wang, D.; Wu, L. P. Nanomaterials for Delivery of Nucleic Acid to the Central Nervous System (CNS). *Mater. Sci. Eng.: C* **2017**, *70*, 1039–1046.

Wang, X.; Cai, J.; Sun, L.; Zhang, S.; Gong, D.; Li, X.; Yue, S.; Feng, L.; Zhang, D. Facile Fabrication of Magnetic Microrobots Based on Spirulina Templates for Targeted Delivery and Synergistic Chemo-Photothermal Therapy. *ACS Appl. Mater. Interf.* **2019,** *11* (5), 4745–4756.

Wicki, A.; Witzigmann, D.; Balasubramanian, V.; Huwyler, J. Nanomedicine in Cancer Therapy: Challenges, Opportunities, and Clinical Applications. *J. Contr. Releas.* **2015,** *200,* 138–157.

Wu, L.; Moghimi, S. M. A Nanoengineered Peptidic Delivery System with Specificity for Human Brain Capillary Endothelial Cells. *Nanomed.: Nanotechnol., Biol. Med.* **2016,** *12* (2), 474–475.

Wu, L. P.; Ficker, M.; Mejlsøe, S. L.; Hall, A.; Paolucci, V.; Christensen, J. B.; Trohopoulos, P. N.; Moghimi, S. M. Poly-(amidoamine) Dendrimers with a Precisely Core Positioned Sulforhodamine B Molecule for Comparative Biological Tracing and Profiling. *J. Contr. Releas.* **2017,** *246,* 88–97.

Wu, L. P.; Wang, D.; Li, Z. Grand Challenges in Nanomedicine. *Mater. Sci. Eng.: C* **2020,** *106,* 110302.

Wu, P. H.; Gambhir, S. S.; Hale, C. M.; Chen, W. C.; Wirtz, D.; Smith, B. R. Particle Tracking Microrheology of Cancer Cells in Living Subjects. *Mater. Today,* **2020.**

Xu, F.; Ma, D.; MacPherson, K. P.; Liu, S.; Bu, Y.; Wang, Y.; Tang, Y.; Bi, C.; Kwok, T.; Chubykin, A. A.; Yin, P. Three-Dimensional Nanoscopy of Whole Cells and Tissues with in Situ Point Spread Function Retrieval. *Nat. Methods* **2020,** *17* (5), 531–540.

Xu, X.; Ho, W.; Zhang, X.; Bertrand, N.; Farokhzad, O. Cancer Nanomedicine: from Targeted Delivery to Combination Therapy. *Trends Mol. Med.* **2015,** *21* (4), 223–232.

Yamamoto, V.; Suffredini, G,; Nikzad, S.; Hoenk, M. E.; Boer, M. S., et al. From Nanotechnology to Nanoneuroscience/Nanoneurosurgery and Nanobioelectronics. A Historical Review of Milestones. In *The Textbook of Nanoneuroscience and Nanoneurosurgery*; Kateb, B., Heiss, J. D., Eds.; CRC Press: Boca Raton, FL, 2014.

Yang, Y.; Asiri, A. M.; Tang, Z.; Du, D.; Lin, Y. Graphene Based Materials for Biomedical Applications. *Mater. Today* **2013,** *16* (10), 365–373.

Zhang, C.; Ni, D.; Liu, Y.; Yao, H.; Bu, W.; Shi, J. Magnesium Silicide Nanoparticles as a Deoxygenation Agent for Cancer Starvation Therapy. *Nat. Nanotechnol.* **2017,** *12* (4), 378.

Zhou, Q.; Shao, S.; Wang, J.; Xu, C.; Xiang, J.; Piao, Y.; Zhou, Z.; Yu, Q.; Tang, J.; Liu, X.; Gan, Z. Enzyme-Activatable Polymer–Drug Conjugate Augments Tumour Penetration and Treatment Efficacy. *Nat. Nanotechnol.* **2019,** *14* (8), 799–809.

Zhu, Y.; Liao, L. Applications of Nanoparticles for Anticancer Drug Delivery: A Review. *J. Nanosci. Nanotechnol.* **2015,** *15* (7), 4753–4773.

Zou, W.; Rohatgi, N.; Brestoff, J. R.; Moley, J. R.; Li, Y.; Williams, J. W.; Alippe, Y.; Pan, H.; Pietka, T. A.; Mbalaviele, G.; Newberry, E. P. Myeloid-specific Asxl2 Deletion Limits Diet-Induced Obesity by Regulating Energy Expenditure. *J. Clin. Invest.* **2020,** *130* (5), 2644–2656.

CHAPTER 2

APPLICATION OF NANOTECHNOLOGY IN DRUG DELIVERY

MUZAFAR AHMAD RATHER[1,2*], SHOWKEEN MUZAMIL BASHIR[1*], SHOWKAT UL NABI[3], SALAHI UDDIN AHMAD[4,5], JIYU ZHANG[4], and MINAKSHI PRASAD[6]

[1] *Division of Veterinary Biochemistry, F. V. Sc. & A.H, SKUAST-Kashmir, Shuhama, J&K, India*

[2] *CSIR-Indian Institute of Integrative Medicine (IIIM), Sanat Nagar Srinagar, J&K, India*

[3] *Division of Veterinary Clinical Medicine, Ethics and Jurisprudence, F. V. Sc. & A.H, SKUAST-Kashmir, Shuhama, J&K, India*

[4] *Lanzhou Institute of Husbandary and Pharmaceutical Sciences, Chinese Academy of Agricultural Sciences, Lanzhou, China*

[5] *Faculty of Veterinary, Animal and Biomedical Sciences, Sylhet Agricultural University, Sylhet, Bangladesh*

[6] *Department of Animal Biotechnology, College of Veterinary Sciences, Lala Lajpat Rai University of Veterinary and Animal Sciences, Hisar, Haryana, India*

Corresponding authors. E-mails: muzafarbiochem.jh@gmail.com; showkeen.muzamil82@gmail.com

ABSTRACT

Nanotechnology has potential to bridge the gaps of biological and physical sciences by applying nanostructures and nanophases at various applications of biomedical science, more explicitly in nanomedicine. Nanomedicine is of major concern for biomedical application owing to fascinating applications in the field of drug delivery and disease diagnosis. Nanomedicine and nano-delivery systems are moderately latest fields but quickly emerging science where substances in the nanorange are exploited to offer a means for diagnostic tools or to carry therapeutically active substances to specific targeted sites in a controlled manner. In this chapter, the benefits of nanotechnology in drug-delivery applications are examined. The transport of therapeutic agents to their desired sites occurs by employing different nanosystems. Over the several years, drug-delivery systems more precisely the nanosystems have emerged into diverse forms of second-generation drug-delivery systems, including nanobots, injectable nanoparticle generator, nanoghosts, niosomes, nanoclews, carbon nanotubes, nanocrystals. A detailed description of these systems has been also included. As nanotechnology offers multiple benefits in treating various human diseases, we have provided an updated review of current advances in the field of nanomedicines and nano-based drug-delivery systems through inclusive inspection of the discovery and application of nanomaterials in improving the efficacy of both new and old drugs.

2.1 INTRODUCTION

The definition of drug-delivery system put forth by the National Institute of Health in the United States is underlined as "Formulation of a device that enables the introduction of therapeutic substances in to the body and improves efficiency and safety by the controlling the rate, time and place of release of drug in the body" (Wanigasekara and Witharana, 2016). This drug delivery spans several sequential steps such as introduction of therapeutically active substance (drug/gene) into the human body utilizing bioacceptable carrier (vehicle) thus contributing a broad range of functional platforms for smart application in biotechnology and medicine, the subsequent transport of the active ingredients to the site of action, and finally the release of the pharmacologically active ingredient from the carrier to induce the biological effect. Drug delivery has a remarkable prospective in the management of several human diseases. Tremendous developments have been in the field of drug delivery to convey pharmacologically active principle to its target location for the management of human diseases. To achieve these goals, several drug-delivery systems were designed and effectively exploited from the recent past; nevertheless, there are still certain challenges that need to be addressed and a cutting-edge innovation should be produced for effective delivery of a medication to its target site. Nanotechnology-based drug-delivery systems are currently being investigated and are expected to assist the advanced system of drug delivery. The application of nanotechnology to achieve the innovation in health care is more commonly defined as nanomedicine. Drug-delivery procedures that utilize nanosystems allow the development of novel platforms for the efficient transport, site specific, controlled release, and target-oriented delivery of drug molecules. Nanomedicine is an interdisciplinary field, where nanoscience, nanoengineering, and nanomedicines are complex engineered nanoscale substances, more commonly used for management and targeting of diseases owing to their enumerable therapeutic advantages. Nanotechnology has gained much interest since its beginning several decades ago both in the pharmaceutical industry and academics. Much attention is focused on the areas of biotechnology and medical sciences from diagnosis, monitoring, control, prevention, and treatment of the diseases (Tinkle et al., 2014). The most common applications of nanotechnology for medical use are schematically represented in Figure 2.1.

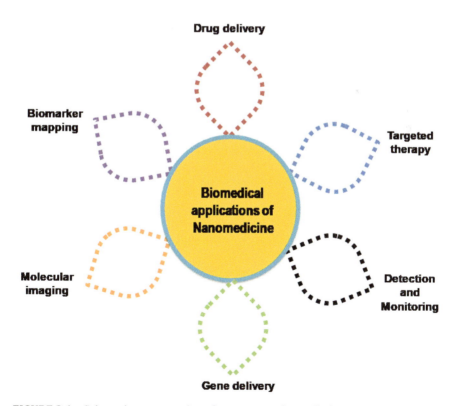

FIGURE 2.1 Schematic representation of nanosystems for medical use.

The interest in the field has led to the inception of new nanotechnology platforms and breakthroughs for medical use (Schäfer-Korting, 2010). Nanomedicine is a fairly latest and rapidly developing field that combines nanotechnology with the biomedical and pharmaceutical sciences (Ventola et al., 2012; Bobo et al., 2016; Caster et al., 2017). According to of European Medicines Agency (EMA) working group, nanomedicines are defined as purposely designed systems for clinical use, with at least one component at the nanorange, that leads to definable properties and characteristics, related to the specific nanotechnology application and characteristics for the intended use (such as the route of administration and dosage) and associated with the expected clinical advantages of nanoengineering (e.g., preferential organ/tissue distribution; Ossa, 2014). Nanotechnology-based drug-delivery systems with different compositions and biological properties exhibiting different applications for development of nanosystems have been extensively explored for drug/gene delivery purposes.

Here, we will discuss the potential advantages of nanomedicine, recent developments in nanotechnology platforms for drug delivery, and the role of nanomedicine in the management of several human diseases.

2.2 POTENTIAL ADVANTAGES OF NANOMEDICINE

Nanotechnology has revolutionized the field of pharmaceutics over the past several years. Nanotechnology can overcome some of the common limitations of the conventional medication. The potential advantages (Fig. 2.2) of nanomedince over the bulk formulation include the following.

i. Nanodrug-delivery improves the safety and efficacy of conventional therapeutics by controlled and site-specific releases, favoring a preferential biodistribution of the drug (e.g., in areas of tumor) and also improves drug transport across biological barriers (Chan, 2006; Zhang et al., 2012; Ossa, 2014).

ii. Nano preparation of a drug improves bioavailability by directly targeting the drug to the site required (Sajja et al., 2009, Galvin et al., 2012)

iii. The large ratio of surface area/volume and greater reactivity of nanodrug reduces the dose as well as dosing frequency, which in turn can improve patient compliance (Bawa, 2011; Galvin et al., 2012; Muzafar et al., 2016)

Application of Nanotechnology in Drug Delivery

iv. Address the unmet medical needs by reducing the toxicity profile of the drug that otherwise could not be used in bulk form owing to high adverse effects (e.g., Mepact).

v. Nanodrugs also possess the large surface area that allows the increase in the dissolution rate (due to increased surface area), saturation of solubility, and drug uptake by cell, thus enhanced the *in vivo* performance (Bawa, 2011; Galvin et al., 2012).

vi. Taking together the drug encapsulation, drug release kinetics, and surface alterations so as to improve the therapeutic targeting or bioavailability could improve the efficacy of therapeutic agent several fold compared with conventional bulk counterparts (Bharali and Mousa, 2010; Sajja, 2010).

vii. Delivery of drugs through a mechanism, called enhanced permeability and retention (EPR) effect (an example of passive drug targeting), can also transport large doses of therapeutic agents into malignant cells while avoiding the damage to healthy normal cells (Bharali and Mousa, 2010; Sajja et al., 2009).

viii. The conjugation or encapsulation of a therapeutically active agent to nanoparticles can promote more desirable pharmacokinetics, including absorption, distribution, metabolism, and elimination (ADME), independent of the molecular structure of the active ingredient (Havel et al., 2016).

ix. Nanodrug delivery exploits multiple mechanisms of action, for instance, nanomag, multifunctional gels (Soares et al., 2018).

x. Nanodrugs can be much inexpensive than the bulk forms of drugs (Bosetti et al., 2013).

xi. Drug release from a nanoformulation can occur at a constant rate over the desired timescale (Karimi et al., 2016).

Nanodrugs significantly expand a scope for drug development opportunities; however, some safety issues have arisen. The physicochemical properties of the nanoformulation that can prompt the pharmacokinetics being altered, in particular the ADME, the prospective for more easily crossing the biological barriers, adverse effects, and their accumulation in the environment and in the human body, are some instances of the issues over the use of the nanomaterials (Bleeker et al., 2013; Tinkle et al., 2014).

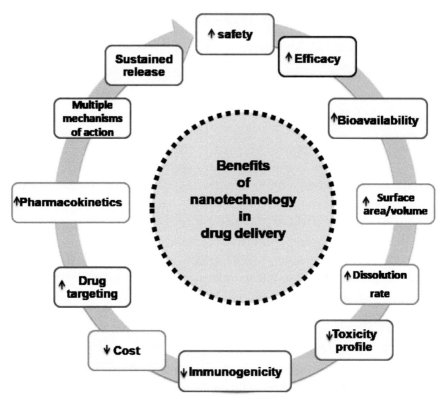

FIGURE 2.2 The potential benefits of nanotechnology-based drug-delivery systems over the conventional drugs.

2.3 NANOCARRIERS FOR DRUG DELIVERY

Nanocarriers have been broadly researched as they indicated promising results in the field of drug delivery. For medical applications, the carrier for nanoformulation should be biodegradable (Masood, 2016). A biodegradable polymer undergoes degradation on in vivo use. The choice for polymer/carrier for nanoparticle constitution should be such that its degradation products need to be bio-acceptable, nonimmunogenic, and nontoxic to induce a minimal inflammatory response. The need for biocompatible and biodegradable nanocarrier is to have a tendency to get self-assembled in nanometric dimensions (Letchford et al., 2007). Biocompatible nanoparticles have been extensively exploited as carrier (vehicles) for drug delivery as they offer a series of benefits, including administration of hydrophobic drugs to protect the unstable compounds against degradation. Drugs can be loaded into the polymer matrix of a nanoparticle by a mechanism involving adsorption, dispersion, or encapsulation. In this context, a noticeable distinction can be drawn between the nanosystems (Fig. 2.3) and can be categorized as nanospheres (matrix system) and nanocapsules (reservoir system), micelle, and polymersomes (Letchford et al., 2007; Rao et al., 2011; Ezhilarasi et al., 2013).

2.3.1 NANOSPHERES

Nanospheres are the matrix system of amorphous or crystalline particles with the size range of 10–200 nm in diameter (Singh et al., 2010). Owing to their small size, these nanoparticles can be administered locally, orally, and systemically. Nanospheres are synthesized with PLA (poly(lactic acid)), PLGA (poly(lactide-*co*-glycolide)), chitosan, alginate, and other polymer combinations with emulsion evaporation technique (Feng *et al.,* 2000). Emulsion polymerization and interfacial polymerization are used for preparation of nanospheres (Letchford and Burt, 2007). The shape of nanospheres needs not to be essentially spherical. The therapeutic agent can be conjugated on the surface (absorbed) of nanosphere or wrapped (entrapped), dispersed (dissolved) within the colloidal matrix both by physical entrapment or chemical bonding. The tissue selectivity of the polymeric nanosphere can be altered using different compounds or proteins to improve the target delivery. These modifications impart the ability to nanoparticle to cross the blood–brain barrier by specific mechanisms for delivery to the central nervous system after administration through various routes (Grumezescu, 2016).

2.3.2 NANOCAPSULES

Nanocapsules are the reservoir system and can be categorized as submicroscopic vesicular structures with an oil-filled cavity jacketed by a narrow polymeric watery or oily envelope, a core–shell structure, with a size ranging from 50 to 300 nm (Grumezescu, 2017). In other words, these nanostructures consist of a solid or liquid core in which the drug molecules are placed into hollow center of the polymer. Nanocapsules are mostly spherical in shape. Both synthetic and natural polymers can be exploited for nanocapsule preparation (Kothamasu et al., 2012). Polymers such as hydroxypropyl-methylcellulose, diacyl-*b*-cyclodextrin, hydroxypropyl methylcellulose, ethyl oleate, poly-(alkylcyanoacrylate), poly-(d,l-lactide), benzyl benzoate, and vegetable or mineral oils have been used for nanocapsule preparation (Kondiah et al., 2018). The common preparation techniques for monomer and polymer nanocapsules include interfacial nanodeposition and interfacial nanodeposition. Nanocapsules possess many advantages, including solubility of poorly soluble drugs gets improved. These nanostructures are suitable for delivery of both hydrophilic and hydrophobic drug molecules; drugs stability gets enhanced by protecting the drug molecule from the enzymatic or chemical degradation. PK profile of the drugs improves, drugs molecule becomes feasible for oral administration and allows controlled release (Erdoğar et al., 2018). Moreover, nanocapsules have high circulation time in the plasma, easy surface modification, and high drug-loading capacity (Kondiah et al., 2018).

2.3.3 NANOMICELLES

Nanomicelles are self-assembled colloidal nanosized dispersions composed of amphiphilic monomers. The constituent monomer consists of hydrophobic head and a long hydrophilic tail. The hydrophobic heads of the monomers interact with the hydrophobic drugs/agents and form a core structure, whereas the hydrophilic tails interact with the water and improve water solubility. This monomer orientation stabilizes the nanomicelles in state of minimum free energy. These nanoconstructs have a size range of 10–50 nm. Nanomicelles have proved to be efficient carriers for drug delivery owing to their advantages, including ability to protect the drug molecule and prevent its degradation, minimize drug toxicity, and improve drug permeation through biological barriers, ultimately resulting in enhanced bioavailability (Gabizon et al., 2003). Nanomicelles have been widely used in healthcare settings for imaging and diagnostics, including magnetic resonance imaging, computed tomography, and near-infrared fluorescent (Trinh et al., 2017). Nanomicelles gained much attention in the scientific community and this interest led to the market approval by Food and Drug Administration (FDA) for Genexol® PM, a nanomicellar formulation of paclitaxel for the treatment of breast, lung, and ovarian cancer in 2007 (Kesharwani et al., 2019). The other advantages of nanomicelle include ease of preparation, economical, enhanced solubility in aqueous media, including unstirred water layer of the intestine and controlled drug release profile (Shakeri et al., 2016).

2.3.4 POLYMERSOMES

Polymersomes are colloidal spherical hollow structures formed by the self-assembly of amphiphilic molecules with an aqueous core surrounded by a polymeric bilayer membrane. These nanostructures are potential candidates for next-generation drug-delivery systems. Polymersomes are synthetic mimics of nature and synthetic analogs to liposomes found in all living cells. Polymersomes can carry hydrophilic drug molecules within the aqueous core of hydrophobic drug molecules within the membrane bilayer or a combination of both and thus providing better treatment effects than a single drug when used alone (Thambi et al., 2012). Polymersomes are rigid, low membrane permeability (Discher et al., 1999), and with least or no immunogenicity (Anajafi and Mallik, 2015). Polymersomes vesicles can be functionalized with specific ligands for specific cell receptors for targeted drug delivery through surface. Stimuli-responsive chemistry can be used to achieve the sustained drug release from polymersome. Owing to these benefits, polymersomes have been used in various medical applications such as drug delivery (Anajafi and Mallik, 2015; Hu et al., 2017), gene and protein delivery (Onaca et al., 2009), imaging (Tanner et al., 2011), and diagnostics (Brinkhuis et al., 2011).

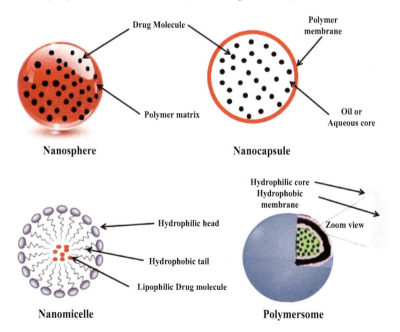

FIGURE 2.3 Schematic diagram showing the types of nanoparticles investigated in nanomedicine

2.4 NOVEL NANOTECHNOLOGY-BASED DRUG-DELIVERY SYSTEMS

Many advanced nanotechnology platforms for the design and development of effective drug-delivery systems came into reality for obtaining the goal of targeted drug delivery. Both biocompatible and inorganic carriers have been explored in these systems and are gaining much scientific attention. These drug-delivery systems improve the solubility of hydrophobic drug molecules, protect drug degradation of the delicate drugs, and promote the controlled release of the drug molecules in response to stimuli. Some of the novel drug-delivery systems are briefly discussed in the following sections.

2.4.1 NANOBOTS

Nanobots are nanomotor robots that carry out a very specific function. These nanodevices are self-driven, made of biodegradable bionanocomponents, and carry cargo to the target precise location in the body. These submicron-sized devices are in the size range of ~50–100 nm (Subramani and Ahmed, 2017). Nanobots are efficient drug-delivery systems. Drug administered in their conventional forms works through the entire body before reaching to the diseased area. However, drugs loaded into the nanobots can be effectively targeted to the desired location (diseased site) thereby improving the drug efficacy and preventing the drug-associated toxicity to the other body sites.

The general assembly nanobot may include motors, power source, sensors, manipulators, and molecular computers. Several nanobots have sensors that are chemical in nature and sense the target molecule. In response, it would emit an electrical signal relative (proportional) to the sensed quantity in the biological system. This electrical impulse would be received by a programmed microprocessor that controls the mechanism, direction, and velocity of the nanobot.

The design of nanbots includes a 5- to 10-atom thick outer carrier wall and ~50- to 100-nm wide inner drug-loaded cell. On in vivo use, the thin wires on the walls of the nanobots emit electrical signals in response to the damage that cause the walls to degrade and drug gets released in a controlled manner and in a precise mechanism in response to electrical signal. Nanobot drug-delivery systems hold a significant promise in curing life-threatening diseases such as cancers in future and in optimizing the drug delivery. Future nanobots could be designed and programmed to repair specific cell damage, similar to those of antibodies. Sensor nanobots implanted beneath the skin

Application of Nanotechnology in Drug Delivery   49

and coated with human molecules onto the microchips can be programmed
to send out an electrical impulse signal in response to blood glucose and may
help in monitoring the blood sugar levels. Elan Pharma has already started
using this technology in their drugs Merck's Emend and Wyeth's Rapamune
(Subramani and Mehta, 2018) (Table 2.1).

TABLE 2.1 List of Nanobots and Their Applications.

Nanobot	Significance	Reference
Merck's Emend	Antinausea drug (aprepitant) approved by the FDA in March 2003 for chemotherapy. It was marketed in the United States in April 2003.	Subramani and Mehta (2018)
Wyeth's Rapamune		Khan et al. (2015)
Respirocytes	Artificial mechanical version of RBC with a capacity of delivering oxygen to cells by 236 fold more than natural RBC, applications comprise isubstitution for blood transfusion; management for perinatal and neonatal disorders, and a variety of lung diseases and conditions and anemia; effective against tumor therapies and diagnostics, certain aggressive cardiovascular and neurovascular procedures; prevents asphyxia; helpful in maintaining artificial breathing in adverse environments; and other sports, veterinary, battlefield, and other applications.	Freitas et al. (1998)
Microbivores	Kills microbial pathogens in the human bloodstream using a digest and discharge protocol.	Freitas (2005)
Clottocytes	Ability to heal the wounds by making fiber mesh upon command at the wound site	Gao et al. (2015)
Chromallocytes	Capable of reversing the effect of genetic diseases by replacing its entire chromosome.	Gao et al. (2015)

2.4.2 NANOGHOSTS

Nanoghost is among the most recent technology aimed for smart drug
delivery of therapeutics (drug or gene). Nanoghosts are basically vesicles
obtained from naturally functionalized mammalian cell surface membranes
of whole biological cells such as mesenchymal stem cells (MSCs) that are
deficient in cytoplasmic cell organelle. These drug-delivery systems have
many benefits over the other drug-delivery systems such as loading issues
are nullified, tumor-specific immune responses are prevented, nanoparticle
stability, and drug release profiles are improved.

2.4.3 NANOCLEWS

Nanoclew or nanococoon is a self-assembled DNA-based bio-acceptable drug-delivery system. The mechanism of rolling circle amplification has recently emerged as a robust technique to generate the regularly patterned single-stranded self-woven structure (Sun and Gu et al., 2016) called DNA nanococoon. The self-assembly of DNA makes a yarn or cocoon- or a clew-like structure by rolling circle amplification. On contact of nanoclew with a cell, it gets absorbed onto cell surface, swallowed and wrapped into an endosome. The positive charge of nanoclews cleaves the endosome membrane and sets nanoclew free that makes its way on its own to target site. For instance, DNA nanoclews have been used for efficient delivery of CRISPR–Cas9 for genome editing (Sun et al., 2015). CRISPR–Cas9 gene-editing tool on its release from DNA nanoclew moves themselves to the nucleus and for loading and delivering anticancer drug (Sun and Gu, 2016).

2.4.4 NANOCLUSTERS

Nanoclusters signify a bunch of self-assembled nanoparticles with at least one dimension and size distribution in nanoscale range. The composition is either the single atom of an element or collection of atoms of different elements in stoichiometric ratios. These atoms are linked with each other by ionic, covalent, metallic, hydrogen bonds or by van der Waals forces (Sinha et al., 2017). Precious metal nanoclusters (typically in the size range from 1 to 3 nm and containing $\sim$ 10 to $\sim$1000 metal atoms) represent a bridge state between small molecules ($<\sim$1 nm) and metal nanoparticles ($>\sim$3 nm) (Fornasiero and Cargnello, 2017). Bulk materials have constant physical properties whereas the physical properties of nanoclusters depend on their size (Sinha et al., 2017). The properties of nanoclusters are particularly interesting and have gained much scientific attention due to their unique physical and chemical properties that depend on size, shape, composition, and environment and influence properties of the cluster. With their size approaching the Fermi wavelength of electrons, metal NCs possess molecule-like properties and excellent fluorescence emission. Owing to their ultrasmall size, strong fluorescence, and excellent biocompatibility, they have been widely studied in environmental and biological fields concerning their applications. The fluorescence properties of nanoclusters (such as silver, gold, or magnetic particles) owing to their molecular-like size have been applied in biomedical applications for biosensing and bio-imaging and also to drug delivery.

Application of Nanotechnology in Drug Delivery 51

2.4.4.1 NANOBUBBLES

Nanobubbles are stable gas-filled cavity in the aqueous solution with several unique physical properties and more often stabilized by polymeric/lipid shells. The reason for their long-lasting property could be explained by the low internal pressure and surface tension, which may be due to the charged gas/liquid interface. Nanobubbles containing a nanoparticulate anticancer drug, doxorubicin, are integrated inside the magnetic poly(lactic-*co*-glycolic acid) bubble filled with perfluorocarbon gas to deliver the particles into the tumor lymph node to induce apoptosis. Antibody-modified nanobubbles using FC region–binding polypeptides have been used for ultrasound imaging (Hamano et al., 2019). Nanobubbles have been investigated for theranostic strategy for contrast-enhanced ultrasound and drug (doxorubicin)-loaded bubbles in osteosarcoma (Kuo et al., 2019).

2.4.5 EXOSOMES

Exosomes are tiny extracellular cell-derived phospholipid nanoparticles that transfer cargo (information) between the cells and thus function as signalo-somes, transporting small amounts of bioactive molecules to specific recipient tissues (Syn et al., 2017). Due to their endogenous nature, they have been used as novel and most promising next-generation nanovesicles for targeted drug/gene/protein/microRNA delivery as well as for disease diagnosis. Exosomes are in the size range of 40–200 nm. Exosomes vesicles are derived from the healthy cells of patients and thus exhibit a unique property of cell tropism (interact with the specific cell types with exceptional ability by displaying receptors in the membranes) toward the originated cells. Exosomes are next-generation drug-delivery system with high drug-carrying capacity (Lv et al., 2015). Exosomes are ideal for designing nonimmunogenic and personalized therapeutic approaches owing to their biological origin and their ability to cross the biological membranes.

2.4.6 INJECTABLE NANOPARTICLE GENERATOR

Injectable nanoparticle generator (iNG) is the novel drug-delivery system developed by Houston Methodist Research Institute in Texas (Xu et al., 2016). The drug is conjugated to the biodegradable polymer particle by means of a pH-sensitive cleavable linker that loads the polymeric drug (pDrug) into iNPG to assemble iNPG-pDrug. This assembly is injected intravenously and the assembly is internalized by host cells by natural tropism. Due to the acidic pH inside the cell, the pDrug gets released from iNPG and pDrug spontaneously forms the pDrug nanoparticles in aqueous solution. Afterwards, pDrug nanoparticles get transported to the perinuclear region and cut into Dox, thereby preventing drug efflux by drug efflux pumps. Xu et al. (2016) created the first iNPG drug with doxorubicin (Dox) with assembly comprising iNPG-pDox that transformed the treatment of metastatic triple-negative breast cancer by creating the first drug comprising doxorubicin conjugated to successfully eliminate lung metastases in mouse models of metastatic breast cancer with cure rate in at least 50% mice.

2.4.7   NANO-TERMINATORS

Investigators at North Carolina State University in collaboration with University of North Carolina at Chapel Hill have developed a new drug-delivery technique comprising biodegradable liquid metal nano-terminators that target cancer cells and improves the effect of anticancer drugs (Lu et al., 2015). The assembly comprises the drug-loaded nanodroplets made of a liquid-phase eutectic gallium–indium core and a thiolated polymeric shell and thiolated hyaluronic produced by sonication. These nanoparticles loaded with Dox with an average size of 107 nm bear the ability to fuse with the plasma membrane of the cancer cell and on internalization subsequently degrade under an acidic condition inside the endosome and Dox is released in acidic endosomes. The formulation is also loaded with hyaluronic acid, a tumor-targeting ligand, leading to enhanced chemotherapeutic inhibition toward the xenograft tumor-bearing mice.

2.4.8   DENDRIMERS

Dendrimers are repetitively branched multivalent nanopolymers radially symmetric globular structures. The origin of the term Dendron is from the Greek word which translates to "tree." The synonymous term "dendrimer" means a cascade molecule. Dendrimers were first introduced in 1985 by Tomalia and Newkome (Fréchet, 2002). The molecular architecture of dendrimers comprises three distinct components: (1) a symmetric core, (2) an inner shell or branches, (3) and an outer shell with terminal functional groups. The functional groups present at the outer surface of the polymeric

Application of Nanotechnology in Drug Delivery

molecule define the nucleic acid complexation or drug entrapment effi-cacy. Dendrimers exist in diverse forms and each form possesses different biological roles such as self-assembling, electrostatic interactions, chemical stability, polyvalency, low cytotoxicity, and solubility (Abbasi et al., 2014). The fascinating properties and unique structures make dendrimers a good choice in nanomedicine.

Two methods have been more commonly adopted for the dendrimers synthesis. In the divergent approach (Tomalia et al., 1985), the growth of the nanopolymer initiates at the core by adding the building blocks to the central core and proceed radially outward toward the dendrimer periphery. In the convergent approach (Hawker and Frechet, 1990), the growth of dendrimer starts at the exterior resulting in the shell synthesis first and then proceeds inward.

2.4.9 LIPOSOMES

Liposomes are artificial spherical vesicles created from one or more phos-pholipid bilayer (Akbarzadeh et al., 2013). Lipids are amphiphilic molecules containing both hydrophilic and hydrophobic parts. When these amphiphilic monomers are placed in an aqueous solution, the nonfavorable interactions of the hydrophobic parts of the molecule with the water force the self-assembly of lipids with an aqueous core surrounded by a lipid bilayer often in the form of spherical liposomes.

Liposomes were first described in the mid-1960s and are still fascinating drug-delivery systems. Despite extensive research from nearly last six decades and the numerous positive results at preclinical and clinical levels of liposome, this drug-delivery system has advanced incrementally (Sercombe et al., 2015).

Liposomes have been used to enhance drug bioavailability, drug absorp-tion and to reduce drug-associated adverse effects. The benefits of using the liposomes as drug-delivery system include ability to encapsulate both hydro-philic and hydrophobic materials and have been used since long for delivery of therapeutic agents. A novel class of photo-triggerable liposomes consisting of dipalmitoyl phosphatidylcholine and photopolymerizable diacetylene phospholipid efficiently released calcein (a water-soluble fluorescent dye) upon UV (254 nm) treatment. The liposome induced the improved killing of a coculture (Raji and MCF-7) following light-triggered release of an encap-sulated anticancer agent (Dox) from photosensitive liposomes (Yavlovichet al., 2011). A liposome-based codelivery system composed of a fusogenic liposome coencapsulating ATP-responsive elements and anticancer drug (Dox) and a liposome containing ATP was designed for ATP-mediated drug release induced by fusion of liposome (Mo et al., 2014). The codelivery system helped in keeping the time bomb for killing the cancer cell and its detonator apart till penetrate the cancer cell, where both combine to destroy the cell.

2.4.10   NIOSOMES

Niosomes are synthetic microscopic vesicles also called nonionic surfactant vesicles with sizes ranging between 20 nm and 50 µm. Niosomes are promising novel drug-delivery systems and exhibit an ability to enhance the solubility and stability of therapeutic molecules (Gharbavi et al., 2018). Niosomes are biodegradable, physiologically acceptable, and nonimmunogenic drug-delivery carriers. Moreover, niosomes are stable, long-lasting and enable the targeted drug delivery at the site required in a controlled manner (Mahale et al., 2012).

Niosomes are formed by self-assembly of nonionic surfactants and cholesterol in an aqueous phase. Self-association of surfactant molecules into vesicular structures was originally reported in the 1970s. Niosomes are prepared upon hydration and the lamellar structures are obtained by combining nonionic surfactant of the alkyl or dialkyl polyglycerol ether class with cholesterol (Malhotra and Jain, 1994), using some energy (e.g., heat/physical agitation) resulting in the formation of a closed bilayer vesicle in aqueous phase. Hydrophobic parts of the bilayer are pointed away from the aqueous phase and the hydrophilic heads interact with the aqueous phase.

Niosomes have exploited for transport of hemoglobin inside the blood (Radha et al., 2013), have been applied as vaccine adjuvants (Brewer et al., 1996), and may be developed as probable anticancer chemotherapy transporter for brain-targeted delivery of the neuropeptide such as dynorphin-B (Bragagni et al., 2014) and doxorubicin (Bragagni et al., 2012).

2.4.11   CARBON NANOTUBES

Carbon nanotubes (CNTs) are composed of carbon atoms covalently bonded to each other in a hexagon with each carbon atom bonded to other three carbon atoms. CNTs are an allotropic form of carbon related to the fullerene

Application of Nanotechnology in Drug Delivery 55

family. The diameter of CNT could be <1 nm and length up to several centimeters and the ratio of length-to-diameter could be > 1,000,000. CNTs are tough like bulky balls but not brittle. CNTs can be bent and they possess an ability to regain their original shape when released.

Three most common approaches for preparation of CNTs include (1) arc discharge (Journet et al., 1997), (2) laser ablation (Thess et al., 1996), and (3) chemical vapor deposition (Kong et al., 1998). The fundamental requirements for synthesizing CNTs include a catalyst, a carbon source and energy. In all these strategies, the supply of energy to a carbon source results in the generation of small fragments (groups or single C atoms) that then recombine to generate CNT. The source of energy for arc discharge is electricity, heat from a furnace (~900°C) for chemical vapor deposition, or the high-intensity light from a laser for laser ablation.

CNTs possess distinct inner and outer surfaces for functionalization and provide improved drug loading capacity over the other nanoparticles. For instance, several anticancer drugs such as Dox, epirubicin, quercetin, cisplatin, methotrexate, and paclitaxel have been conjugated with functionalized CNTs and successfully tested in vitro and in vivo (Li et al., 2010; Lay et al., 2011; Madani et al., 2011; Elhissi et al., 2012). The experimental data obtained so far on cytotoxicity CNTs is lacking in comparability, and sometimes there is controversy about it (Zhu and Li, 2008), thereby demanding more research on toxicity evaluation of CNTs.

2.4.12 *GRAPHENE*

Graphene is an allotrope of carbon in the form of monolayer of atoms tightly bound in a two-dimensional hexagonal lattice in which one atom forms each vertex. Generally, three major approaches have been used to synthesize graphene/nanoparticle composites, including pregraphenization, syngraphenization (usually called the one-pot method), and postgraphenization (Kakaei et al., 2019). Graphene and its modified nanostructures have emerged novel nanomaterial and attracted much scientific attention in recent past for nanomedicine applications due to ease scale-up and cost-effective preparation and other physical and chemical features, including biocompatibility, high specificity, large surface area, and ease of conjugation to antibodies/peptides/proteins (Tabish, 2018). The other novel application includes grapheme-based sensors for the detection of cancer cells (Gu et al., 2019) or protein biomarkers (Xu et al., 2019).

2.5 NANOMEDICINE IN CLINICS

The approval procedure for nanomedicines for human use is currently governed by the US FDA and is principally same as that for any bulk medication. Following the discovery/invention of the material, its evaluation from preclinical phase (in animal) to clinical phase (in humans) to demonstrate efficacy, safety, toxicity profile, and to elucidate suitable dosage, the whole process is estimated to take nearly 10–15 years and approximately cost of $1 billion/new drug. Drug efficacy and safety have been the two essential criteria for a nanoformulation to reach the market. Yet, the greater part of the nanomedicines fails to achieve these prerequisites and consequently their fruitful commercialization (Kaur et al., 2014). From last few decades, quite a number of nanomedicine products have been developed and have received the approval by the US FDA and the European Medicines Agency (EMA) for clinical practice for a number of indications and there are also a number of nanomedicines in clinical trials (Anselmo and Mitragotri, 2016; Sainz et al., 2015; Ragelle et al., 2016), enabling more effective therapeutic interventions reaching the market (Table 2.2).

2.6 CONCLUSION

Drug-delivery systems based on nanotechnologies are being designed and eventually the nanodrug development has advanced significantly in the recent past owing to many common reasons; nanotechnology improves the pharmacokinetics and therapeutic properties of bulk drug counterparts. Nanodrugs offer a novel solution in drug delivery by reducing the adverse effects associated with the highly toxic drugs and improve the transport of low selectivity drug across the biological barriers as well as to their target sites without their accumulation in the body. Additionally, nanosystems improve bioavailability of drugs, reduce the cost of products, improve patient compliance, and reduce the drug resistance (e.g. microbial infectious diseases), which are some of the common reasons that nanotechnology is used for drug delivery. A number of nanosystems/nanotools, including nanocapsules, nanospheres, polymersomes, dendrimers, micelles, niosomes, liposomes, iNG dendrimers, CNTs, nanobubbles, and nanoclews, have been developed to achieve these goals and a quiet a number of nanodrugs (nanomedicines) have received the approval by US FDA, and many more are in early stages of development

TABLE 2.2 US FDA Approved Nanomedicines for Different Clinical Indications have been Listed from the Perspective of the Class of Nanomaterial and their Therapeutic Application.

Trade name	Manufacturer	Therapeutic agent	Indication(s)	Benefit	Month and year of approval	Reference
Liposome nanoparticles						
Doxil	Janssen	Doxorubicin HCl liposome injection	Used for treatment of acquired immunodeficiency syndrome-related Kaposi's sarcoma, ovarian cancer, breast cancer, and other solid tumors	Increased delivery to disease site, decreased systemic toxicity of free drug	1995	Barenholz (2012)
Curosurf	Chiesi Farmaceutici	Poractant alfa	Treatment of premature newborn with acute respiratory distress syndrome	Increased delivery with smaller volume, decreased toxicity	November, 1999	Chao et al. (2018)
Abelcet	Sigma-Tau	Equimolar mixture of amphotericin B complexed with two lipids	Treatment of invasive mycoses	Abelcet® was clearly demonstrated to be less toxic than amphotericin B desoxycholate	1995 2005 2008	Lister et al. (1996)
AmBIsome	Gilead Sciences	Liposomal amphotericin B	Treatment of server underlying fungal and protozoan (leishmaniasis) infection	Decreased nephrotoxicity	July, 1997	Meunier et al. (1991)
Visudyne	Bausch + Lomb	Liposomal verteporfin	Used for the treatment of patients with predominantly classic subfoveal choroidal neovascularization	Increased delivery to site of diseased vessels, photosensitive release	April, 2000	Bressler and Bressler (2005)
DepoDur	Pacira Pharmaceuticals	Liposomal morphine sulphate	Postoperative analgesic effect	Extended release	May, 2004	Gambling et al. (2005)
DepoCyt	Sigma-Tau	Sustained-release liposomal formulation of the chemotherapeutic agent cytarabine	Treatment of neoplastic meningitis in patients with	improved delivery to tumor site, decreased adverse effects	DepoCyt was received accelerated approval in 1999 and was granted full approval in 2007 by US FDA	Craig (2000)
Marqibo	Spectrum Pharmaceuticals	Sphingomyelin and cholesterol-based nanoparticle formulation of vincristine	Treatment of adult patients with Philadelphia chromosome-negative (Ph-) acute lymphoblastic leukemia (ALL).	Optimize delivery to target tumor lesions and facilitate dose intensification without increasing toxicity, decreased systemic toxicity	2012	Silverman et al. (2013)

TABLE 2.2 (Continued)

Trade name	Manufacturer	Therapeutic agent	Indication(s)	Benefit	Month and year of approval	Reference
Onivyde	Ipsen Biopharmaceuticals	Liposomal irinotecan	Used to treat colon cancer, small cell lung cancer, and metastatic pancreatic cancer	Liposomal formulation improved the pharmacokinetics and therapeutic index of chemotherapeutic agents	October, 2015	US FDA press release (2015)
Vyxeos	Jazz Pharmaceuticals	Liposomal formulation of daunorubicin and cytarabine	Treatment of adults with newly diagnosed acute myeloid leukemia with myelodysplasia-related changes	Increased efficacy through synergistic delivery of coencapsulated agents	August, 2017	Deutsch et al. (2018)
Polymer nanoparticles						
Adagen	Leadiant Biosciences	Pegademase bovine	Severe combined immunodeficiency disease	Longer circulation time, reduced immunogenicity	March 1990	Shams and Kobrynski (2019)
PegIntron	Merck	Pegylated interferon alfa-2b	Hepatitis C and melanoma	Improved stability of protein	August, 2001	Bukowski et al. (2002)
Pegasys	Genentech	Pegylated interferon alfa-2a	Hepatitis B, hepatitis C	Greater protein stability	October, 2002	Foser et al. (2003)
Neulasta	Amgen	Pegfilgrastim, PEGylated form of the recombinant human granulocyte colony-stimulating factor analog filgrastim	Decrease the incidence of infection, as manifested by febrile neutropenia, in patients with nonmyeloid malignancies	Greater protein stability	January, 2002	Crawford (2003)
Somavert	Pfizer	Pegvisomant, a growth hormone receptor antagonist	Acromegaly	Greater protein stability	March, 2003	Van der et al. (2011)
Macugen	Bausch and Lomb	Pegaptanib sodium, a selective vascular endothelial growth factor (VEGF) antagonist	Neovascular age-related macular degeneration	Greater aptamer stability	December, 2004	Tobin (2006)
Oncaspar	Baxalta US	Pegaspargase is L-asparaginase (L-asparagine amidohydrolase) that is covalently conjugated to monomethoxypolyethylene glycol	Treatment of acute lymphoblastic leukemia	Greater protein stability	July, 2006	Dimndorf et al. (2007)

TABLE 2.2 (*Continued*)

Trade name	Manufacturer	Therapeutic agent	Indication(s)	Benefit	Month and year of approval	Reference
Renvela and Renagel	Genzyme	Sevelamer carbonate; and Sevelamer HCl, phosphate binder helps to prevent hypocalcemia (low levels of calcium in the body) caused by elevated phosphorus	Chronic kidney disease	Longer circulation time and therapeutic delivery	October, 2007	Calkins and Harris (2017)
Krystexxa	Horizon	Pegloticase, a PEGylated uric acid specific enzyme indicated for the treatment of chronic gout	Chronic gout	Greater protein stability	September, 2010	US FDA (2010)
Plegridy	Biogen	Pegylated interferon beta-1a	Multiple sclerosis	Greater protein stability	2014	Idec (2015)
Eligard	Tolmar	leuprolide, a luteinizing hormone- agonist, leuprolide acetate and polymer	Decrease testosterone to suppress tumor growth in patients with hormone-responsive prostate cancer	Sustained release	December, 2014	Sartor (2003)
Adynovate	Shire	Full-length coagulation factor VIII molecules linked to polyethylene glycol	Treatment of hemophilia A in children and in adults	Greater protein stability, longer half-life	December 22, 2016	Turecek et al. (2016)
Rebinyn	Novo Nordisk	Coagulation Factor IX (Recombinant), glycol polyethylene glycolated	Hemophilia B	Longer half-life, greater drug levels between infusions	June, 2017	Davis et al. (2019)
Copaxone	Teva	Glatimer acetate which is a combination of four amino acids (proteins) that affect the immune system	Multiple sclerosis to reduce the frequency of relapses, but not for reducing the progression of disability	Controlled clearance	October, 2017	Ross et al. (2017)
Zilretta	Flexion Therapeutics	Triamcinolone acetonide	Osteoarthritis-related knee pain	Extended release	October, 2017	Barlas (2018)
Mircera	Vifor	Methoxy polyethylene glycol-epoetin beta	Anemia associated with chronic kidney disease dosing	Greater aptamer stability	June, 2018	Hayashi et al. (2019)
Cimzia	UCB	PEGylated anti-TNF (tumor necrosis factor)	Crohn's disease, rheumatoid arthritis, psoriatic arthritis and ankylosing spondylitis	Longer circulation time, increased in vivo stability	March 2019 for adults with active nonradiographic axial spondyloarthritis	Lang (2008)

TABLE 2.2 *(Continued)*

Trade name	Manufacturer	Therapeutic agent	Indication(s)	Benefit	Month and year of approval	Reference
Micelle nanoparticles						
Estrasorb	Novavax	Micellar estradiol is an estrogen steroid hormone and the major female sex hormone	Treatment of moderate to severe vasomotor symptoms due to menopause	Controlled delivery	October, 2003	Simon (2006)
Nanocrystal nanoparticles						
Avinza	Pfizer	Morphine sulfate	An opioid associated with pain medication	Greater drug loading and bioavailability, ER	March, 2002	Portenoy et al. (2002)
Tricor	AbbVie	Fenofibrate	Hyperlipidemia	Greater bioavailability simplifies administration	September, 2001 (February, 1998)	Junghanns and Müller (2008)
Zanaflex	Acorda	Tizanidine HCl	Used to treat muscle spasticity due to spinal cord injury or multiple sclerosis	Greater drug loading and bioavailability	July, 2002 (initial U.S. *Approval,*1996)	Spiller et al. (2004)
Vitoss	Stryker	Calcium phosphate	Mimics bone structure allowing cell adhesion and growth	Mimics bone structure	2003	Van et al. (2012)
Ritalin LA	Novartis	Methylphenidate HCl, is a stimulant medication	Used to treat attention deficit hyperactivity disorder and narcolepsy	Extended release, greater drug loading, and bioavailability	June, 2002	Lyseng-Williamson and Keating (2002)
OsSatura	IsoTis Orthobiologics	Hydroxyapatite nanocrystals	Bone substitute	Mimics bone structure	2003	Odiba et al. (2017)
Focalin	Novartis	Dexmethylphenidate HCl	Mild stimulant to the central nervous system. It affects chemicals in the brain that contribute to hyperactivity and impulse control. **it** is used to treat attention deficit hyperactivity disorder	Greater drug loading and bioavailability	May, 2005	Focalin (2007)

Application of Nanotechnology in Drug Delivery

TABLE 2.2 *(Continued)*

Trade name	Manufacturer	Therapeutic agent	Indication(s)	Benefit	Month and year of approval	Reference
Megace ES	Par Pharmaceuticals	Megestrol acetate, a progestin of the 17α-hydroxyprogesterone group	Anorexia, cachexia, or an unexplained, significant weight loss in patients with a diagnosis of acquired immunodeficiency syndrome	Lower dosing	July, 2005	Narang et al. (2007)
NanOss	Pioneer Surgical Technology	Filler, Bone Void, Calcium Compound	Bone substitute	Mimics bone structure	August, 2008	Epstein (2015)
EquivaBone	Zimmer Biomet	Proprietary combination of osteoinductive demineralized bone matrix (DBM) and ETEX's osteoconductive nanocrystalline calcium phosphate	Fills bone voids or defects of the skeletal system (including extremities, pelvis, and spine)	Mimics bone structure	April, 2009	Voor e al. (2011)
Invega Sustenna	Janssen	Paliperidone Palmitate extended-release injectable suspension	Treatment of schizoaffective disorder in adults as monotherapy and as an adjunct to mood stabilizers or antidepressants	sustained and improved solubility of the drug	July, 2009	OMJ (2009)
Rapamune	Wyeth Pharmaceuticals	Sirolimus	Immunosuppressant	Greater bioavailability	September, 2010 (Initial U.S. Approval: 1999)	Wyeth (2011)
Ryanodex	Eagle Pharmaceuticals	Dantrolene sodium	Malignant hypothermia	More rapid rate of administration at higher doses	July, 2014	McAvoy et al. (2019)
Emend	Merck	Aprepitant	Prevents chemotherapy-induced nausea and vomiting and to prevent postoperative nausea and vomiting	Greater absorption and bioavailability	May, 2018	Giagnuolo et al. (2019)
Ostim	Heraeus Kulzer	Hydroxyapatite	Treatment of traumatic bone defects of radius, calcaneus and tibia	Mimics bone structure	—	Huber et al. (2006)
Inorganic nanoparticle						
Dexferrum	American Regent	Iron Dextran Injection	Iron deficiency in chronic kidney disease	Allows increased dose	1957	Bobo et al. (2016)

TABLE 2.2 *(Continued)*

Trade name	Manufacturer	Therapeutic agent	Indication(s)	Benefit	Month and year of approval	Reference
Venofer	American Regent	Iron sucrose	Iron deficiency in chronic kidney disease	Increased dose	October, 2005	Charytan et al. (2005)
Feraheme	AMAG Pharmaceuticals	Ferumoxytol	Iron deficiency in chronic kidney disease	Prolonged, steady release thereby reducing the dosing frequency	June, 2009	Mandeville et al. (2010)
Ferrlecit	Sanofi-Aventis	Elemental iron as the sodium salt of a ferric ion carbohydrate complex in an alkaline aqueous solution with 20% sucrose in water for injection	Iron deficiency in chronic kidney disease	Increased dose	2011 (September, 1999)	Wu et al. (2011)
Infed	Actavis Pharma	Iron dextran, complex of ferric hydroxide and dextran	Iron deficiency in chronic kidney disease	Allows increased dose	2016 (1992)	Anselmo and Mitragotri (2019)
Protein nanoparticle						
Ontak	Eisai	An engineered protein combining interleukin-2 and diphtheria toxin	CD25-directed cytotoxin indicated for the treatment of patients with persistent or recurrent cutaneous T-cell lymphoma	Targeted T-cell specificity, lysosomal escape	1999	Manoukian and Hagemeister (2009)
Abraxane	Celgene	Protein-bound paclitaxel, also known as nanoparticle albumin–bound paclitaxel or nab-paclitaxel	Breast cancer, late-stage pancreatic cancer, non–small cell lung cancer (NSCLC)	Greater solubility, improved delivery to tumor	July 2005 for breast cancer, October 2012 for NSCLC and September 2013 for pancreatic cancer	Chapman et al. (2018)

Application of Nanotechnology in Drug Delivery

or in clinical trials. While many challenges face nanodrug development, it may only be a matter of time until these agents provide unique solutions for unmet medical needs and greatly alter clinical practice. Most nanomedicines in clinics are based on conventional therapeutics that were already granted approval and are mostly composed of simple nanoparticles. The quantum of effort that is occurring in the field of nanomedicine predicts that many new nanodrugs will eventually be available for clinical use.

KEYWORDS

- **drug delivery**
- **nanomedicine**
- **nanoparticles**
- **second-generation drug-delivery systems**

REFERENCES

Abbasi, E.; Aval, S. F.; Akbarzadeh, A.; Milani, M.; Nasrabadi, H. T.; Joo, S. W.; Hanifehpour, Y.; Nejati-Koshki, K.; Pashaei-Asl, R. Dendrimers: Synthesis, Applications, and Properties. *Nanosc. Res. Lett.* **2014,** *9* (1), 247.

Ahmad, U.; Faiyazuddin, M.; Hussain, M. T.; Ahmad, S.; Alshammari, T. M.; Shakeel, F. Silymarin: An Insight to Its Formulation and Analytical Prospects. *Acta Physiol. Plant.* **2015,** 37 (11), 253.

Akbarzadeh A.; Rezaei-Sadabady R.; Davaran S.; Joo SW.; Zarghami N.; Hanifehpour Y.; Samiei M.; Kouhi M.; Nejati-Koshki K. Liposome: Classification, Preparation, and Applications. *Nanosc. Res. Lett.* **2013,** 8 (1), 102.

Anajafi T.; Mallik S. Polymersome-Based Drug-Delivery Strategies for Cancer Therapeutics. *Therap. Deliv.* **2015,** *6* (4), 521–534.

Anselmo, A. C.; Mitragotri, S. Nanoparticles in the Clinic. *Bioeng. Transl. Med.* **2016,** *1* (1), 10–29.

Anselmo, A. C.; Mitragotri, S. Nanoparticles in the Clinic: An Update. *Bioeng. Transl. Med.* **2019,** *4* (3), e10143.

Barenholz, Y. C. Doxil®—the First FDA-Approved Nano-Drug: Lessons Learned. *J. Control. Release.* **2012,** *160* (2), 117–134.

Barlas S. Congress Moves Toward Opioid Response 2.0: Bills Are Wide Ranging and Well Intentioned But Mostly Lightweight. *Pharm. Therap.* **2018,** *43* (6), 332.

Bawa, R. Regulating Nanomedicine—Can the FDA Handle It? *Curr. Drug Deliv.* **2011,** *8* (3), 227–234.

Bharali, D. J.; Mousa, S. A. Emerging Nanomedicines for Early Cancer Detection and Improved Treatment: Current Perspective and Future Promise. *Pharmacol. Ther.* **2010**, *128* (2), 324–335.

Bleeker, E. A.; de Jong, W. H.; Geertsma, R. E.; Groenewold, M.; Heugens, E. H..; Koers-Jacquemijns, M. et al. Considerations on the EU Definition of a Nanomaterial: Science to Support Policy Making. *Regul. Toxicol. Pharmacol.* **2013**, *65*, 119–125.

Bobo, D.; Robinson, K. J.; Islam, J.; Thurecht, K. J.; Corrie, S. R. Nanoparticle-Based Medicines: A Review of FDA-Approved Materials and Clinical Trials to Date. *Pharma. Res.* **2016** Oct 1, *33* (10), 2373–2387.

Bosetti, R.; Marneffe, W.; Vereeck, L. Assessing the Need for Quality-Adjusted Cost–Effectiveness Studies of Nanotechnological Cancer Therapies. *Nanomedicine* **2013**, *8* (3), 487–497.

Bragagni, M.; Mennini, N.; Furlanetto, S.; Orlandini, S.; Ghelardini, C.; Mura, P. Development and Characterization of Functionalized Niosomes for Brain Targeting of Dynorphin-B. *Eur. J. Pharma. Biopharma.* **2014**, *87* (1), 73–79.

Bragagni, M.; Mennini, N.; Ghelardini, C.; Mura, P. Development and Characterization of Niosomal Formulations of Doxorubicin Aimed at Brain Targeting. *J. Pharm. Pharma. Sci.* **2012**, *15* (1), 184–196.

Bressler, N. M.; Bressler, S. B. Photodynamic Therapy with Verteporfin (Visudyne): Impact on Ophthalmology and Visual Sciences. *Invest. Ophthalmol. Visual Sci.* **2000**, *41* (3), 624–628.

Brewer, J. M.; Roberts, C. W.; Conacher, M.; McColl, J.; Blarney, B. A.; Alexander, J. An Adjuvant Formulation That Preferentially Induces T Helper Cell Type 1 Cytokine and CD8+ Cytotoxic Responses Is Associated with Up-Regulation of IL-12 and Suppression of IL-10 Production. *Vaccine Res.* **1996**, *5* (2), 77–89.

Brinkhuis, R. P.; Rutjes, F. P.; van Hest, J. C. Polymeric Vesicles in Biomedical Applications. *Polym. Chem.* **2011**, *2* (7), 1449–1462.

Bukowski RM.; Tendler C.; Cutler D.; Rose E.; Laughlin MM.; Statkevich P. Treating Cancer with PEG Intron: Pharmacokinetic Profile and Dosing Guidelines for an Improved Interferon-Alpha-2b Formulation. *Cancer: Interdiscip. Int. J. Am. Cancer Soc.* **2002**, *95* (2), 389–3896.

Calkins, T.; Harris, L. Genzyme: The Renvela Launch Decision; Kellogg School of Management Cases, 2017 Jan 20.

Caster, J. M.; Patel, A. N.; Zhang, T.; Wang, A. Investigational Nanomedicines in 2016: A Review of Nanotherapeutics Currently Undergoing Clinical Trials. *Wiley Interdiscip. Rev. Nanomed. Nanobiotechnol.* **2017**, *9* (1).

Chan, V. S. Nanomedicine: An Unresolved Regulatory Issue. *Regul. Toxicol. Pharmacol.* **2006**, *46*, 218–224.

Chao, Y. S.; Grobelna, A. Curosurf (*poractant alfa*) for the Treatment of Infants at Risk For or Experiencing Respiratory Distress Syndrome: A Review of Clinical Effectiveness, Cost-Effectiveness, and Guidelines, 2018.

Chapman, B. C.; Gleisner, A.; Rigg, D.; Messersmith, W.; Paniccia, A.; Meguid, C.; Gajdos, C.; McCarter, M. D.; Schulick, R. D.; Edil, B. H. Perioperative and Survival Outcomes Following Neoadjuvant FOLFIRINOX versus Gemcitabine Abraxane in Patients with Pancreatic Adenocarcinoma. *JOP* **2018**, *19* (2), 75.

Charytan, C.; Qunibi, W.; Bailie, G. R. Venofer Clinical Studies Group. Comparison of Intravenous Iron Sucrose to Oral Iron in the Treatment of Anemic Patients with Chronic Kidney Disease Not on Dialysis. *Nephron. Clin. Pract.* **2005**, *100* (3), c55–c62.

Application of Nanotechnology in Drug Delivery 65

Chiappini, C.; Almeida, C. Silicon Nanoneedles for Drug Delivery. In *Semiconducting Silicon Nanowires for Biomedical Applications*; Woodhead Publishing, 2014; pp 144–167.

Craig, C. Current Treatment Approaches for Neoplastic Meningitis: Nursing Management of Patients Receiving Intrathecal DepoCyt. *Oncol. Nurs. Forum* **2000**, *27* (8), 1225–1230.

Crawford, J. Once-Per-Cycle Pegfilgrastim (Neulasta) for the Management of Chemotherapy-Induced Neutropenia. *Semin. Oncol.* **2003**, *30*, 24–30.

Davis, J.; Yan, S.; Matsushita, T.; Alberio, L.; Bassett, P.; Santagostino, E. Systematic Review and Analysis of Efficacy of Recombinant Factor IX Products for Prophylactic Treatment of Hemophilia B in Comparison with rIX-FP. *J. Med. Econ.* **2019**, *22* (10), 1014–1021.

Deutsch, Y. E.; Presutto, J. T.; Brahim, A.; Raychaudhuri, J.; Ruiz, M. A.; Sandoval-Sus, J.; Fernandez, H. F. Safety and Feasibility of Outpatient Liposomal Daunorubicin and Cytarabine (Vyxeos) Induction and Management in Patients with Secondary AML. *Blood* **2018**, *132* (Suppl 1), 3559.

Dinndorf, P. A.; Gootenberg, J.; Cohen, M. H.; Keegan, P.; Pazdur, R. Regulatory Issues: FDA. *Oncologist* **2007**, *12*, 991–998.

Discher, B. M.; Won, Y. Y.; Ege, D. S.; Lee, J. C.; Bates, F. S.; Discher, D. E.; Hammer, D. A. Polymersomes: Tough Vesicles Made from Diblock Copolymers. *Science* **1999**, *284* (5417), 1143–1146.

Elhissi, A.; Ahmed, W.; Hassan, I. U.; Dhanak, V.; D'Emanuele, A. Carbon Nanotubes in Cancer Therapy and Drug Delivery. *J. Drug Deliv.* **2012**, 2012.

Epstein, N. E. Preliminary Documentation of the Comparable Efficacy of Vitoss versus NanOss Bioactive as Bone Graft Expanders for Posterior Cervical Fusion. *Surg. Neurol. Int.* **2015**, *6* (Suppl 4):S164.

Erdoğar, N.; Akkın, S.; Bilensoy, E. Nanocapsules for Drug Delivery: An Updated Review of the Last Decade. *Recent Patents Drug Deliv. Formulation* **2018**, *12* (4), 252–266.

Ezhilarasi, P. N.; Karthik, P.; Chhanwal, N.; Anandharamakrishnan, C. Nanoencapsulation Techniques for Food Bioactive Components: A Review. *Food Bioprocess Technol.* **2013**, *6*, 628–647.

Feng, Q. L.; Wu, J..; Chen, G. Q..; Cui, F. Z..; Kim, T. N..; Kim, J. O. A Mechanistic Study of the Antibacterial Effect of Silver Ions on *Escherichia coli* and *Staphylococcus aureus*. *J. Biomed. Mater. Res.* **2000**, *52*, 662–668.

Focalin, X. R. Drug Approval Package, Drugs@ FDA, Approval Date: May 26, 2005, date created, 2007.

Fornasiero, P.; Cargnello, M., Ed. *Morphological, Compositional, and Shape Control of Materials for Catalysis*; Elsevier, 2017

Foser, S.; Schacher, A.; Weyer, K. A.; Brugger, D.; Dietel, E.; Marti, S.; Schreitmüller, T. Isolation, Structural Characterization, and Antiviral Activity of Positional Isomers of Monopegylated Interferon α-2a (PEGASYS). *Protein Express. Purif.* **2003**, *30* (1), 78–87.

Fréchet, J. M. Dendrimers and Supramolecular Chemistry. *Proc. Natl. Acad. Sci.* **2002**, *99* (8), 4782–4787.

Freitas, Jr. R. A. A Mechanical Artificial Red Cell: Exploratory Design in Medical Nanotechnology. *Arti. Cells Blood Substit. Immobil. Biotechnol.* **1998**, *26*, 411–430.

Freitas, Jr. R. A. Microbivores: Artificial Mechanical Phagocytes Using Digest and Discharge Protocol. *J. Evol. Technol.* **2005**, *14*, 1–52.

Gabizon, A.; Shmeeda, H.; Barenholz, Y. Pharmacokinetics of Pegylated Liposomal Doxorubicin. *Clinical Pharmacokinet.* **2003**, *42* (5), 419–436.

Galvin, P.; Thompson, D.; Ryan, K. B. et al. Nanoparticle-Based Drug Delivery: Case Studies for Cancer and Cardiovascular Applications. *Cell Mol. Life Sci.* **2012,** *69* (3), 389–404.

Gambling, D.; Hughes, T.; Martin, G.; Horton, W.; Manvelian, G. A Comparison of Depodur™, a Novel, Single-Dose Extended-Release Epidural Morphine, with Standard Epidural Morphine for Pain Relief after Lower Abdominal Surgery. *Anesthesia Analgesia.* **2005,** *100* (4), 1065–1074.

Gao, W.; Dong, R.; Thamphiwatana, S.; Li, J.; Gao, W.; Zhang, L.; Wang, J. Artificial Micromotors in the Mouse's Stomach: A Step toward in Vivo Use of Synthetic Motors. *ACS Nano.* **2015,** *9* (1), 117–123.

Gharbavi, M.; Amani, J.; Kheiri-Manjili, H.; Danafar, H.; Sharafi, A. Niosome: A Promising Nanocarrier for Natural Drug Delivery through Blood-Brain Barrier. *Adv. Pharmacol. Pharma. Sci.* **2018,** 2018./*8.

Giagnuolo, G.; Buffardi, S.; Rossi, F.; Petruzziello, F.; Tortora, C.; Buffardi, I.; Marra, N.; Beneduce, G.; Menna, G.; Parasole, R. Single Center Experience on Efficacy and Safety of Aprepitant for Preventing Chemotherapy-Induced Nausea and Vomiting (CINV) in Pediatric Hodgkin Lymphoma. *PLoS One* **2019,** *14* (4).

Grumezescu, A. M., Ed. *Nanobiomaterials in Drug Delivery: Applications of Nanobiomaterials*; William Andrew, 2016 Apr 26.

Grumezescu, A. M. *Nano-and Microscale Drug Delivery Systems: Design and Fabrication*; William Andrew, 2017.

Gu, H.; Tang, H.; Xiong, P.; Zhou, Z. Biomarkers-Based Biosensing and Bioimaging with Graphene for Cancer Diagnosis. *Nanomaterials* **2019,** *9* (1), 130.

Hamano, N.; Kamoshida, S.; Kikkawa, Y.; Yano, Y.; Kobayashi, T.; Endo-Takahashi, Y.; Suzuki, R.; Maruyama, K.; Ito, Y.; Nomizu, M.; Negishi, Y. Development of Antibody-Modified Nanobubbles Using Fc-Region-Binding Polypeptides for Ultrasound Imaging. *Pharmaceutics* **2019,** *11* (6), 283.

Havel H.; Finch G.; Strode P. et al. Nanomedicines: From Bench to Bedside and Beyond. *AAPS J.* **2016,** *18* (6), 1373–1378.

Hawker, C. J.; Frechet, J. M. Preparation of Polymers with Controlled Molecular Architecture. A New Convergent Approach to Dendritic Macromolecules. *J. Am. Chem. Soc.* **1990,** *112* (21), 7638–7647.

Hayashi, T.; Uemura, Y.; Kumagai, M.; Kimpara, M.; Kanno, H.; Ohashi, Y.; MIRACLE-CKD Study Group. Effect of Achieved Hemoglobin Level on Renal Outcome in Non-Dialysis Chronic Kidney Disease (CKD) Patients Receiving Epoetin Beta Pegol: MIRcerA CLinical Evidence on Renal Survival in CKD patients with renal anemia (MIRACLE-CKD Study). *Clin. Exp. Nephrol.* **2019,** *23* (3), 349–361.

Hu X.; Zhang Y.; Xie Z.; Jing X.; Bellotti A.; Gu Z. Stimuli-Responsive Polymersomes for Biomedical Applications. *Biomacromolecules* **2017,** *18* (3), 649–673.

Huber, F. X.; Belyaev, O.; Hillmeier, J.; Kock, H. J.; Huber, C.; Meeder, P. J.; Berger, I. First Histological Observations on the Incorporation of a Novel Nanocrystalline Hydroxyapatite Paste OSTIM® in Human Cancellous Bone. *BMC Musculoskeletal Disorders.* **2006,** *7* (1), 50.

Idec, B. Biogen Idec's Plegridy (Peginterferon Beta-1a) Approved in the US for the Treatment of Multiple Sclerosis. Press release, 2014.

Journet, C.; Maser, W. K.; Bernier, P.; Loiseau, A.; de La Chapelle, M. L.; Lefrant, D. S.; Deniard, P.; Lee, R.; Fischer, J. E. Large-Scale Production of Single-Walled Carbon Nanotubes by the Electric-Arc Technique. *Nature* **1997,** *388* (6644), 756–758.

Junghanns, J. U.; Müller, R. H. Nanocrystal Technology, Drug Delivery and Clinical Applications. *Int. J. Nanomed.* **2008,** *3* (3), 295.

Kakaei, K.; Esrafili, M. D.; Ehsani, A. Graphene and Anticorrosive Properties. *Interf. Sci. Technol.* **2019,** *27,* 303–337.

Karimi, M.; Eslami, M.; Sahandi-Zangabad, P.; Mirab, F.; Farajisafiloo, N.; Shafaei, Z.; Ghosh, D.; Bozorgomid, M.; Dashkhaneh, F.; Hamblin, M. R. pH-Sensitive Stimulus-Responsive Nanocarriers for Targeted Delivery of Therapeutic Agents. *Wiley Interdiscip. Rev. Nanomed. Nanobiotechnol.* **2016,** *8* (5), 696–716.

Kaur, I. P.; Kakkar, V.; Deol, P. K.; Yadav, M.; Singh, M.; Sharma, I. Issues and Concerns in Nanotech Product Development and Its Commercialization. *J. Control. Release* **2014,** *193,* 51–62.

Kesharwani, S. S.; Kaur, S.; Tummala, H.; Sangamwar, A. T. Multifunctional Approaches Utilizing Polymeric Micelles to Circumvent Multidrug Resistant Tumors. *Colloids Surf. B Biointerf.* **2019,** *173,* 581–590.

Khan, S. B.; Asiri, A. M.; Akhtar, K. *Development and Prospective Applications of Nanoscience and Nanotechnology*; Nanomaterials and their Fascinating Attributes, Eds.; Bentham Science Publishers Ltd: Sharjah, UAE, 2015; p 261.

Kondiah, P. P.; Choonara, Y. E.; Kondiah, P. J.; Marimuthu, T.; Kumar, P.; du Toit, L. C.; Modi, G.; Pillay, V. Nanocomposites for Therapeutic Application in Multiple Sclerosis. In *Applications of Nanocomposite Materials in Drug Delivery*; Woodhead Publishing, 2018; pp 391–408.

Kong, J.; Cassell, A. M.; Dai, H. Chemical Vapor Deposition of Methane for Single-Walled Carbon Nanotubes. *Chem. Phys. Lett.* **1998,** *292* (4–6), 567–574.

Kothamasu, P.; Kanumur, H.; Ravur, N.; Maddu, C.; Parasuramrajam, R.; Thangavel, S. Nanocapsules: The Weapons for Novel Drug Delivery Systems. *BioImpacts: BI* **2012,** *2* (2), 71.

Krishnamurthy, S.; Vijayasarathy, S. Role of Nanomaterials in Clinical Dentistry. In *Nanobiomaterials in Dentistry*; William Andrew Publishing; pp 211–240.

Kuo, T. T.; Wang, C. H.; Wang, J. Y.; Chiou, H. J.; Fan, C. H.; Yeh, C. K. Concurrent Osteosarcoma Theranostic Strategy Using Contrast-Enhanced Ultrasound and Drug-Loaded Bubbles. *Pharmaceutics* **2019,** *11* (5), 223.

Lang, L. FDA Approves Cimzia to Treat Crohn's Disease. *Gastroenterology* **2008,** *134* (7), 1819.

Lay, C. L.; Liu, J.; Liu, Y. Functionalized Carbon Nanotubes for Anticancer Drug Delivery. *Expert Rev. Med. Devices* **2011,** *8* (5), 561–566.

Letchford, K.; Burt, H. A Review of the Formation and Classification of Amphiphilic Block Copolymer Nanoparticulate Structures: Micelles, Nanospheres, Nanocapsules and Polymersomes. *Eur. J. Pharm. Biopharm.* **2007,** *65* (3), 259–269.

Li, R.; Wu, R. A.; Zhao, L.; Wu, M.; Yang, L.; Zou, H. P-Glycoprotein Antibody Functionalized Carbon Nanotube Overcomes the Multidrug Resistance of Human Leukemia Cells. *ACS Nano* **2010,** *4* (3), 1399–1408.

Lister, J. Amphotericin B Lipid Complex (Abelcet®) in the Treatment of Invasive Mycoses: The North American Experience. *Eur. J. Haematol.* **1996,** *56* (S57), 18–23.

Lu, Y.; Hu, Q.; Lin, Y.; Pacardo, D. B.; Wang, C.; Sun, W.; Ligler, F. S.; Dickey, M. D.; Gu, Z. Transformable Liquid-Metal Nanomedicine. *Nat. Commun.* **2015,** *6* (1), 1-0.

Lv, L.; Zeng, Q.; Wu, S.; Xie, H.; Chen, J.; Guo, X. J.; Hao, C.; Zhang, X.; Ye, M.; Zhang, L. Exosome-Based Translational Nanomedicine: The Therapeutic Potential for Drug Delivery. In *Mesenchymal Stem Cell Derived Exosomes*; Academic Press, 2015 Jan 1; pp. 161–176.

Lyseng-Williamson, K. A.; Keating, G. M. Extended-Release Methylphenidate (Ritalin® LA). *Drugs* **2002,** *62* (15), 2251–2259.

Madani, S. Y.; Naderi, N.; Dissanayake, O.; Tan, A.; Seifalian, A. M. A New Era of Cancer Treatment: Carbon Nanotubes as Drug Delivery Tools. *Int. J. Nanomed.* **2011,** *6,* 2963.

Mahale, N. B.; Thakkar, P. D.; Mali, R. G.; Walunj, D. R.; Chaudhari, S. R. Niosomes: Novel Sustained Release Nonionic Stable Vesicular Systems—An Overview. *Adv. Colloid Interf. Sci.* **2012,** *183,* 46–54.

Malhotra, M.; Jain, N. K. Niosomes: A Controlled and Novel Drug Delivery System. *Indian Drugs* **1994,** *31* (3), 81–86.

Mandeville, J. B.; Srihasam, K.; Vanduffel, W.; Livingston, M. S. Evaluating Feraheme as a Potential Contrast Agent for Clinical IRON fMRI. *Proc. Int. Soc. Magn. Reson. Med*; Stockholm, 2010; p 1110.

Manoukian, G.; Hagemeister, F. Denileukin Diftitox: A Novel Immunotoxin. *Expert Opin. Biol. Ther.* **2009,** *9* (11), 1445–1451.

Masood, F. Polymeric Nanoparticles for Targeted Drug Delivery System for Cancer Therapy. *Mater. Sci. Eng. C.* **2016, 60,** 569–578.

McAvoy, J. C.; Brodsky, J. B.; Brock-Utne, J. Pennywise and a Pound Foolish: The Advantage of Dantrolene Nanosuspension (Ryanodex) in the Treatment of Malignant Hyperthermia. *Anesthesia Analgesia* **2019,** *129* (6), e201–e202.

Meunier, F.; Prentice, H. G.; Ringden, O. Liposomal Amphotericin B (AmBisome): Safety Data from a Phase II/III Clinical Trial. *J. Antimicrob. Chemother.* **1991,** *28* (Suppl_B), 83–91.

Mo, R.; Jiang, T.; Gu, Z. Enhanced Anticancer Efficacy by ATP-Mediated Liposomal Drug Delivery. *Angew. Chem. Int. Ed.* **2014,** *53* (23), 5815–5820.

Narang, A. S.; Delmarre, D.; Gao, D. Stable Drug Encapsulation in Micelles and Microemulsions. *Int. J. Pharm.* **2007,** *345* (1–2), 9–25.

Odiba, A.; Ottah, V.; Ottah, C.; Anunobi, O.; Ukegbu, C.; Edeke, A.; Uroko, R.; Omeje, K. Therapeutic Nanomedicine Surmounts the Limitations of Pharmacotherapy. *Open Med.* **2017,** *12* (1), 271–287.

OMJ, P. Invega Sustenna (Paliperidone Palmitate) Extended Release Injectable Suspension: Prescribing Information, 2009.

Onaca, O.; Enea, R.; Hughes, D. W.; Meier, W. Stimuli-Responsive Polymersomes as Nanocarriers for Drug and Gene Delivery. *Macromol. Biosci.* **2009,** *9* (2), 129–139.

Ossa, D. Quality Aspects of Nano-Based Medicines SME Workshop: Focus on Quality for Medicines Containing Chemical Entities London. http://www.ema.europa.eu/docs/en_GB/document_library/Presentation/2014/04/WC500165444.pdfn

Portenoy, R. K.; Sciberras, A.; Eliot, L.; Loewen, G.; Butler, J.; Devane, J. Steady-State Pharmacokinetic Comparison of a New, Extended-Release, Once-Daily Morphine Formulation, Avinza™, and a Twice-Daily Controlled-Release Morphine Formulation in Patients with Chronic Moderate-to-Severe Pain. *J. Pain Symp. Manage.* **2002,** *23* (4), 292–300.

Radha, G. V.; Rani, T. S.; Sarvani, B. A Review on Proniosomal Drug Delivery System for Targeted Drug Action. *J. Basic Clin. Pharm.* **2013,** *4* (2), 42.

Ragelle, H.; Danhier, F.; Preat, V.; Langer, R.; Anderson, D. G. Nanoparticle-Based Drug Delivery Systems: A Commercial and Regulatory Outlook as the Field Matures. *Expert. Opin. Drug. Deliv.* doi:10.1080/17425247.2016.1244187 (2016) (Epub ahead of print).

Rao, J. P.; Geckeler, K. E. Polymer Nanoparticles: Preparation Techniques and Size-Control Parameters. *Prog. Polym. Sci.* **2011,** *36,* 887–913.

Application of Nanotechnology in Drug Delivery

Rather, M. A.; Amin, S.; Maqbool, M.; Bhat, Z. S.; Gupta, P. N.; Ahmad, Z. Preparation and In Vitro Characterization of Albumin Nanoparticles Encapsulating an Anti-Tuberculosis Drug-Levofloxacin. Advanced Science. *Eng. Med.* **2016,** *8* (11), 912–917.

Ross, C. J.; Towfic, F.; Shankar, J.; Laifenfeld, D.; Thoma, M.; Davis, M.; Weiner, B.; Kusko, R.; Zeskind, B.; Knappertz, V.; Grossman, I. A Pharmacogenetic Signature of High Response to Copaxone in Late-Phase Clinical-Trial Cohorts of Multiple Sclerosis. *Genome Med.* **2017,** *9* (1), 50.

Sainz, V.; Conniot, J.; Matos, A. I.; Peres, C.; Zupanŏiŏ, E.; Moura, L.; Silva, L. C.; Florindo, H. F.; Gaspar, R. S. Regulatory Aspects on Nanomedicines. *Biochemical Biophys. Res. Commun.* **2015,** *468* (3), 504–510.

Sajja, H. K.; East, M. P.; Mao, H.; et al. Development of Multifunctional Nanoparticles for Targeted Drug Delivery and Noninvasive Imaging of Therapeutic Effect. *Curr. Drug Discov. Technol.* **2009,** *6* (1), 43–51.

Sartor, O. Eligard: Leuprolide Acetate in a Novel Sustained-Release Delivery System. *Urology* **2003,** *61* (2), 25–31.

Schäfer-Korting, M., Ed. *Drug Delivery*; Springer Science & Business Media, Ed..; Springer: Berlin, Germany, 2010; pp 55–86.

Sercombe, L.; Veerati, T.; Moheimani, F.; Wu, S. Y.; Sood, A. K.; Hua, S. Advances and Challenges of Liposome Assisted Drug Delivery. *Front. Pharmacol.* **2015,** *6,* 286.

Shakeri, A.; Sahebkar, A. Opinion Paper: Nanotechnology: A Successful Approach to Improve Oral Bioavailability of Phytochemicals. *Recent Pat. Drug Deliv. Formul.* **2016,** *10,* 4–6.

Shams, M.; Kobrynski, L. Management of ADA-Deficient SCID Patient on Adagen during Pregnancy. *J. Clin. Immunol.* **2019,** *39* (8), 846–848.

Silverman, J. A.; Deitcher, S. R. Marqibo®(Vincristine Sulfate Liposome Injection) Improves the Pharmacokinetics and Pharmacodynamics of Vincristine. *Cancer Chemother. Pharmacol.* **2013,** *71* (3), 555–564.

Simon, J. A. ESTRASORB Study Group: Estradiol in Micellar Nanoparticles: The Efficacy and Safety of a Novel Transdermal Drug-Delivery Technology in the Management of Moderate to Severe Vasomotor Symptoms. *Menopause* **2006,** *13,* 222–231.

Singh, A.; Garg, G.; Sharma, P. K. Nanospheres: A Novel Approach for Targeted Drug Delivery System. *Int. J. Pharm. Sci. Rev. Res.* **2010,** *5* (3), 84–88.

Sinha, N.; Kulshreshtha, N. M.; Dixit, M.; Jadhav, I.; Shrivastava, D.; Bisen, P. S. Nanodentistry: Novel Approaches. *Nanostruct. Oral Med.* **2017,** 751–776.

Soares, S.; Sousa, J.; Pais, A.; Vitorino, C. Nanomedicine: Principles, Properties, and Regulatory Issues. *Front. Chem.* **2018,** *6,* 360.

Spiller, H. A.; Bosse, G. M.; Adamson, L. A. Retrospective Review of Tizanidine (Zanaflex®) Overdose. *J. Toxicol. Clin. Toxicol.* **2004,** *42* (5), 593–596.

Subramani, K.; Ahmed, W., Ed. *Emerging Nanotechnologies in Dentistry*; William Andrew, 2017.

Subramani, K.; Mehta, M. Nanodiagnostics in Microbiology and Dentistry. In *Emerging Nanotechnologies in Dentistry*; William Andrew Publishing, 2018; pp 391–419.

Sun, W.; Gu, Z. DNA Nanoclews for Stimuli-Responsive Anticancer Drug Delivery. In *Rolling Circle Amplification (RCA)*; Springer: Cham, 2016; pp 141–150.

Sun, W.; Ji, W.; Hall, J. M.; Hu, Q.; Wang, C.; Beisel, C. L.; Gu, Z. Self-Assembled DNA Nanoclews for the Efficient Delivery of CRISPR–Cas9 for Genome Editing. *Angew. Chem. Int. Ed.* **2015,** *54* (41), 12029–12033.

Syn, N. L.; Wang, L.; Chow, E. K.; Lim, C. T.; Goh, B. C. Exosomes in Cancer Nanomedicine and Immunotherapy: Prospects and Challenges. *Trends Biotechnol.* **2017,** *35* (7), 665–676.

Tabish, T. A. Graphene-Based Materials: The Missing Piece in Nanomedicine? *Biochem. Biophys. Res. Commun.* **2018,** *504* (4), 686–689.

Tanner P.; Baumann P.; Enea R.; Onaca O.; Palivan C.; Meier W. Polymeric vesicles: from drug carriers to nanoreactors and artificial organelles. Accounts of chemical research. **2011,** *44* (10), 1039–1049.

Thambi, T.; Deepagan, V. G.; Ko, H.; Lee, D. S.; Park, J. H. Bioreducible Polymersomes for Intracellular Dual-Drug Delivery. *J. Mater. Chem.* **2012,** *22* (41), 22028–22036.

Thess, A.; Lee, R.; Nikolaev, P.; Dai, H.; Petit, P.; Robert, J.; Xu, C.; Lee, Y. H.; Kim, S. G.; Rinzler, A. G.; Colbert, D. T. Crystalline Ropes of Metallic Carbon Nanotubes. *Science* **1996,** *273* (5274), 483–487.

Tinkle, S.; McNeil, S. E.; Mühlebach, S.; Bawa, R.; Borchard, G.; Barenholz, Y. C. et al. Nanomedicines: Addressing the Scientific and Regulatory Gap. *Ann. N. Y. Acad. Sci.* **2014,** *1313*, 35–56.

Tobin, K. A. Macugen Treatment for Wet Age-Related Macular Degeneration. *Insight (Am. Soc. Ophthal. Reg. Nurses)* **2006,** *31* (1), 11–14.

Tomalia, D. A.; Baker, H.; Dewald, J.; Hall, M.; Kallos, G.; Martin, S.; Roeck, J.; Ryder, J.; Smith, P. A New Class of Polymers: Starburst-Dendritic Macromolecules. *Polym. J.* **1985,** *17* (1), 117–132.

Trinh, H. M.; Joseph, M.; Cholkar, K.; Mitra, R.; Mitra, A. K. Nanomicelles in Diagnosis and Drug Delivery. In *Emerging Nanotechnologies for Diagnostics, Drug Delivery and Medical Devices*; Elsevier, 2017; pp 45–58.

Turecek, P. L.; Romeder-Finger, S.; Apostol, C.; Bauer, A.; Crocker-Buque, A.; Burger, D. A.; Schall, R.; Gritsch, H. A World-Wide Survey and Field Study in Clinical Haemostasis Laboratories to Evaluate FVIII: C Activity Assay Variability of ADYNOVATE and OBIZUR in Comparison with ADVATE. *Haemophilia* **2016,** *22* (6), 957–965.

US Food and Drug Administration. FDA Approves New Treatment for Advanced Pancreatic Cancer. Press Release, October 22, 2015.

US Food and Drug Administration. FDA Labelling Information—Krystexxa (Pegloticase). FDA Website, 2010.

Van, C. H.; Samora, J. B.; Griesser, M. J.; Crist, M. K.; Scharschmidt, T. J.; Mayerson, J. L. Effectiveness of Ultraporous β-Tricalcium Phosphate (Vitoss) as Bone Graft Substitute for Cavitary Defects in Benign and Low-Grade Malignant Bone Tumors. *Am. J. Orthoped.* (Belle Mead, NJ). **2012,** *41* (1), 20–23.

Van der Lely, A. J.; Lundgren, F.; Biller, B. M.; Brue, T.; Cara, J.; Ghigo, E.; Hadavi, J. H.; Rajicic, N.; Saller, B.; Sanocki, J.; Strasburger, C. Long-Term Treatment of Acromegaly with Pegvisomant (Somavert): Cross-Sectional Observations from ACROSTUDY, a Post-Marketing, International, Safety, Surveillance, Study. In *13th European Congress of Endocrinology*; BioScientifica, 2011; p 26.

Ventola, C. L. The Nanomedicine Revolution: Part 1: Emerging Concepts. *P T* **2012,** *37* (9), 512–525.

Voor, M. J.; Yoder, E. M.; Burden, Jr. R. L. Xenograft Bone Inclusion Improves Incorporation of Hydroxyapatite Cement into Cancellous Defects. *J. Orthopaed. Trauma* **2011,** *25* (8), 483–487.

Wanigasekara, J.; Witharana, C. Applications of Nanotechnology in Drug Delivery and Design-an Insight. *Curr. Trends Biotechnol. Pharm.* **2016,** *10* (1), 78–91.

Wu, M.; Sun, D.; Tyner, K.; Jiang, W.; Rouse, R. Comparative Evaluation of US Brand and Generic Intravenous Sodium Ferric Gluconate Complex in Sucrose Injection: In Vitro Cellular Uptake. *Nanomaterials* **2017,** *7* (12), 451.

Wyeth, L. *Rapamune (Sirolimus) Oral Solution and Tablets*; Philadelphia, PA, 2011.

Xu, L.; Wen, Y.; Pandit, S.; Mokkapati, V. R.; Mijakovic, I.; Li, Y.; Ding, M.; Ren, S.; Li, W.; Liu, G. Graphene-Based Biosensors for the Detection of Prostate Cancer Protein Biomarkers: A Review. *BMC Chem.* **2019,** *13* (1), 112.

Xu, R.; Zhang, G.; Mai, J.; Deng, X.; Segura-Ibarra, V.; Wu, S.; Shen, J.; Liu, H.; Hu, Z.; Chen, L.; Huang, Y. An Injectable Nanoparticle Generator Enhances Delivery of Cancer Therapeutics. *Nat. Biotechnol.* **2016,** *34* (4), 414.

Yavlovich, A.; Singh, A.; Blumenthal, R.; Puri, A. A Novel Class of Photo-Triggerable Liposomes Containing DPPC: DC8, 9PC as Vehicles for Delivery of Doxorubcin to Cells. *Biochim. Biophys. Acta (BBA)-Biomembr.* **2011,** *1808* (1), 117–126.

Zhang, X. Q..; Xu, X..; Bertrand, N..; Pridgen, E..; Swami, A..; Farokhzad, O. C. Interactions of Nanomaterials and Biological Systems: Implications to Personalized Nanomedicine. *Adv. Drug Deliv. Rev.* **2012,** *64*, 1363–1384.

Zhu, Y.; Li, W. Cytotoxicity of Carbon Nanotubes. *Sci. China Ser. B Chem.* **2008,** *51* (11), 1021–1029.

CHAPTER 3

APPLICATION OF NANOPARTICLES IN BIOMEDICAL IMAGING

AFODUN ADAM MOYOSORE*

Department of Anatomy, Faculty of Biomedical Sciences, Kampala International University, Kampala, Uganda

Department of Medical Imaging, Ultrasound and Doppler Unit, Crystal Specialist Hospital, Dopemu-Akowonjo, Lagos, Nigeria

*E-mail: aadam.afodun@kiu.ac.ug; afodunadam@yahoo.com

ABSTRACT

The importance of nanotechnology cannot be overemphasized because some traditional imaging approaches do not measure up to patients' requirements due to limitations and lack of specificity, thus, the need for microimage-based drug delivery, detection, and metastatic ablation. Molecular imaging involves the movement, calibration, visualization, and quantification of specific biological mechanisms at the "tracer level" in living systems. We report on specific nanoparticles (NPs) used (with various potency potentials: quantum dots, gold NPs, copper gadolinium, and fluorescent dyes) as labeling markers and molecules for biological screening.

At the current NPs research advancement rate, contrast product with the necessary attributes can be manufactured for any wanted application. There are promising prospects for improved efficacy, compatibility, detection, and isolation of disease targets. Summarily, nanotechnology will further produce particles that harbor interesting properties, which will eventually be explored in the formulations of novel contrast agents to be used in medical imaging. The combination of nanotechnology modalities (with radiological means: CT, ultrasonography, X-rays, magnetic resonance imaging, positron emission tomography, and single-photon emission computed tomography) and telecommunication will improve the accuracy of the theragnostic procedure,

74 Diverse Applications of Nanotechnology in the Biological Sciences

which eventually influences healthcare. With the advent of multipurpose NPs (at the nanometer-size-range) undergoing clinical trials, we are ushering in the era of "controlled imaging," producing a comprehensive, smart, customizable, and specific end-user (radiologist/oncologist/imaging-scientist) advantage. In this chapter, we discussed various conditions and factors when considering the design of NP transducers and expatiated examples of the most developed types and fluorescence techniques

3.1 GENERAL INTRODUCTION

Nanoparticles (NPs) are beginning to be employed in different branches of science. The accelerating interests in nanomedicine are evidence in literature publication with dates mostly falling in a 10-year period. An attribute common to most NPs is that they are primarily biocompatible, for example, gold and silver—others (NPs) with magnetic features required "artificial shells" to become biocompatible. Not only are NPs used to treat and target tumors or cancerous growths, but they also increase visibility in sonography and magnetic resonance imaging (MRI).

The use of NPs in malignancy investigations has provided a glimmer of hope to the medical society for the production of antimalignant drugs. Cancers of the gastrointestinal tract account for greater than 55% mortality associated with malignant growths (Song, 2007). The importance of nanotechnology cannot be overemphasized because some traditional imaging approaches do not measure up to patients' requirements due to limitations and lack of identity, thus, the need for microimage-based drug-delivery, detection, and metastatic ablation.

Nanoimaging methods require altering or manufacturing materials on a specific nanometer scale by firing up from an isolated group of atoms or by "cutting" heavy substances to "usable" NPs (Jabir et al., 2012). A range of NP are used for therapeutic or diagnostic (theragnostic) purposes in different cancer manifestations. Nanodevices developed for oncologic use include gold GNPs, quantum dots, liposomes, and carbon nanotubes (CNTs) embedded through MRI contrast for sought specificity protein and DNA detection (Ferrari, 2005; Jamieson et al., 2007). These innovative advances offer a revolutionary bridge between molecular profiles, protein, and genetic biomarkers to treat cancerous masses.

Nanogels: These are hydrogels consisting mainly of polymeric-linked chains crossed by noncovalent interactions or strong covalent bonds (Kabanov and

Vinogradov, 2009; Sasaki et al., 2010). Nanogels react to outer nociception changes by adjusting their shape, volume, and equilibrium in a spatial elastic manner (Eichenbaum et al., 1998). Cheng et al. (2010) gave detailed evidence of nanogel use in targeting umbilical-cord vein endothelial cells (HUVECs); as alternate potent capabilities of conjugated carboxybetaine methacrylate become apparent.

Nanopyramids: Different particle types, including rod-like dendritic nanostructural gold, sometimes are combined with extra additives such as system and Pb^{4+}. The electrochemical lattice scaffold of pyramidal structures is mostly for a single use and may be less costly. Scientists at Northwestern Higher Institution of learning in the United States have constructed gold nanopyramids fixable on silicon metallic pedestals (Sweeny et al., 2011). When kindled with photons, the anisotropic nanopyramids produce thermal energy. When nanopyramids become absorbed by cancerous cells with damaged cell walls, the "leaky" streaming after heat change (through IR/infrared) illumination results in cell death.

3.2 THERAGNOSTIC DENDRIMERS

Dendrimers are big, complicated polymers with a well-circumscribed border and a nucleic-like inner core. The mass, shape, diameter, and physiologic features of a dendrimer can be channeled toward working NPs (Svenson and Tomaila, 2005). Most often technologically developed dendrimer usage has been in the open aspect of NIR/MIR coupling materials in oncology. Some dendrimer variants rely on multivalent property conjugation on rallied or attracted antibodies (such as CD14 and prostate-targeting membranous antigen), providing lucent contrast substances. These nanomaterials are studied with microscopic aids and velocity tracking in flow cytometry (Tomalia et al., 2007).

3.3 NANOWIRES AND ITS USE IN DIAGNOSTICS

Real-time identification of three cancer markers (mucin 1, prostate-targeting antigen, and carcino-embryologic antigen) using SiNW bioinformatic sensors has been reported (Zhancy and Ning, 2012). The concurrent high reactive dissections of many biomarkers can enhance malignancy's early discovery (Etzioni et al., 2003; Stern et al., 2010). A study described ZnO synthesis using vapor–solid (mass transformation) with a demonstration by a scanning electron microscope (SEM). Nanowires incorporated with silicon (Si) have a feature of surface receptors fixed and incorporated into arrayed columns (Zheng et al., 2005). Part of the spectroscopic and monochromatic electrical properties exhibited by single-walled CNTs (SWCNTs) is a forceful resonance scattering and infrared photoluminescence that is valuable in medical imaging applications (Sun et al., 2002).

The combination of numerous modalities and energy strength is a distinct feature of NPs that makes them valuable for therapy and diagnosis of a broad range of ailments as MRI developed at a geometric rate during the last decades and now widely used for anatomical and functional medical imaging. The primary nanoparticulate material that allows for molecular imaging is superparamagnetic iron oxide (SPIO) NP (SPION) used for biological contrast enhancement with MRI machines (Kircher and Willmann, 2012). Ferumoxytol has displayed practical use in different aspects like noninvasive recognition of type I diabetes, a glucose metabolic disease (Gaglia et al., 2015). Worldwide therapeutic and diagnostic deliveries are increasingly relying on NPs as a technological bedrock. Due to the different chemical contents of NPs, they are an emerging "game-changer" in nanomedicine. Positron emission tomography (PET) and single-photon emission computed tomography (SPECT) are noninvasive radiological methods, cardiology, and the fields of oncology (Lee et al., 2011; Pratt et al., 2011) with radioactive nuclides fixed with small molecular antibodies [^{18}F-fluorodeoxyglucose shortened as (^{18}F-FDG)]. Fluorescent NPs (FNPs) combined with fluorescent dyes are being researched as an emerging modality due to minute molecular diameter. The mechanism of action by aligning active or passive specificity targeting (Hill and Mohs, 2016; Kamila et al., 2016) increased turnaround time (by avoiding kidney clearance and immune detection) and smart activation (through acidity or alkalinity).

There have been marked advances in techniques to identify and observe diseased tissue in real time in modern times. Pathophysiologic changes can be seen through tissue morphology and the monitoring of active cell functional sites (Minchin and Martin, 2010). Molecular imaging involves the movement, calibration, visualization, and quantification of specific biological mechanisms at the "tracer level" in living systems. Molecular imaging techniques involve a device and transducer, chemical changes during metabolic "buildup or breakdown," underlining blood flow, or oxygen transport has been visualized with various imaging modalities (SNM, 2010). Nanoagents allow for more exact marker-specific imaging to assist visualize

Application of Nanoparticles in Biomedical Imaging 77

and diagnose disease before STAGE I manifestation. There is a prospect to improve treatment options and minimize unwanted side effects through nano/microstructuring that allow "live" tracking when patients are on sickbeds or hospital couches (Veiseh et al., 2010).

3.4 COMPUTER AXIAL TOMOGRAPHY AND NANOPARTICLES

Nanoparticle-based particles and contrast agents are innovative and transformative media for a wide range of roentgenologic modalities: MRI, CAT, and SPECT for diagnostic reporting. CT/CAT photo-slides usually appear in grayscale depending on the forms, wave-reverberate coefficient of different tissue thickness, and texture. On direct AP or internal-view X-rays, bone displays an efficient echogenic attenuation due to density, forming a whitish or dark gray image. At the same time, soft tissues (muscle mass) appear black or dark gray in appearance (Cho et al., 2010; Cormode et al., 2014). Neutral-state intrinsic contrast exists between surrounding tissues and bone; however, there is a limit in isolating typical organ/tumors with various attenuation coefficients.

In recent years, nonparticulate contrast agents have been introduced for CT scans to surmount this inefficiency. They absorb far higher contrast substance content and better X-ray immersion and refraction relative to fluoride- or iodine-based contrast agents, with the overall advantage of lesser patient exposure to radiation. There are distinct from other available agents in terms of pharmacokinetic features and would display viewing over a broader turn-around time and show improved window perfusion (Mukundan et al., 2006; Naha et al., 2014). In a study, Myoung et al. have used bioinert tantalum oxide NPs as contrast and coupling materials (Rand et al., 2011). Iodinated materials demonstrate adverse side effects with reduced in vivo timing, thus its limited use in CT imaging laboratories. Therefore, the need for nanomaterials long-circulating half-life, the ability to remotely modify passive (nonactive) targeting, and surface modifications are ever more on demand.

3.5 CONTRAST USE IN X-RAY TECHNOLOGY

Contrasts X-rays are obtained from the gap in density and solid attenuation between two biological surfaces or mass. Bones absorb more X-rays because they are denser, albeit with a more significant atomic number. Therefore, a decrease in the enthalpy of X-ray produces increased contrast between the two tissue masses as the photoelectric phenomenon occurs at decreased levels less than (<50 KVp) (Carlton and Adler, 2013). Iodine is the most commonly used radiopaque element because of its high atomic number relative to its surrounding biotissue. Iodine is commonly grouped along with triiodobenzene compounds in either dimer or monomeric lattice structure; iodine-based CA is often linked to DNA application and damage during CT scans (Piechowiak et al., 2015). Most iodinated chemical mediums are in watery solutions with dilution ranging from $0.4 \rightarrow 1.0$ mol/L (corresponding to $150 \rightarrow 400$ mgI/mL, with reports of dimeric iodine-based CA showing retention in the renal column).

3.6 ROLE OF THE MAGNETIC NANOPARTICLE

NPs can be used as labeling markers and molecular units for biological screening. Paramagnetic NPs are used for all tracking, and calcium detection made potential when MRI modality is combining with powerful SPIO—[a versatile calcium-sensing protein that doubles as an NP contrast medium]. A 2- to 3-nm-sized SPIO coupled with MRI reveals minute previously undetected lymph node metastases. Dextran-plated iron oxide NP improves MRI imaging of cranial/cephalic tumors for over a day—refractory period, same applications for visualizing cerebral ischemic lesions (Rajasundari and Illamurugu, 2010).

3.7 GOLD NANOPARTICLE AND X-RAY IMAGING EQUATIONS II

When traveling through the human body, X-rays change attenuation caused by photoelectric diffraction or scattering and loss of beam intensity. This phenomenon is cited by Beer–Lambert's Law: $I = I_0 e^{-\mu x}$ (Szabo et al., 2004).

μ = absorption coefficient
x = tissue depth
I_0 = first beam intensity.

3.8 GOLD NANOPARTICLE AND ULTRASOUND

Sonography uses sound waves above > 20 kHz more significant than the human hearing range. The latest technology in ultrasonographic contrast agents rests with the utility of microbubbles applied in Doppler techniques during perfusion analysis of blood velocity and oxygenated/deoxygenated

Application of Nanoparticles in Biomedical Imaging

status (Lentacker et al., 2009). Scientists producing US/CT contrast agents combine gold particles with microbubbles, as stated by Ke et al. (2014). It has been discovered that perfluoroalkyl bromide (PFOB) contains elevated acoustic impedance than gas, making it suitable for a sonographic contrast agent. Jin et al. (2013) invented a microcapsule of polylactic acid (with graphene oxide or gold contents) to successfully imaging rat liver, spleen, and kidneys. The physics of graphene oxide generates harmonic mode created by horizontal scattering irrespective of environmental impedance. Capsules can also deliver theranostic payloads when microcapsules are programmed or "wired" to shield pancreatic islet cells in mice. Regulation of pancreatic cells by the capsule was maintained in a 6- to 8-week period after injection in diabetic mice for the optimal maintenance of regular glucose bar (Arifin et al., 2011). Postintroductory tracking of particles was attained by CT and US modalities, respectively. The encapsulation procedure utilizes many controversial methods that reduce CT signal strength.

3.9 PHOTOACOUSTIC IMAGING

This is a modern method that uses optical beams and properties as fluorescence imaging. Near-infrared light or visible (light) irradiate tissues, which results in adiabatic expansion. Pressure waves generated are calibrated (in automation) and used to construct an image, partially depending on the biological tissues' thermal and optical characteristics. Contrast materials can be used when low contrast exists between tissues (e.g., muscle vs. myoma) or when penetration depth is low (Wu et al., 2014). The Prussian blue staining dye has been used as a fluorescent component to improve and generate photoacoustic signaling. Cheheltani et al. (2016) used GNPs (AuNP) embedded into polydi(carboxylatophenoxy)phosphazene (PCPP) nanopheres, producing passive, bright contrast on CT with variable adjustable wavelength altered based on AuNP size.

3.10 GOLD NANOPARTICLES MRI/CT MODALITY

Gold particles could be reengineered with other imaging media apart from ultrasound, mainly CT and MRI. X-rays are considered poor for soft tissue observation due to reduced attenuation. The discussion mechanism of operation MRI is beyond the scope of this chapter. Gadolinium-based contrast substance (GBCA) is ideal for MRI contrast imaging due to spiked paramagnetism. Nine isotopic forms of GBCA are medically approved, with some controversial toxicity (Marckmann et al., 2006). The literature documents alternate use of reactive oxygen species initiated by SPIOs; therefore, adjusted alterations to formulations will be needed for successful compatibility (Luo et al., 2016). Two compounds can be mixed to form amalgamated heterogeneous compound in unified core–shell fashion. Research implored Au nanorods instead of circular particles plated with a polypyrrole material covering and internally embedded with iron oxide crystals. Particles are text run in vitro by opposing properties (T_2 and T_1); phase relaxation was introduced as CT contrast. The relaxavity constants were 7.99 mM^{-1} s^{-1} and 128.57 nM^{-1} s^{-1} with $T_2 > T_1$. Researchers noted that future in vivo tests are needed to evaluate imaging prowess and internal metabolic distribution since CT values are linked to being directly parallel with concentration status (Feng et al., 2015).

Kim et al. (2011) plated AuNP/Fe$_3$O heterogeneous particle with an amphiphilic polymer (DMA-r-mPEGMa-r-MA) consisting of a PEG moiety. Ethically, a guinea pig modeled experiment was developed to check if liver tumor can be differentiated from hepatic parenchyma. The resulting CT images were inconclusive; however, the authors noted 1.5 times enhancement phenomenon (24 h) after contrast introduction. MRI contrast was densely more massive than that of CA—Resovist®, under similar concentrations (Kim et al., 2011). Molecular isolation is being used as a particle formation strategy for MRI/CAT modality. An alternate procedure uses thiolated Fe$_3$O$_4$ "building-blocks" via the autoassembly of amphiphilic block laced with cross-links by 3-mercaptopropyl trimethoxysilane.

AuNPs utilized in medicated growth adjusting molar ratio at the rate of 2:1 (T_2 relaxivity is 92.67 s^{-1} mM^{-1}) on both MR and CT radiological modality, respectively. A study on biodistribution localized significant accumulation in the spleen and liver (Cai et al., 2015). PEI PEGylated, conditional as a chemical lattice support to reduce AuNP, can be used to vary contrast differentiation and abilities in mouse models. The hetero-particle ratio of gold to iron oxide (3:1) despite high T_2 relaxivity (221.92 s^{-1} mM^{-1}) gives excellent contrast with both CT and MRI, with CT showing ignorable differences (Sun et al., 2016). Wen et al. (2013) manufactured dendrimers with gadolinium ions, gold clusters, and PEG before intravenous (IV) introduction in mice tail and focused imaging. Graphene is a reliable encapsulator in aerosol medium to create greater than >100 nm graphene particles covering Fe$_3$O$_4$ and AuNP. See Tables 3.1 and 3.2.

The brand particle strength was assessed for porosity, pore volume, and material property, vis-à-vis its zenith conjugation with CT imaging. Coughlin et al. (2014) used 120-nm silica particles with the gold shell to amalgamate PEG and gadolinium, thus making Au a resourceful platform for making multimodal particles. Since the relativity of T_1 is more than that of commercially available dimeglumine gadopentetate (Zeng et al., 2014), coupling AuNP (31 nm) plus gadolinium (1.9 nm) before imaging at 30 min proved CT had the most particle (gold) accumulation in the kidneys, with sediments seen in situ urinary bladder by MRI (Alric et al., 2008). To have a positive effect, NPs have to be created for a specific reason, one of which is targeted treatment in cancer management. Since NPs do not select a particular tissue, intravenous injection traveling in the bloodstream before using ultrasound, MRI, or CT can be "reference umpired" with or without contrast content, perhaps by using antibodies to improve selectively.

Nonetheless, current exogenous gel agents' immense importance for supplying main diagnostic information or data; certain side effects like nonspecific distribution, rapid clearance, reduced blood half-life, dangers of kidney toxicity, and poor resolution in obese patients with huge habitus should be noted. Biomedical imaging is an essential component in the diagnosis and management of diseased conditions. Modern technology has led to advances in imaging modalities such as MRI (McRobbie, 2007), sonography (Szabo, 2009), position emission tomography (PET) (NUH, 2007), and CAT (Hsieh, 2015). "Suspended colloidal gold" was first documented by Faraday in 1857 (Faraday, 1857). Au NPs were earlier introduced as an X-ray field block by Hainfeld, who imaged viscera and blood vessels of laboratory rats after AUNP (IV) injections (Hainfeld et al., 2006a,b).

NPs have great potency as therapeutic components, as they are engineered to deliver genes and drugs. Ultrasound can promote the accumulation of NPs and drug release in pathologic masses through the process of cavitation, with bridging of NPs reported to enter the blood–brain barrier (Timbie et al., 2014; Afodun et al., 2017). GNPs with high radiofrequency waves can be used for photodynamic (PDT) therapy, before luminous activation of transferring-coated TiO_2 photosensitizer and tumor mediation by a free-radical generation with immune cell infiltration (Kotagiri et al., 2015). One of the most intriguing modern applications of NPs in nanotechnology is the criteria for stratification into negative or positive contrast in MRI.

TABLE 3.1 Imaging Methods Derived Using Gold Nanoparticles.

Radiological method	Formulation/affiliation
CT	AuNP
X-ray	AuNP
Sonography	Acoustic impedance
MRI	Heavy metals (iron oxide, gadolinium)
Nuclear Medicine	[111]In, [64]Cu
Photoacoustic	Fluorescent dye

3.11 MICELLES

Micelles are chemical surfactant molecules scattered or dissolved in a colloidal liquid. They are of two forms: (1) inverse micelle and (2) typical micelle.

Typical micelle: also named normal phase micelle (oil in H_2O micelle); aggregate formation in aqueous solution with the proximity of hydrophilic caudal region in proximity with enveloped solvent.

Inverse micelle: also known as H_2O in oil micelle; it has its head region in proximity to the center point with protruding tails externally. All micelles are divided according to their shape, diameter, molecular affinity, and phase. Micelles have different medical applications in the form of drug delivery (Kataoka et al., 2012), imaging (Dubertret et al., 2002), and especially anticarcinogenic preparations. A research performed by Kim et al. (2014), through the use of stationed MRI, designed a cancer-marking contrast (CmR-CAs) by using pH-responsive polymer micelles, demonstrating the potency of CR-CA's for malignant diagnosis.

3.12 CORE–SHELL NPS

Core–shell NPs are particle nucleoid hybrid systems with resilient semi-conductivity, magnetism, and metal conductivity (Ghosh et al., 2011). In drug delivery and medical imaging core–shell, NPs are increasingly being used (Karamipour et al., 2015; Zhou et al., 2015). Single-bodied core–shell consists of bicovalent-bonding polymers separated by the accommodative groups at the shell and core. The core–shell functional mechanism is designed to merge conjugating Gd3+ at the rest-phase nucleoid with encapsulating doxorubicin (Dox) at the circumferential border.

Application of Nanoparticles in Biomedical Imaging

Previous literature has shown core–shelled agents promising a diagnostic delivery, as a superior MRI contrast agent (Lee et al., 2010; Zhu et al., 2013) for brain mapping (Yu et al., 2010) hepatic conditions (Ratanajanchai et al., 2014) and toward tumor growth (Ho et al., 2015) image analysis.

3.13 MACROMOLECULAR DENDRIMERS

Dendrimers are three-dimensional macromolecules with well-branched angled architecture, consisting of end groups, core, and branches (Crooks et al., 2001). Dendrimers can encapsulate different molecules in between their bonds for focused tissue-specific targeting purposes. Among the commercially available dendrimers are polypropylene imine (PPI) and polyamidoamine dendrimers (PAMAM) (Esumi et al., 2004; Majoros et al., 2006; Omidi et al., 2005). PEGylated dendrimers (plated with polyethylene glycol coating) are a sorted class that attracts scientists due to minimal toxicity in "heme"/blood circulation, less accumulation in visceral organs, and prolonged circulation time (Duncan and Izzo, 2015).

Medical imaging is a viable application of dendrimers (Zheng and Dickson, 2002) as well as controlled drug payload delivery, since the multipurpose nature of NPs [nanoprobes specifically for CT imaging (Zhu, 2015) cancer cells in vitro] optimizes some opaque contrast agents. PAMAM dendrimers have been recently used as delivery vehicles for MRI contrast agents, with gadolinium complexes dispatched to determine imaging properties of the synthesized compound in biological in vivo studies. Recent innovations in isomeric NPs are prepared with PEGylated polyethyleneimine-entrapped gold adjusted with folic acid or ferric for tumor CT imaging. Results from the dendrimer pilot study showed that this elaborate preparation holds excellent prospects to be used as a nanotransducer for CAT imaging suspicious masses (Zhout et al., 2016).

3.14 NANOPARTICLES IN MRI OPERATIONS

MRI is known to be the most advanced noninvasive body-viewing modality in radiology, as its mechanism of operation is by linear magnetic field concentrated on hydrogen protons in the body, which become excited and aligned before production of high-resolution images of tissue function, structures, and body cavities.

A type of MRI contrast correlates with remarkable changes in the bandwidth characteristics of MRI signal, which involves a transverse and a longitudinal relaxation capped time constant; TA and TB or $\mathbf{T_1}$ **and** $\mathbf{T_2}$ simultaneously. Depending on the application's target use, both types of resource techniques can be implemented on any MR system. Longitudinal relaxation time (T_1) causes positive contrast enhancement while supplied by (Gd3+) paramagnetic ion complexes. The finding shows toxicity fallback related to metallic ion (by damaging contrast agents) that rely on magnetic iron oxide NPs (SPIONs); which results in darker colored T_2—a weighted image representing nonparticulate agents (Fang and Zhang, 2009; Khalkhali et al., 2015).

Of recent, the most widely used contrast agents for MRI are small molecular weighted Gd3+ complexes and paramagnetic couplers that increase contrast by a nonspecific increase in H_2O proton relaxation rates within the "heme" in vessels. Examples are Gd-DTPA about product commercial name, Magnevist®, which is excreted by micturition.

3.15   Fe OXIDE NANOPROTEINS

Gadolinium, a nontoxic compound that chelates with the massive magnetic moment and unpaired electrons, is used in medical imaging. Park et al. (2014) conjugated Gd-DOTA (tetraazacyclododecane tetraacetic acid) from a tri-amino acid sequence, arginine-glycine-aspartate (RGD)-peptide precursor to fetch an MRI contrast agent with tumor-targeting capacity. The use of gadolinium in NP formulations can better understand tissue sensitivity to magnetic fields during imaging procedures (Helm, 2010). A widely available contrast for MRI is SPION.

SPIONs increase pixel quality indirectly by shortening the T_2 relaxation time of H_2O protons. T_2 relativity (SPION-based) is greater than that of gadolinium agents (Szpak et al., 2013). Surface plating, volume, size, and electrical charge can influence metabolism, biological distribution, and stability. Surface charging is a significant determinant particulate nature of NPs, with similar effects on plasma protein binding and in vivo clearance from the kidneys (Thomas et al., 2013). Furthermore, iron oxide deemed nonharmful is metabolized into hemoglobin (Gypta and Gupta, 2005; Laurent et al., 2008) with large iron elements increasing the possibility of toxicity (Qiao et al., 2009). It is vital to moderate SPIONs with the magnetic saturation field to reduce iron loading during in vivo procedures. SPIONs coupled with gadolinium substrates increase MRI contrast (Santra et al.,

2012), but NPs are channeled toward regional anatomy in the body (Zhang et al., 2016; Ni et al., 2016). Some biocompatible materials conjugated with marker moieties like chitosan (Szpak et al., 2014), glycosaminoglycan (Yang et al., 2014) antibodies (Wan et al., 2016), and thermal-sensitive proteins (Shevtsov et al., 2016) have been attached to SPIONs surface to seek (hexagonal-shaped) hepatocytes, cardiac molecules, macrophages, endothelial cells, tumors and minimize nonspecific uptake to improve biocompatibility.

3.16 NANOBIOPSY

Brain tumors are challenging and difficult to diagnose in the human body. Unlike other viscera in the body, the cerebrum's delicate nature, cerebellum, and protective calvaria make obtaining a biopsy sample dangerous. The decreased invasiveness enabled by nanotechnology offers a profound solution to this organ/faculty of coordination's unique specificity. A new technique based on a nano-controlled pen developed to fetch proteins by surface adhesion exists to map brain tumors. An endoscopic pen can be channeled to the brain through the nasal cavity, where "excised" biomolecular can be tapped without removing brain tissue. However, extreme caution must be exhibited due to the brain's delicate nature; accurate placement of endoscopic pen ball head by alternate routes through stereotaxic methods is necessary without damaging normal tissues around it.

3.17 NANOSHELLS FOR TREATMENT USES

Nanoshells are small beads coated superficially with (Ag) gold. Due to their diameter, they can be safely infused into rats or guinea pigs as a test site for cancer-specific lesioned localization, named enhanced permeation and retention (EPR). In the end-use dynamics of energy (radio, optical, motion, or mechanical) to cellular units, there is the subsequent pickup by nanoshells. Postnanoshell excitation causes relaxation to the base state forming spiked local heating and selective destruction of tumor cells without negatively affecting healthy cells' anatomy.

Targeted delivery of a nanotheragnostic medium through chronologic ablation of colorectal cancer (CRC) was described in the literature by Fortina et al. (2007). Surface-bound guanylyl cyclase C (GCC) still manifests on all intestinal epithelial cells, often as CRC growths.

3.18 REVIEW ON NANOPARTICLE CONTRAST DESIGN FOR IMAGING

Mulder et al. (2007) cited a multifunctional NP for diseased imaging arteries in atherosclerosis. The methodology employed uses a multistep synthesis of molecular leverage containing fluorophore contrast agent and targeting group. Micelle agent coating enclosing a quantum dot, polyethylene glycol (PEG), and modified phospholipids for MRI contrasts targets moiety in lipid embedding. A group of scientists from Washington University researched the use of microemulsions as contrast agents horned at atherosclerotic dots like fibrin (Winter et al., 2006), collagen III (Cyrus et al., 2006) or $\alpha_1 v_3$—integrin (Winter et al., 2008). The perfluorocarbon core allows MR imaging modality comprising "F" central nucleoid (Ahrens et al., 2005; Bulte, 2005) and the possibility of a hydrophilic inclusion drug. When MR images are recovered 3.5 h after the first injection, there was a rise in aortic combined pressure due to contrast agent accumulation, ascribed to anti-angiogenic effect of fumagillin injection before imaging.

3.19 THE USE OF QUANTUM DOTS IN NANOTECHNOLOGY

Quantum dots are semiconductor micro–nanocrystals with a neutral feature between discrete molecules and bulk semiconductors, with a diameter ranging between 2 and 10 nm. They display size variable fluorescent properties and charged energy levels (Larson et al., 2003). Part conductor NPs may collect at a target site by their characteristic permeability and concentration at a tumor site. Targeted clusters of QDs have low scientifically pro*vivo* using xenographic graft model involving human prostate cancer cells implanted in rats (Gao et al., 2004).

DNA-associated QDs are incubated with unknown nucleic acid samples to hybridize portions correlating to the oncogene. Photoillumination is the process of specific barcode generation after initial ignition by QDs prior sequence identification and isolation (Majumdar et al., 2002). The band wave of quantum dots enables the formation of unique tags that can be used to mark several deoxyribonucleic acid areas concurrently. A known hallmark of quantum dots–based protocols is its role in cancer detection due to dumping various independent DNA changes within a cell. Summarily, the (minimally) noninvasive nature of QDs makes it biopsy friendly. However, further investigations are needed in the area of its contested toxicologic side effects (Tang et al., 2010; Yan et al., 2011; Chen et al., 2012; Akbarzadeh et al., 2012; Navarro et al., 2012).

The energy level difference between the conduction and valence layer is called bandgap. This varies depending on the nature and features of the compound semiconductor. As the size of a quantum dot becomes smaller, the bandgap ratio becomes more massive, thereby exciting and initiating smaller particles (e.g., light with shorter particles-wavelength) (Belykh et al., 2015; Yang et al., 2015). The excitation commences in a wide variety of wavelengths, with the width and bilateral edges of the emission spectrum of the quantum dots becoming narrow. Since the wavelengths from (UV to infrared) are adjustable (Lodahl et al., 2015; Rabouw et al., 2015), it makes the use of quantum dots in medical imaging particularly attractive because of stability and high-brightness tone.

3.20 THE POTENCY OF NANOPARTICLE TYPE

A wide baserange of NPs have been suggested for use as contrast agents with different imaging mechanism requiring unique biochemical properties for each type. Gadolinium NPs such as micelles (Amirbekian et al., 2006), viruses (Anderson et al., 2006), CNTs (Sitharaman et al., 2005), and lipids are adaptable for providing contrast for T_1-weighted MRI through magnetic control fields. Iron-based superparamagnetic oxides have been modeled to generate T_2-weighted differences for MRI (Corot et al., 2006). Van Tilborg et al. (2006) found out that iron oxide NPs were generally more sensitive in imaging than gadolinium chelates. Gadolinium chelates most often produce (+ve) contrasts that early accumulate in the targeted tissue. Whereas iron oxides have (−ve) contrast, signal reduction in MRI imaging (with the exception) of those caused by iron oxide accumulation may be challenging to trace the source of the flow disruption. New image acquisition cycles generate positive contrast from iron oxide to level this difficulty (Briley-Saebo et al., 2007; Lipinski et al., 2008).

Quantum dots provide a reliable contrast medium for fluorescent imaging. Its characteristic varied excitation phase (Leatherdale et al., 2002); they glow with high efficiency and not subjectable to photobleaching (Sukhanova et al., 2004). When the fluorescent label is applied to NPs in imaging, proximity to the infrared band range of between 650 and 900 nm is advised, especially in tissue sensitivity (Sosnovik and Weissleder, 2005). The application is for fluorescence tomography where their tissue slices are needed; for light to penetrate tissues processed for confocal microscopy, about 10 µm of specimen depth is necessary. NPs involved in the use with CT contrast agents are based on their atomic numbers and electron density, like bismuth (Rabin et al., 2006), gold (Hainfeld et al., 2006), or iodine (Hyafil et al., 2007). GNPs are an indispensable choice for CT (Kim et al., 2007) due to quality X-ray attenuation during clinical trials. The ultrasound scanning system needs contrast agents that can provide sound waves to the transducer/probe, mostly achieved by oscillating micron-sized bubbles of decafluorobutane (a water-insoluble gas) (Villanueva et al., 1998). The bubbles' stability is achieved by external superficial coating, a hallmark of particle flexibility to yield a usable contrast of quality. Multilayer vesicles have been employed as sonographic contrast agents; however, the echogenic response generated entraps gaseous reverberations. Similarly, microemulsions have been experimented with to be used as sonographic contrast agents (Huang et al., 2002).

3.21 TYPES OF COATING

Raw substances used to generate contrast for molecular imaging usually having particular characteristics, such as reduced biocompatibility, fast excretion, decrease stability, and toxicity. Concerted efforts are being made to make the material biofriendly incorporating materials like polyvinylpyr-rolidone (Liu et al., 2007), phospholipids (Van Tilborg, 2006), silicon (Shang et al., 2002). External coating by the process of PEGylation has been widely used to improve half-life turnaround and particle ejection by the reticuloen-dothelial system (RES) (Otsuch et al., 2005).

An alternative to artificially manufactured coatings is the use of naturally existing NPs, such as lipoproteins or viruses, to evade the body's lymphatic system. A natural NP's external surface can be adjusted to contain contrast-producing dyes and ions (Sitharaman et al., 2005). Another approach from the former involves a blend of natural materials and inorganic substances at the core of the lipoprotein or virus (Huang et al., 2007; Sun et al., 2007; Cormode et al., 2008). The iron oxide, gold HDL, and quantum dots were collectively given the acronyms FeO–HDL, Au–HDL, and QD–HDL simul-taneously. Fluorescent and paramagnetic phospholipids were inducted into particles so that each provided contrast for MRI techniques and procedures and the indispensable use of Au–HDL in CTs. These NPs are absorbed by macrophage cells in vitro using a rat replica of induced atherosclerotic disease with CT and MRI modalities.

3.22 SYNTHETIC AND TARGETING STRATEGIES

There is better control of contrast properties and size by synthesized organic phased NPs (Gupta and Gupta, 2005; Michalet et al., 2005). The organic

Application of Nanoparticles in Biomedical Imaging 89

phase's key hallmark is that NPs synthesized in it are not soluble in H_2O and require a double coating to make it applicable. There are several other productive methods for doing this using phospholipid molecule or amphiphilic polymers (Kim et al., 2005; Yu et al., 2007).

A significant paradox before synthesis is how additional functions such as fluorophores and targeting groups would be attracted to the external membrane. Cross-linked iron oxides have a dextran coating modified with amine groups (McCarthy et al., 2007), which can be chemically targeted by groups of chelates, metal ions, fluorophores, or even conjugated (Nahrendorf et al., 2008). It should be noted that when there is marked angiogenesis in tissues, pathological conditions like atherosclerosis (Sanz and Fayad, 2008), cancer, or any other compromised cellular internal environs; leaky blood vessels lead to an elevated load of materials with extended half-life distribution caused by spiked retention and enhanced permeability (EPR) effect (Mulder et al., 2006).

A nonactive form of biotargeting involves the use of dextran as a particle coating compound, with macrophages actively taking up and absorbing iron-based NPs plated with dextran (Ruehm et al., 2001). Several types of molecular combination can be joined to NPs to produce antibody fragments (Kang et al., 2002), peptides (Nahrendorf et al., 2006), proteins (Schellenberger et al., 2004), sugars (Villaneva et al., 2007), small molecules (Zhang et al., 2005), and aptamers (Javier et al., 2005). The source targeting ligand must show a high chance of selectivity, but other factors must also be considered. In another instance, antibodies usage is expensive, highly specific, and their bonding may increase the particulate size by about 10 nm (Frangioni, 2008).

Techniques developed by Weissleder et al. (2005) where a cluster of small molecules targeted NPs where separated against several cell forms for identifying prospective ligands. The biotin–streptavidin pathway has been most used for NP channeling, but the immunologic reaction is experienced by some patients (Serda et al., 2007).

3.23 NANOPARTICLES AND NUCLEAR IMAGING

Nuclear medical imaging consists of two different 3D modalities, PET and SPECT, which both works on the same principle using a of variety contrast agents. The two depend on patient internalization of a radiotracer composed of a biological targeting ray and an isotope. The targeting element of the radiotracer generates relevant biological information, such as the metabolism of a compound.

A variety of isotopes used as nuclear imaging radiotracers can be added to GNPs to produce imaging PEGylated. The gold nanosphere has been used in research to attach ^{64}Cu to create PET imaging (He et al., 2014). The intravenous channel of positive tumor mice was used as the route of particle introduction with iodinated oil to improve tumor uptake, making CAT scans to produce high contrast after 60 min in the mass region (Tian et al., 2013), see Table 3.1. T-cell electroporation loaded with ^{64}Cu-labeled AuNP introduced intravenously can be used for mice imaging (Weissleder et al., 1989). In ^{64}C labeled, cells were injected into the dorsal vein and monitored using CAT imaging over a period of over 17 h (Li et al., 2014) (Fig. 3.1).

TABLE 3.2 Comparison of Different Selected Nanoparticle Properties.

Radiological modality	Form	Tissue biocompatibility	Structure	Size range (nm)	Material component	Selectivity
MRI	Diagnostic imaging	Shell required	Spherical	<4	Iron oxide	N/A
Sonography	Diagnostic imaging	Not removed (real time)	Spherical	500–3500	Perfluorocarbon emulsion	N/A
Laser explosion	Therapy	Natural	Spherical	15–40	Gold (Au)	Marker or viruses

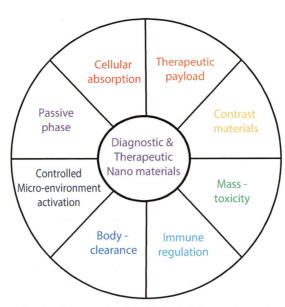

FIGURE 3.1 Schematic diagram showing biological interactions of nanoparticles with (purpose) setup mechanism and external environment correlation.

Application of Nanoparticles in Biomedical Imaging 91

3.24 ALIMENTARY TRACT DISTRIBUTION OF ANTI-INFLAMMATORY NANOPARTICLES USED IN TREATING INFLAMMATORY DISEASE

The larger intestinal caliber in humans consists of the ascending colon, transverse colon, descending colon, and sigmoid colon, while the first part of the ascending colon is the cecum. The colon is mainly the "recipient" organ in inflammatory bowel disease, Crohn's disease, and ulcerative enteritis. Drugs can be loaded into NPs before commencing the nanoimaging procedure. Micromolecules such as siRNA (Laroni et al., 2011), tripeptides, or proteins (Theiss et al., 2011) (in antibody or hormonal form) may be shelled in a saturated way inside an NP. The concurrent formation, extraction, and loading with anti-inflammatory elements were immediately followed by delivery to the colon. A recovery method was invented to target NPs to regions of the alimentary canal tract, using a soluble hydrogel bound through electrostatic interactions between negative polysaccharides and +ive ions. The static dual cross-link of alginate and chitosan regulated by SO_4^{2-} and Ca^{2+} in addition to rat GIT via double gavage formed gel agents. Drugs coupled with NPs are designed to withstand the "hostile" environmental pH and physiologic of the bowel mucosa. The combination of nanostructures and hydrogels (biomaterials) enabled dose reduction and progressive load and bulk addition to the large intestine, where the drug reduces inflammation (Laroui et al., 2012).

3.25 NANOPARTICLES AND BOWEL IMAGING

At present, the diseased condition of a colorectal carcinoma can be reliably evaluated by histopathology of the resected specimen. The above imaging applications use nanostructured to target and treat colorectal tumors (Weissleder et al., 2005). Nanostructure-based MRI contrasts show active potency when channeled for in vivo viewing and for differential diagnosis of CRC.

Molecular Imaging in Nanotechnology and Theranostics (MINT) was established in 2015. The unique size of NPs, their adaptable, functionalization abilities, and molecular structure are being "rewired" for use in biomedicine. Specified gene and drug-delivery strategies versus stimuli-responsive therapies are actively developed in pilot-experimental tests (Anselmo and Mitragotri, 2016). NPs can be remodified to suit specific imaging outcomes; for example, photoacoustic imaging (PAI) through a contrast medium ignites a mechanical response ultimately converted to ultrasound (Lemaster and Jokerst, 2016). The sonogram obtained is a noninvasive procedure and can penetrate deep inside tissues and with better pixel resolution compared to techniques based solely on optical imaging methods.

Plasmonic (Yang et al., 2013; Jing et al., 2014; Liu et al., 2015; Dixon et al., 2015), polymeric (Li and Liu, 2014; Lu et al., 2016; Yan et al., 2016; Xie et al., 2017) nanodevices like cantilevers have been integrated into a highly sensitive marker for disease diagnostics and staple-identifiers overprogrammed time frame to display performance properties. NP-based drug-delivery systems protect drugs from degradation; after that, eliminating dose and frequency of use, which ultimately improves patient comfort and downsizing cost of treatment and elevated signal-key intensity (Wang et al., 2014). Developments in our understanding of fluorogenic laced with FNPs later produce aggregate-induced emissions (AIE) (Yan et al., 2016). Subsequently, due to the fluorogenic broad emission wave, it is not suitable for multiplexing through Förster resonance energy transfer (Geng et al., 2014).

Nanofluorescence imaging can be overloaded by increased false-positive rates (Tummers et al., 2015); developments in bond specificity have been reported in the literature, leading to the higher occurrence of total tumor regression (Stummer et al., 2006). NP, when combined with positive target antibodies, surface-enhanced (resonance) Raman scattering NPs (SErRSNPs), and peptides were reported to decimate tumors (postsurgically) and intraoperatively (Kircher et al., 2012). Other applications of the latter are malignant tumor detection (through trace markers) and accurate diagnosis for a variety of visceral metastasis and foci glioblastoma (Karabeber et al., 2014; Harmsen et al., 2015; Nayak et al., 2016; Huang et al., 2016; Oseldehyk et al., 2017). In financial and logical terms, it is reasonable to produce supplementary agents to achieve the potential of current radiological devices in diagnosis/detection and medical intervention in pathological conditions. Naturally, there is a gulf distinctive contrast (in humans under anatomically "normal" conditions) between bone and surrounding soft tissues, with or without nano-aided imaging radiologically (Cormode et al., 2010; He et al., 2015). There are shortfalls of clinical procedures in body image enhancement, such as on the accurate gagging of renal toxicity, blood half-life, nonspecific biodistribution, and lipid clearance in different human body habitus (Razi et al., 2015).

NP agents are a promising strategy for safe, noninvasive diagnosis; mostly compatible with traditional contrast agents due to the following:

1. The presence of surface area, which can bond, functionalize with targeted molecules at different densities.
2. Flexible physicochemical property and shortened plasma movement time.
3. Drugs and agents can be added at predetermined ratios internally or on the surface (Janib et al., 2010).

3.26 PHYSIOLOGY OF CLEARANCE ROUTES

Most nanomaterials are manufactured for use as contrast agents in medical imaging and are still hampered by ultrafiltration in the kidney loop of Henlé. Furthermore, there is a limitation by the process of nonspecific stalling and accumulation in the mononuclear phagocyte system (MPS) found in the lymphatic drainage, liver, and spleen tissue (Mohammadian et al., 2016; Sadeghzadeh et al., 2017).

NPs with greater (hydrodynamic) diameters of 200 nm and above are meshed by splenic MPS while smaller NPs <10 nm are porous to the RES, with traces found in the hepatocytes (Amirsaadat et al., 2016). Irrespective of size, surface properties of NPs play vital roles in RES uptake. It has been discovered that neutral (apolar) and hydrophilic surfaces are not susceptible to opsonization or absorption (Sum et al., 2008; Farayzadeh et al., 2017). Certain products degenerate into very toxic materials during catabolism in vivo and are considered unsafe for use in nanotechnology (Lee et al., 2012).

3.27 LIMITATIONS IN THE USE OF NP-BASED CONTRASTS

An iron oxide-containing NP in some settings has been introduced into clinical trials (Kiessling et al., 2014) and put to use; however, there are challenges in obtaining wanted pharmacokinetic properties and substance homogeneity. Other concerns about nephrologists, radiologists, oncologists, and internal physicians are tracking biodegradation, elimination, and toxicity. To alleviate these limitations, the most commonly used method is the alternate stereological stabilization of NP by polyethylene glycol coating (Jokerst et al., 2011). Therefore, advances in optimal features associated with rapid body clearance time and tumor-specific contrast targeting are needed to design NPs.

3.28 REDUCING THE EFFECTS OF TOXICITY

NPs are physiologically handled in the aspect of toxicity of drugs or contrast agents in humans. Larger molecular-weighted particles initially remain in the spleen and liver before been broken down (Van Schooneveld, 2008).

Shaw et al. (2008) used different in vitro assays to expand NPs' outcomes on various cellular processes. Some NPs degenerate into nontoxic materials and are considered nonharmful. It should be noted that the human body has limited ability to process iron; therefore, if the iron oxide is injected in a crystalline form, it is nontoxic, rapidly splits to release Fe^- iron ions at a

94   Diverse Applications of Nanotechnology in the Biological Sciences

determined rate. In cases where particles break into toxic substances like quantum dots (cadmium is ejected), doubts will emerge by a physician on its relevance for clinical use (Mancini et al., 2008).

3.29 FUTURE DIRECTIONS

NPs are becoming indispensable imaging agents, projecting multimodal radiology into the limelight (Chen et al., 2015; Rieffel et al., 2015). Future developments promise to predict NP distribution before administration, mainly due to the high variability of the EPR effect. Such a process may take the form of microcosmic computer models (van-de-Ven et al., 2015) or companion NPs (Miller et al., 2015). Artificially made materials biochemically derived and synthesized NPs obtained using "green chemistry" or cells (Wu et al., 2013; Sharma et al., 2012; Kikuchi et al., 2016) can overlap the limits of biocompatibility and become covert to the body's immune system through biomimicry (Parodi et al., 2013). Biocompatible advancement could enable selective chemical manipulation of NPs in vivo (Broun et al., 2014). The morphology of NPs from the laboratory to the clinic gives room for modern projected applications like image-guided oncosurgery (Kim et al., 2016) and the use of iron oxide reprogram to attack tumor cells (Zanganeh et al., 2016).

3.30 CONCLUSION

Nanotechnology and nanostructures are under sustained development toward specialized micro-imaging. Since NPs have a size range of 1–100 nm, it is functionalized to display atomic, molecular, and cellular properties at the electronic level. Nanomaterials are now frequently used in therapeutic, diagnostic, and targeted drug channel delivery. The combination of NP technological modalities and telecommunication will improve the theragnostic procedure's accuracy, which eventually affects healthcare. In the future (SSS), sensitive sensor systems would allocate biomarkers (analytes) to nanotools (quantum dots and liposomes) with nanodevices (biochips and nanowire), which would allow innovative measurements such as electromagnetic transmission and biomagnetism for motion detection in tissues.

Since organic dyes have been used as contrast agents in radiologic imaging, NPs' multipurpose properties mirror a promising strategy for noninvasive diagnosis. However, most are still being tested in vitro/in vivo

Application of Nanoparticles in Biomedical Imaging

(pilot groups) with laboratory animals due to poisoning concerns. Medically efficient imaging contrast agents can be manufactured using NPs as their bedrock reagent. At the current NP research advancement rate (NPs) contrast, agents with the required attributes can be synthesized for any desired application. There are promising prospects for significant developments in efficacy, detection, and specificity of disease targets. Summarily, nanotechnology will further produce particles that harbor exciting properties, which will eventually be explored in a novel contrast agent's formulations to be used in medical imaging.

With the advent of multipurpose theragnostic NPs undergoing clinical trials, we are ushering in the era of "controlled imaging," producing a comprehensive, smart, customizable, and specific end-user (radiologist/oncologist/imaging-scientist) advantage. This review discussed various conditions and factors when considering the design of an NP probe and expatiate examples of the most advanced types and fluorescence techniques. There is also a need to be cautious about placing the premature burden on speculative hopes or concerns about nanotechnologies documented ahead of evidence, since it carries the risk of public engagement and worries over a developmental issue that is ongoing. Nanostructures can be used to adjust traditional contrast agents like gadolinium or imaging agents like iron oxide to enhance the diagnostic potency of clinical imaging (Harisinghani et al., 2003; Perez et al., 2003; Mahmood and Weissleder, 2003; Kobayashi et al., 2005).

KEYWORDS

- **nanoparticles**
- **imaging**
- **theragnostic**
- **molecular**

REFERENCES

Afodun, A. M.; Eze, E. D.; Bakare, A. A. et al. Comparative Ultrasound Review of Free Intra-Peritoneal Fluid (Ascites). *Open J. Med. Imag.* **2017,** *7,* 229–236.

Ahrens, E.T.; Flores, R.; Xu, H.; Morel, P. A. In Vivo Imaging Platform for Tracking Immunotherapeutic Cells. *Nat. Biotech* **2005,** *23,* 983–987.

Akbarzadeh, A.; Mikaeili, H.; Zarghami, N. et al. Preparation and in Vitro Evaluation of Doxorubicin-Loaded Fe_3O_4 Magnetic Nanoparticles Modified with Biocompatible Copolymers. *Int. J. Nanomed.* **2012,** *7,* 511–526.

Alric, C.; Taleb, J.; Le, G.; Duc et al. Gadolinium Chelate Coated Gold Nanoparticles as Contrast Agents for Both X-ray Computed Tomography and Magnetic Resonance Imaging. *J. Am. Chem. Soc.* **2008,** *130* (18), 5908–5915.

Amirbekian, V.; Lipinski, M. J.; Briley-Saebo, K. C.; Amirbekian, S.; Aguinaldo, J. G.; Weinreb, D. B.; Vucic, E.; Frias, J. C.; Hyafil, F.; Mani, V.; Fisher, E. A.; Fayad, Z. A. Detecting and Assessing Macrophages in Vivo to Evaluate Atherosclerosis Noninvasively Using Molecular MRI. *Proc. Natl. Acad. Sci. U. S. A.* **2007,** *104,* 961–966. [PubMed: 17215360].

Amirsaadat, S.; Pilehvar-Soltanahmadi, Y.; Zarghami, F. et al. Silibinin-Loaded Magnetic Nanoparticles Inhibit hTERT Gene Expression and Proliferation of Lung Cancer Cells. *Artif. Cells Nanomed. Biotechnol. Forthcoming.* [cited Jan 12, 2017]. doi: 10.1080/21691401.2016.1276922.

Anderson, E. A.; Isaacman, S.; Peadbody, D. S.; Wang, E. Y.; Canary, J. W.; Kirshenbaum, K. Viral Nanoparticles Donning a Paramagnetic Coat: Conjugation of MRI Contrast Agents to the MS2 Capsid. *Nano Lett.* **2006,** *6,* 1160–1164. [PubMed: 16771573].

Anselmo, A. C.; Mitragotri, S. Nanoparticles in the Clinic. *Bioeng. Transl. Med.* **2016,** *1,* 10–29.

Arifin, D. R.; Long, C. M.; Gilad, A. A. et al. Trimodal Gadolinium-Gold Microcapsules Containing Pancreatic Islet Cells Restore Normoglycemia in Diabetic Mice and Can Be Tracked by Using US, CT, and Positive-Contrast MR Imaging. *Radiology* **2011,** *260* (3), 790–798.

Belykh, V.; Yakovlev, D.; Schindler, J. et al. Large Anisotropy of Electron and Hole G Factors in Infrared-Emitting InAs/InAlGaAs Selfassembled Quantum Dots. *Phys. Rev.* **2015,** *93,* 125302.

Braun, G. B.; Friman, T.; Pang, H. B. et al. Etchable Plasmonic Nanoparticle Probes to Image and Quantify Cellular Internalization. *Nat. Mater.* **2014,** *13,* 904–911. [PubMed: 24907927].

Briley-Saebo, K. C.; Mulder, W. J. M.; Mani, V.; Hyafil, F.; Amirbekian, V.; Aguinaldo, J. G. S.; Fisher, E. A.; Fayad, Z. A. Magnetic Resonance Imaging of Vulnerable Atherosclerotic Plaques: Current Imaging Strategies and Molecular Imaging Probes. *J. Magn. Reson. Imag.* **2007,** *26,* 460–479. [PubMed: 17729343].

Bulte, J. Hot spot MRI Emerges from the Background. *Nat. Biotech.* **2005,** *23,* 945–946.

Cai, H.; Li, K.; Li, J. et al. Dendrimer-Assisted Formation of Fe 3O4/Au Nanocomposite Particles for Targeted Dual Mode CT/MR Imaging of Tumors. *Small* **2015,** *11* (35), 4584–4593.

Carlton, R. R.; Adler, A. M. *Principles of Radiographic Imaging: An Art and A Science*; Delmar/Cengage Learning: Clifton Park, NY, 2013.

Cheheltani, R.; Ezzibdeh, R. M.; Chhour, P. et al. Tunable, Biodegradable Gold Nanoparticles as Contrast Agents for Computed Tomography and Photoacoustic Imaging. *Biomaterials* **2016,** *102,* 87–97.

Chen, F.; Rieffel, J.; Chen, G. et al. Hexamodal Imaging in Vivo with Nanoparticles. *J. Nucl. Med.* **2015,** *56,* 56–56. [PubMed: 25525184].

Application of Nanoparticles in Biomedical Imaging

Chen, N.; He, Y.; Su, Y. et al. The Cytotoxicity of Cadmium-Based Quantum Dots. *Biomaterials* **2012,** *33,* 1238–1244.

Cheng, G.; Mi, L.; Cao, Z. Q. et al. Functionalizable and Ultrastable Zwitterionic Nanogels. *Langmuir* **2010,** *26,* 6883–6886.

Cho, E. C.; Glaus, C.; Chen, J. et al. Inorganic Nanoparticle-Based Contrast Agents for Molecular Imaging. *Trends Mol. Med.* **2010,** *16,* 561–573.

Cormode, D. P.; Jarzyna, P. A.; Mulder, W. J. et al. Modified Natural Nanoparticles as Contrast Agents for Medical Imaging. *Adv. Drug Deliv. Rev.* **2010,** *62,* 329–338.

Cormode, D. P.; Naha, P. C.; Fayad, Z. A. Nanoparticle Contrast Agents for Computed Tomography: A Focus on Micelles. *Contrast Media Mol. Imag.* **2014,** *9,* 37–52.

Cormode, D. P.; Skajaa, T.; van Schooneveld, M. M.; Koole, R.; Jarzyna, P.; Lobatto, M. E.; Calcagno, C.; Barazza, A.; Gordon, R. E.; Zanzonico, P.; Fisher, E. A.; Fayad, Z. A.; Mulder, W. J. M. Nanocrystal Core High-Density Lipoproteins: A Multimodal Molecular Imaging Contrast Agent Platform. *Nano Lett* **2008,** *8,* 3715–3723. [PubMed: 18939808].

Corot, C.; Robert, P.; Idee, J. M.; Port, M. Recent Advances in Iron Oxide Nanocrystal Technology for Medical Imaging. *Adv. Drug Deliv. Rev.* **2006,** *58,* 1471–1504. [PubMed: 17116343].

Coughlin, A. J.; Ananta, J. S.; Deng, N.; Larina, I. V.; Decuzzi, P.; West, J. L. Gadolinium-Conjugated Gold Nanoshells for Multimodal Diagnostic Imaging and Photothermal Cancer Therapy. *Small* **2014,** *10* (3), 556–565.

Crooks, R. M.; Zhao, M.; Sun, L. et al. Dendrimer-Encapsulated Metal Nanoparticles: Synthesis, Characterization, and Applications to Catalysis. *Acc. Chem. Res.* **2001,** *34,* 181–190.

Cyrus, T.; Abendschein, D. R.; Caruthers, S. D.; Harris, T. D.; Glattauer, V.; Werkmeister, J. A.; Ramshaw, J. A. M.; Wickline, S. A.; Lanza, G. M. MR Three-Dimensional Molecular Imaging of Intramural Biomarkers with Targeted Nanoparticles. *J. Cardiovasc. Magn. Reson.* **2006,** *8,* 1–7.

Dixon, A. J.; Hu, S.; Klibanov, A. L.; Hossack, J. A. Oscillatory Dynamics and In Vivo Photoacoustic Imaging Performance of Plasmonic Nanoparticle-Coated Microbubbles. *Small* **2015,** *11,* 3066–3077. [PubMed: 25703465].

Dubertret, B.; Skourides, P.; Norris, D. J. et al. In Vivo Imaging of Quantum Dots Encapsulated in Phospholipid Micelles. *Science.* **2002,** *298,* 1759–1762.

Duncan, R.; Izzo, L. Dendrimer Biocompatibility and Toxicity. *Adv. Drug Deliv. Rev.* **2005,** *57,* 2215–2237.

Eichenbaum, G. M.; Kiser, P. F.; Simon, S. A. et al. pH and Ion-Triggered Volume Response of Anionic Hydrogel Microspheres. *Macromolecules* **1998,** *31,* 5084–5093.

Esumi, K.; Isono, R.; Yoshimura, T. Preparation of PAMAM-and PPImetal (Silver, Platinum, and Palladium) Nanocomposites and Their Catalytic Activities for Reduction of 4-nitrophenol. *Langmuir* **2004,** *20,* 237–243.

Etzioni, R.; Urban N, Ramsey S, et al. The Case for Early Detection. *Nat. Rev. Cancer* **2003,** *3,* 243–252.

Fang, C.; Zhang, M. Multifunctional Magnetic Nanoparticles for Medical Imaging Applications. *J. Mater. Chem.* **2009,** *19,* 6258–6266.

Faraday, M. The Bakerian Lecture: Experimental Relations of Gold (and Other Metals) to Light. *Phil. Trans. R. Soc. A: Math. Phys. Eng. Sci.* **1857,** *147,* 145–181.

Farajzadeh, R.; Pilehvar-Soltanahmadi, Y.; Dadashpour, M. et al. Nano-Encapsulated Metformin-Curcumin in PLGA/PEG Inhibits Synergistically Growth and hTERT Gene

Expression in Humanbreast Cancer Cells. *Artif. Cells Nanomed. Biotechnol.* **2017,** Forthcoming. [cited 2017 Jul 5]. doi: 10.1080/21691401.2017.1347879.

Feng, W.; Zhou, X.; Nie, W. et al. Au/polypyrrole@Fe_3O_4 Nanocomposites for MR/CT Dual-Modal Imaging Guidedphotothermal Therapy: An in Vitro Study. *ACS Appl. Mater. Interf.* **2015,** *7* (7), 4354–4367.

Ferrari, M. Cancer Nanotechnology: Opportunities and Challenges. *Nat. Rev. Cancer* **2005,** *5*, 161–171.

Fortina, P.; Kricka, L. J.; Graves, D. J. et al. Applications of Nanoparticles to Diagnostics and Therapeutics in Colorectal Cancer. *Trends Biotechno.* **2007,** *25*, 145–152.

Frangioni, J. V. New Technologies for Human Cancer Imaging. *J. Clin. Oncol.* **2008,** *26*, 4012–4021. [PubMed: 18711192].

Gaglia, J. L.; Harisinghani, M.; Aganj, I, et al. Noninvasive Mapping of Pancreatic Inflammation in Recent-Onset Type-1 Diabetes Patients. *Proc. Natl. Acad. Sci. USA.* **2015,** *112*, 2139–2144. [PubMed:25650428].

Gao, X. H.; Cui, Y. Y.; Levenson, R. M. et al. In Vivo Cancer Targeting and Imaging with Semi-Conductor Quantum Dots. *Nat. Biotechnol.* **2004,** *22*, 969–976.

Geng, J.; Zhu, Z.; Qin, W. et al. Near-Infrared Fluorescence Amplified Organic Nanoparticles with Aggregation-Induced Emission Characteristics for in Vivo Imaging. *Nanoscale* **2014,** *6*, 939–945. [PubMed: 24284804].

Ghosh, Chaudhuri R.; Paria, S. Core/Shell Nanoparticles: Classes, Properties, Synthesis Mechanisms, Characterization, and Applications. *Chem. Rev.* **2011,** *112*, 2373–2433.

Gupta, A. K.; Gupta, M. Synthesis and Surface Engineering of Iron Oxide Nanoparticles for Biomedical Applications. *Biomaterials* **2005,** *26*, 3995–4021.

Gupta, A. K.; Gupta, M. Synthesis and Surface Engineering of Iron Oxide Nanoparticles for Biomedical Applications. *Biomaterials* **2005,** *26*, 3995–4021. [PubMed: 15626447].

Hainfeld, J. F.; Slatkin, D. N.; Smilowitz, H. M. The Use of Gold Nanoparticles to Enhance Radiotherapy in Mice. *Phys. Med. Biol.* **2004,** *49* (18), N309 –N315.

Hainfeld, J. F.; Slatkin, D. N.; Focella, T. M.; Smilowitz, H. M. Gold Nanoparticles: A New X-ray Contrast Agent. *Br. J. Radiol.* **2006a,** *79* (939), 248–253.

Hainfeld, J. F.; Slatkin, D. N.; Focella, T. M.; Smilowitz, H. M. Gold Nanoparticles: A New X-ray Contrast Agent. *Br. J. Radiol.* **2006b,** *79*, 248–253. [PubMed: 16498039].

Harisinghani, M. G.; Barentsz, J.; Hahn, P. F. et al. Noninvasive Detection of Clinically Occult Lymph-Node Metastases in Prostate Cancer. *New Engl. J. Med.* **2003,** *348*, 2491–U5.

Harmsen, S.; Bedics, M. A.; Wall, M. A.; Huang, R.; Detty, M. R.; Kircher, M. F. Rational Design of a Chalcogenopyrylium-Based Surface-Enhanced Resonance Raman Scattering Nanoprobe with Attomolar Sensitivity. *Nat. Commun.* **2015,** *6*, 6570. [PubMed: 25800697].

He, W.; Ai. K.; Lu, L. Nanoparticulate X-ray CT contrast agents. *Sci China Chem.;* 58, 753–760.

He, X.; Liu, F.; Liu, L.; Duan, T.; Zhang, H.; Wang, Z. Lectinconjugated Fe_2O_3@ Au Core@ Shell Nanoparticles as Dual Mode Contrast Agents for in Vivo Detection of Tumor. *Mol. Pharm.* **2014,** *11* (3), 738–745.

Helm, L. Optimization of Gadolinium-Based MRI Contrast Agents for High Magnetic-Field Applications. *Future Med. Chem.* **2010,** *2*, 385–396.

Hill, T. K.; Mohs, A. M. Image-Guided Tumor Surgery: Will There Be a Role for Fluorescent Nanoparticles? Wiley Interdisciplinary Reviews. *Nanomed. Nanobiotechnol.* **2016,** *8*, 498–511. [PubMed: 26585556].

Ho, L. C.; Hsu, C. H.; Ou, C. M. et al. Unibody Core–Shell Smart Polymer as a Theranostic Nanoparticle for Drug Delivery and MR Imaging. *Biomaterials*. **2015**, *37*, 436–446.

Hsieh, J. *Computed Tomography: Principles, Design, Artifacts, and Recent Advances*; SPIE Press, 3rd ed., 2015.

Huang, R.; Harmsen, S.; Samii, J. M. et al. High Precision Imaging of Microscopic Spread of Glioblastoma with a Targeted Ultrasensitive SERRS Molecular Imaging Probe. *Theranostics*. **2016**, *6*, 1075–1084. [PubMed: 27279902].

Huang, S. L.; Hamilton, A. J.; Pozharski, E.; Nagaraj, A.; Klegerman, M. E.; McPherson, D. D.; MacDonald, R. C. Physical Correlates of the Ultrasonic Reflectivity of Lipid Dispersions Suitable as Diagnostic Contrast Agents. *Ultrasound Med. Biol.* **2002**, *28*, 339–348. [PubMed: 11978414].

Huang, X.; Bronstein, L. M.; Retrum, J.; Dufort, C.; Tsvetkova, I.; Aniagyei, S.; Stein, B.; Stucky, G.; McKenna. B.; Rennes, N.; Baxter, D.; Kao, C. C.; Dragnea, B. Self-Assembled Virus-Like Particles with Magnetic Cores. *Nano Lett.* **2007**, *7*, 2407–2416. [PubMed: 17630812].

Hyafil, F.; Cornily, J. C.; Feig, J. E.; Gordon, R.; Vucic, E.; Amirbekian, V.; Fisher, E. A.; Fuster, V.; Feldman, L. J.; Fayad, Z. A. Noninvasive Detection of Macrophages Using a Nanoparticulate Contrast Agent for Computed Tomography. *Nat. Med.* **2007**, *13*, 636–641. [PubMed: 17417649].

Jabir, N. R.; Tabrez, S.; Ashraf, G. M. et al. Nanotechnology-Based Approaches in Anticancer Research. *Int. J. Nanomed.* **2012**, *7*, 4391–408.

Jamieson, T.; Bakhshi, R.; Petrova, D, et al. Biological Applications of Quantum Dots. *Biomaterials* **2007**, *8*, 4717–4732.

Janib, S. M.; Moses, A. S.; MacKay, J. A. Imaging and Drug Delivery Using Theranostic Nanoparticles. *Adv. Drug Deliv. Rev.* **2010**, *62*, 1052–1063.

Javier, D. J.; Nitin, N.; Levy, M.; Ellington, A.; Richards-Kortum, R. Aptamer-Targeted Gold Nanoparticles as Molecular-Specific Contrast Agents for Reflectance Imaging. *Bioconjugate Chem.* **2008**, *19*, 1309–1312.

Jin, Y.; Wang, J.; Ke, H.; Wang, S.; Dai, Z. Graphene Oxide Modified PLA Microcapsules Containing Gold Nanoparticles for Ultrasonic/CT Bimodal Imaging Guided Photothermal Tumor Therapy. *Biomaterials* **2013**, *34* (20), 4794–4802.

Jing, L.; Liang, X.; Deng, Z. et al. Prussian Blue Coated Gold Nanoparticles for Simultaneous Photoacoustic/CT Bimodal Imaging and Photothermal Ablation of Cancer. *Biomaterials*. **2014**, *35*, 5814–5821. [PubMed: 24746962].

Jokerst, J. V.; Lobovkina, T.; Zare, R. N. et al. Nanoparticle PEGylation for Imaging and Therapy. *Nanomedicine* **2011**, *6*, 715–728.

Jost, G.; Pietsch, H.; Lengsfeld, P.; Hutter, J.; Sieber, M. A. The Impact of the Viscosity and Osmolality of Iodine Contrast Agents on Renal Elimination. *Investig. Radiol.* **2010**, *45* (5), 255–261.

Kabanov, A. V.; Vinogradov, S. V. Nanogels as Pharmaceutical Carriers: Finite Networks of Infinite Capabilities *Angew. Chem. Int. Ed.* **2009**, *48*, 5418–5429.

Kamila, S.; McEwan, C.; Costley, D. et al. Diagnostic and Therapeutic Applications of Quantum Dots in Nanomedicine. *Top Curr. Chem.* **2016**, *370*, 203–224. [PubMed: 26589510].

Kang, H. W.; Weissleder, R.; Bogdanov, A. Targeting of MPEG-Protected Polyamino Acid Carrier to Human E-selectin in Vitro. *Amino Acids* **2002**, *23*, 301–308. [PubMed: 12373551].

Karabeber, H.; Huang, R.; Iacono, P. et al. Guiding Brain Tumor Resection Using Surface-Enhanced Raman Scattering Nanoparticles and a Hand-Held Raman Scanner. *ACS Nano.* **2014,** *8,* 9755–9766. [PubMed: 25093240].

Karamipour, S.; Sadjadi, M. S.; Farhadyar, N. Fabrication and Spectroscopic Studies of Folic Acid-Conjugated Fe_3O_4@Au Core–Shell for Targeted Drug Delivery Application. *Spectrochim Acta A: Mol. Biomol. Spectrosc.* **2015,** *148,* 146–155.

Kataoka, K.; Harada, A.; Nagasaki, Y. Block Copolymer Micelles for Drug Delivery: Design, Characterization and Biological Significance. *Adv. Drug Deliv. Rev.* **2012,** *64,* 37–48.

Ke, H.; Yue, X.; Wang, J. et al. Gold Nanoshelled Liquid Perfluorocarbon Nanocapsules for Combined Dual Modal Ultrasound/CT Imaging and Photothermal Therapy of Cancer. *Small* **2014,** *10* (6), 1220–1227.

Khalkhali, M.; Rostamizadeh, K.; Sadighian, S. et al. The Impact of Polymer Coatings on Magnetite Nanoparticles Performance as MRI Contrast Agents: A Comparative Study. *DARU J. Pharm. Sci.* **2015,** *23,* 1.

Kiessling, F.; Mertens, M. E.; Grimm, J. et al. Nanoparticles for Imaging: Top or Flop? *Radiology* **2014,** *273,* 10–28.

Kikuchi, F.; Kato, Y.; Furihata, K. et al. Formation of Gold Nanoparticles by Glycolipids of Lactobacillus casei. *Sci. Rep.* **2016,** *6,* 34626. [PubMed: 27725710].

Kim, B. S.; Qiu, J. M.; Wang, J. P.; Taton, T. A. Magnetomicelles: Composite Nanostructures from Magnetic Nanoparticles and Crosslinked Amphiphilic Block Copolymers. *Nano Lett.* **2005,** *5,* 1987–1991. [PubMed: 16218723].

Kim, D.; Park, S.; Lee, J. H.; Jeong, Y. Y.; Jon, S. Antibiofouling Polymer-Coated Gold Nanoparticles as a Contrast Agent for in Vivo X-ray Computed Tomography Imaging. *J. Am. Chem. Soc.* **2007,** *129,* 7661–7665. [PubMed: 17530850].

Kim, D.; Yu, M. K.; Lee, T. S.; Park, J. J.; Jeong, Y. Y.; Jon, S. Amphiphilic Polymer-Coated Hybrid Nanoparticles as CT/MRIdual Contrast Agents. *Nanotechnology* **2011,** *22* (15), ID 155101.

Kim, K. S.; Park, W.; Hu, J. et al. A Cancer-Recognizable MRI Contrast Agents Using pH-Responsive Polymeric Micelle. *Biomaterials* **2014,** *35,* 337–343.

Kim, S. E.; Zhang, L.; Ma, K. et al. Ultrasmall Nanoparticles Induce Ferroptosis in Nutrient-Deprived Cancer Cells and Suppress Tumour Growth. *Nat. Nano.* **2016,** *11,* 977–985.

Kircher, M. F.; de la Zerda, A.; Jokerst, J. V. et al. A Brain Tumor Molecular Imaging Strategy Using a New Triple-Modality MRI-Photoacoustic-Raman Nanoparticle. *Nat. Med.* **2012,** *18,* 829–834. [PubMed: 22504484].

Kircher, M. F.; Willmann, J. K. Molecular Body Imaging: MR Imaging, CT, and US. Part I. Principles. *Radiology* **2012,** *263,* 633–643. [PubMed: 22623690].

Kobayashi, H.; Kawamoto, S.; Brechbiel, M. W. et al. Detection of Lymph Node Involvement in Hematologic Malignancies Using Micromagnetic Resonance Lymphangiography with a Gadolinum-Labeled Dendrimer Nanoparticle. *Neoplasia* **2005,** *7,* 984–991.

Kotagiri, N.; Sudlow, G. P.; Akers, W. J.; Achilefu, S. Breaking the Depth Dependency of Phototherapy with Cerenkov Radiation and Low-Radiance-Responsive Nanophotosensitizers. *Nat Nano.* **2015,** *10,* 370–379.

Lanza, G. M.; Winter, P. M.; Caruthers, S. D.; Hughes, M. S.; Cyrus, T.; Marsh, J. N.; Neubauer, A. M.; Partlow, K. C.; Wickline, S. A. Nanomedicine Opportunities for Cardiovascular Disease with Perfluorocarbon Nanoparticles. *Nanomedicine* **2006,** *1,* 321–329. [PubMed: 17716162].

Laroui, H.; Dalmasso, G.; Nguyen, H. T. et al. Drug-Loaded Nanoparticles Targeted to the Colon with Polysaccharide Hydrogel Reduce Colitis in a Mouse Model. *Gastroenterology* **2010**, *138*, 843–853 e1–2.

Laroui, H.; Sitaraman, S. V.; Merlin, D. Gastrointestinal Delivery Of Anti-Inflammatory Nanoparticles. *Methods Enzymol.* **2012**, *509*, 101–125.

Larson, D. R.; Zipfel, W. R.; Williams, R. M. et al. Water-Soluble Quantum Dots for Multiphoton Fluorescence Imaging in Vivo. *Science* **2003**, *300*, 1434–1436.

Laurent, S.; Forge, D.; Port, M. et al. Magnetic iron Oxide Nanoparticles: Synthesis, Stabilization, Vectorization, Physicochemical Characterizations, and Biological Applications. *Chem. Rev.* **2008**, *108*, 2064–2110.

Leatherdale, C. A.; Woo, W. K.; Mikulec, F. V.; Bawendi, M. G. On the Absorption Cross Section of CdSe Nanocrystal Quantum Dots. *J. Phys. Chem. B* **2002**, *116*, 7619–7622.

Lee, D. S.; Im, H. J.; Lee, Y. S. Radionanomedicine: Widened Perspectives of Molecular Theragnosis. Nanomedicine: Nanotechnology. *Biol. Med.* **2016**, *11*, 795–810.

Lee, J. H.; Park, G.; Hong, G. H. et al. Design Considerations for Targeted Optical Contrast Agents. *Quant. Imag. Med. Surg.* **2012**, *2*, 266–273.

Lee, P. W.; Hsu, S. H.; Wang, J. J. et al. The Characteristics, Biodistribution, Magnetic Resonance Imaging and Biodegradability of Superparamagnetic Core–Shell Nanoparticles. *Biomaterials* **2011**, *31*, 1316–1324.

Lemaster, J. E.; Jokerst, J. V. What Is New in Nanoparticle-Based Photoacoustic Imaging? *Wiley Interdiscipli. Rev.: Nanomed. Nanobiotechnol.* **2016**, n/a-n/a.

Lenhard, D. C.; Pietsch, H.; Sieber, M. A. et al. The Osmolality of Nonionic, Iodinated Contrast Agents as an Important Factor for Renal Safety. *Investig. Radiology* **2012**, *47* (9), 503–510.

Lentacker, I.; De Smedt, S. C.; Sanders, N. N. Drug Loaded Microbubble Design for Ultrasound Triggered Delivery. *Soft Matter* **2009**, *5* (11), 2161–2170.

Li, H.; Diaz, L.; Lee, D.; Cui, L.; Liang, X.; Cheng, Y. In Vivo Imaging of T Cells Loaded with Gold Nanoparticles: A Pilot Study. *La Radiologia Medica* **2014**, *119* (4), 269–276.

Li, K.; Liu, B. Polymer-Encapsulated Organic Nanoparticles for Fluorescence and Photoacoustic Imaging. *Chem. Soc. Rev.* **2014**, *43*, 6570–6597. [PubMed: 24792930].

Lipinski, M. J.; Briley-Saebo, K. C.; Mani, V.; Fayad, Z. A. Positive Contrast Inversion-Recovery with Oxide Nanoparticles-Resonant Water Suppression Magnetic Resonance Imaging: A Change for the Better. *J. Am. Coll. Cardiol.* **2008**, *52*, 492–494. [PubMed: 18672171].

Liu, H. L.; Ko, S. P.; Wu, J. H.; Jung, M. H.; Min, J. H.; Lee, J. H.; An, B. H.; Kim, Y. K. One-Pot Polyol Synthesis of Monosize PVP-Coated Sub-5nm Fe3O4 Nanoparticles for Biomedical Applications. *J. Magnet. Magnet. Mater* **2007**, *310*, E815–E817.

Liu, Y.; He, J.; Yang, K. et al. Folding Up of Gold Nanoparticle Strings into Plasmonic Vesicles for Enhanced Photoacoustic Imaging. *Angew. Chem.* **2015**, *127*, 16035–16038.

Lodahl, P.; Mahmoodian, S.; Stobbe, S. Interfacing Single Photons and Single Quantum Dots with Photonic Nanostructures. *Rev. Mod. Phys.* **2015**, *87*, 347.

Lu, H. D.; Wilson, B. K.; Heinmiller, A.; Faenza, B.; Hejazi, S.; Prud'homme, R. K. Narrow Absorption NIR Wavelength Organic Nanoparticles Enable Multiplexed Photoacoustic Imaging. *ACS Appl. Mater. Interf.* **2016**, *8*, 14379–14388. [PubMed: 27153806].

Luo, C.; Li, Y.; Yang, L.; Wang, X.; Long, J.; Liu, J. Superparamagnetic Iron Oxide Nanoparticles Exacerbate the Risks of Reactive Oxygen Species-Mediated External Stresses. *Arch. Toxicol.* **2015**, *89* (3), 357–369.

Mahmood, U.; Weissleder, R. Near-Infrared Optical Imaging of Proteases in Cancer. *Mol. Cancer Therap.* **2003,** *2,* 489–496.

Majoros, I. J.; Myc, A.; Thomas, T. et al. PAMAM Dendrimer-Based Multifunctional Conjugate for Cancer Therapy: Synthesis, Characterization, and Functionality. *Biomacromolecules* **2006,** *7,* 572–579.

Majumdar, A. Bioassays Based on Molecular Nanomechanics. *Dis. Markers* **2002,** *18,* 167–174.

Mancini, M. C.; Kairdolf, B. A.; Smith, A. M.; Nie, S. Oxidative Quenching and Degradation of Polymerencapsulated Quantum Dots: New Insights into the Long-Term Fate and Toxicity of Nanocrystals in Vivo. *J. Am. Chem. Soc.* **2008,** *130,* 10836–10837. [PubMed: 18652463].

Marckmann, P.; Skov, L.; Rossen, K. et al. Nephrogenic Systemic Fibrosis: Suspected Causative Role of Gadodiamide Used for Contrast-Enhanced Magnetic Resonance Imaging. *J. Am. Soc. Nephrol.,* *17* (9), 2359–2362.

McCarthy, J. R.; Kelly, K. A.; Sun, E. Y.; Weissleder, R. Targeted Delivery of Multifunctional Magnetic Nanoparticles. *Nanomedicine,* *2,* 153–167. [PubMed: 17716118].

McRobbie, D. W. *MRI from Picture to Proton.* Cambridge: University Press, Cambridge, 2007.

Michalet, X.; Pinaud, F. F.; Bentolila, L. A.; Tsay, J. M.; Doose, S.; Li, J. J.; Sundaresan, G.; Wu, A. M.; Gambhir, S. S.; Weiss, S. Quantum Dots for Live Cells, in Vivo Imaging, and Diagnostics. *Science* **2005,** *307,* 538–544. [PubMed: 15681376].

Miller, M. A.; Gadde, S.; Pfirschke, C. et al. Predicting Therapeutic Nanomedicine Efficacy Using a Companion Magnetic Resonance Imaging Nanoparticle. *Sci. Transl. Med.* **2015,** *7,* 314ra183–314ra183.

Minchin, R. F.; Martin, D. J. Minireview: Nanoparticles for Molecular Imaging—An Overview. *Endocrinology* **2020,** *15,* 474–481.

Mohammadian, F.; Pilehvar-Soltanahmadi, Y.; Mofarrah, M. et al. Down Regulation of miR-18a, miR-21 and miR-221 Genes in Gastric Cancer Cell Line by Chrysin-Loaded PLGA-PEG Nanoparticles. *Artif. Cells Nanomed. Biotechnol.* **2016,** *44,* 1972–1978.

Mukundan, Jr S.; Ghaghada, K. B.; Badea, C. T. et al. A Liposomal Nanoscale Contrast Agent for Preclinical CT in Mice. *AJR Am. J. Roentgenol.* **2006,** *186,* 300–307.

Mulder, W. J. M.; Strijkers, G. J.; Briley-Saeboe, K. C.; Frias, J. C.; Aguinaldo, J. G. S.; Vucic, E.; Amirbekian, V.; Tang, C.; Chin, P. T. K.; Nicolay, K.; Fayad, Z. A. Molecular Imaging of Macrophages in Atherosclerotic Plaques Using Bimodal PEG-Micelles. *Magn. Res. Med.* **2007,** *58,* 1164–1170.

Mulder, W. J. M.; Douma, K.; Koning, G. A.; van Zandvoort, M. A.; Lutgens, E.; Daemen, M. J.; Nicolay, K.; Strijkers, G. J. Liposome-Enhanced MRI of Neointimal Lesions in the apoE-KO Mouse. *Magn. Res. Med.* **2006,** *55,* 1170–1174.

Naha, P. C.; Al Zaki, A.; Hecht, E. et al. Dextran Coated Bismuth–Iron Oxide Nanohybrid Contrast Agents for Computed Tomography and Magnetic Resonance Imaging. *J. Mater. Chem.* **2014,** *2,* 8239–8248.

Nahrendorf, M.; Jaffer, F. A.; Kelly, K. A.; Sosnovik, D. E.; Aikawa, E.; Libby, P.; Weissleder, R. Noninvasive Vascular Cell Adhesion Molecule-1 Imaging Identifies Inflammatory Activation of Cells in Atherosclerosis. *Circulation* **2006,** *114,* 1504–1511. [PubMed: 17000904].

Nahrendorf, M.; Zhang, H.; Hembrador, S.; Panizzi, P.; Sosnovik, D. E.; Aikawa, E.; Libby, P.; Swirski, F. K.; Weissleder, R. Nanoparticle PET-CT Imaging of Macrophages in Inflammatory Atherosclerosis. *Circulation* **2008,** *117,* 379–387. [PubMed: 18158358].

Navarro, D. A.; Bisson, M. A.; Aga, D. S. Investigating Uptake of Water-Dispersible CdSe/ZnS Quantum Dot Nanoparticles by Arabidopsis Thaliana Plants. *J. Hazard. Mater.* **2012,** *211–212,* 427–435.

Nayak, T. R.; Andreou, C.; Oseledchyk, A. et al. Tissue Factor-Specific Ultra-Bright SERRS Nanostars for Raman Detection of Pulmonary Micrometastases. *Nanoscale.* **2019,** *9*(3), 1110–1119.

Ni, D.; Zhang, J.; Bu, W. et al. PEGylated NaHoF4 Nanoparticles as Contrast Agents for Both X-ray Computed Tomography and Ultrahigh Field Magnetic Resonance Imaging. *Biomaterials.* **2016,** *76,* 218–225.

Nutt, R. The History of Positron Emission Tomography. *Mol. Imag. Biol.* **2002,** *4* (1), 11–26.

Omidi, Y.; Hollins, A. J.; Drayton, R. et al. Polypropylenimine Dendrimer-Induced Gene Expression Changes: The Effect of Complexation with DNA, Dendrimer Generation and Cell Type. *J. Drug Target.* **2005,** *13,* 431–443.

Oseledchyk, A.; Andreou, C.; Wall, M. A.; Kircher, M. F. Folate-Targeted Surface-Enhanced Resonance Raman Scattering Nanoprobe Ratiometry for Detection of Microscopic Ovarian Cancer. *ACS Nano.* **2017,** *11*(2), 1488–1497.

Otsuka, H.; Nagasaki, Y.; Kataoka, K. PEGylated Nanoparticles for Biological and Pharmaceutical Applications. *Adv. Drug Deliv.* **2003,** *55,* 403–419.

Park, J. A.; Lee, Y. J.; Ko, I. O. et al. Improved Tumor-Targeting MRI Contrast Agents: Gd(DOTA) Conjugates of a Cycloalkane-Based RGD Peptide. *Biochem. Biophys. Res. Commun.* **2014,** *455,* 246–250.

Parodi, A.; Quattrocchi, N.; van de Ven, A. L. et al. Synthetic Nanoparticles Functionalized with Biomimetic Leukocyte Membranes Possess Cell-Like Functions. *Nat. Nanotechnol.* **2013,** *8,* 61–68. [PubMed: 23241654].

Perez, J. M.; Simeone, F. J.; Saeki, Y. et al. Viral-Induced Self-Assembly of Magnetic Nanoparticles Allows the Detection of Viral Particles in Biological Media. *J. Am. Chem. Soc.* **2003,** *125,* 10192–10193.

Piechowiak, E. I.; Peter, J. F. W.; Kleb, B.; Klose, K. J.; Heverhagen, J. T. Intravenous Iodinated Contrast Agents Amplify DNA Radiation Damage at CT. *Radiology* **2015,** *275* (3), 692–697.

Pratt, E. C.; Shaffer, T. M.; Grimm, J. Nanoparticles and Radiotracers: Advances Toward Radionanomedicine. *Wiley Interdiscipl. Rev. Nanomed. Nanobiotechnol.* **2016,** *8,* 872–890. [PubMed: 27006133].

Qiao, R.; Yang, C.; Gao, M. Superparamagnetic Iron Oxide Nanoparticles: From Preparations to in Vivo MRI Applications. *J. Mater. Chem.* **2009,** *19,* 6274–6293.

Rabin, O.; Perez, J. M.; Grimm, J.; Wojtkiewicz, G.; Weissleder, R. An X-ray Computed Tomography Imaging Agent Based on Long-Circulating Bismuth Sulphide Nanoparticles. *Nat. Mater* **2006,** *5,* 118–122. [PubMed: 16444262].

Rabouw, F. T.; Kamp, M.; van Dijk-Moes. R. J. et al. Delayed Exciton Emission and Its Relation to Blinking in CdSe Quantum Dots. *Nano Lett.* **2015,** *15,* 7718–7725.

Rajasundari, K.; Ilamurugu. K. Nanotechnology and Its Applications in Medical Diagnosis. *J. Basic Chem. Appl. Chem.* **2010,** *1,* 26–32.

Rand, D.; Ortiz, V.; Liu, Y. et al. Nanomaterials for X-ray imaging: gold nanoparticle enhancement of X-ray scatter imaging of hepatocellular carcinoma. *Nano Lett.* **2011,** *11,* 2678–2683.

Ratanajanchai, M.; Lee, D. H.; Sunintaboon, P. et al. Photocured PMMA/PEI Core/Shell Nanoparticles Surface-Modified with Gd-DTPA for T1 MR Imaging. *J. Colloid Interf. Sci.* **2014,** *415,* 70–76.

Razi, M.; Dehghani, A.; Beigi, F. et al. The Peep of Nanotechnology in Reproductive Medicine: Amini-Review. *Int. J. Med. Lab.* **2015,** *2,* 1–15.

Rieffel, J.; Chen, F.; Kim, J. et al. Hexamodal Imaging with Porphyrin-Phospholipid-Coated Upconversion Nanoparticles. *Adv. Mater. (Deerfield Beach, Fla)* **2015,** *27,* 1785–1790.

Ruehm, S. G.; Corot, C.; Vogt, P.; Kolb, S.; Debatin, J. F. Magnetic Resonance Imaging of Atherosclerotic Plaque with Ultrasmall Superparamagnetic Particles of Iron Oxide in Hyperlipidemic Rabbits. *Circulation* **2001,** *103,* 415–422. [PubMed: 11157694].

Sadeghzadeh, H.; Pilehvar-Soltanahmadi, Y.; Akbarzadeh, A. et al. The Effects of Nanoencapsulated Curcumin-Fe_3O_4 on Proliferation and hTERT Gene Expression in Lung Cancer Cells. *Anticancer Agents Med. Chem.* **2017,** *17.* doi: 10.2174/1871520617666170 213115756.

Santra, S.; Jativa, S. D.; Kaittanis, C. et al. Gadolinium-Encapsulating Iron Oxide Nanoprobe as Activatable NMR/MRI Contrast Agent. *ACS Nano.* **2012,** *6,* 7281–7294.

Sanz, J.; Fayad, Z. A. Imaging of Atherosclerotic Cardiovascular Disease. *Nature* **2008,** *451,* 953–957. [PubMed: 18288186].

Sasaki, Y.; Akiyoshi, K. Nanogel Engineering for New Nanobiomaterials: From Chaperoning Engineering to Biomedical Applications. *Chem. Record* **2010,** *10,* 366–376.

Schellenberger, E. A.; Sosnovik, D.; Weissleder, R.; Josephson, L. Magneto/Optical Annexin V, a Multimodal Protein. *Bioconjugate Chem.* **2004,** *15,* 1062–1067.

Selvan, S. T.; Tan, T. T.; Ying, J. Y. Robust, Non-Cytotoxic, Silica-Coated CdSe Quantum Dots with Efficient Photoluminescence. *Adv. Mater.* **2005,** *17,* 1620–1625.

Serda, R. E.; Adolphi, N. L.; Bisoffi, M.; Sillerud, L. O. Targeting and Cellular Trafficking of Magnetic Nanoparticles for Prostate Cancer Imaging. *Mol. Imag.* **2007,** *6,* 277–288. [PubMed: 17711783].

Sharma, N.; Pinnaka, A. K.; Raje, M.; Fnu, A.; Bhattacharyya, M. S.; Choudhury, A. R. Exploitation of Marine Bacteria for Production of Gold Nanoparticles. *Microb. Cell Fact.* **2012,** *11,* 86. [PubMed: 22715848].

Shevtsov, M. A.; Nikolaev, B. P.; Ryzhov, V. A. et al. Detection of Experimental Myocardium Infarction in Rats by MRI Using Heat Shock Protein 70 Conjugated Superparamagnetic Iron Oxide Nanoparticle. *Nanomedicine.* **2016,** *12,* 611–621.

Sitharaman, B.; Kissell, K. R.; Hartman, K. B.; Tran, L. A.; Baikalov, A.; Rusakova, I.; Sun, Y.; Khant, H. A.; Ludtke, S. J.; Chiu, W.; Laus, S.; Toth, W.; Helm, L.; Merbach, A. E. Superparamagnetic Gadonanotubes are Highperformance MRI Contrast Agents. *Chem. Commun.* **2005,** 3915–3917.

SNM, Centre for Molecular Imaging, Innovation and Translation. Fact Sheet: What Is Molecular Imaging 2010. http://www.snm.org/mi (accessed Jan 11, 2013).

Song, S. Y. Future Direction of Nanomedicine in Gastrointestinal Cancer. *Korean J. Gastroenterol.* **2007,** *49,* 271–279.

Sosnovik, D.; Weissleder, R. Magnetic Resonance and Fluorescence Based Molecular Imaging Technologies. *Progress Drug Res.* **2005,** *62,* 86–114.

Stern, E.; Vacic, A.; Rajan, N. K. et al. Label-Free Biomarker Detection from Whole Blood. *Nat. Nanotechnol.* **2010,** *5,* 138–142.

Application of Nanoparticles in Biomedical Imaging

Stummer, W.; Pichlmeier, U.; Meinel, T. et al. Fluorescence-Guided Surgery with 5-Aminolevulinic Acid for Resection of Malignant Glioma: A Randomised Controlled Multicentre Phase III Trial. *Lancet Oncol.* **2006**, *7*, 392–401. [PubMed: 16648043].

Sukhanova, A.; Devy, M.; Venteo, L.; Kaplan, H.; Artemyev, M.; Oleinikov, V.; Klinov, D.; Pluot, M.; Cohen, J. H. M.; Nabiev, I. Biocompatible Fluorescent Nanocrystals for Immunolabeling of Membrane Proteins and Cells. *Anal. Biochem.* **2004**, *324*, 60–67. [PubMed: 14654046].

Sun, C.; Lee, J. S.; Zhang, M. Magnetic Nanoparticles in MR Imaging and Drug Delivery. *Adv. Drug Deliv. Rev.* **2008**, *60*, 1252–1265.

Sun, J.; DuFort, C.; Daniel, M-C.; Murali, A.; Chen, C.; Gopinath, K.; Stein, B.; De, M.; Rotello, V. M.; Holzenburg, A.; Kao, C. C.; Dragnea, B. Core-Controlled Polymorphism in Virus-Like Particles. *Proc. Natl. Acad. Sci. USA.* **2007**, *104*, 1354–1359. [PubMed: 17227841].

Sun, L.; Joh, D. Y.; Al-Zaki, A. et al. Theranostic Application of Mixed Gold and Superparamagnetic Iron Oxide Nanoparticle Micelles in Glioblastoma Multiforme. *J. Biomed. Nanotechnol.* **2016**, *12* (2), 347–356.

Sun, Y. P.; Fu, K. F.; Lin, Y. et al. Functionalized Carbon Nanotubes: Properties and Applications. *Acc. Chem. Res.* **2002**, *35*, 1096–1104.

Svenson, S.; Tomalia, D. A. Dendrimers in Biomedical Applications – Reflections on the Field. *Adv. Drug Deliv. Rev.* **2005**, *57*, 2106–2129.

Sweeney, C. M.; Stender, C. L.; Nehl, C. L. et al. Optical Properties of Tipless Gold Nanopyramids. *Small* **2011**, *7*, 2032–2036.

Szabo, T. L. *Diagnostic Ultrasound Imaging: Inside Out.* Academic Press: Cambridge, MA, USA, 2004.

Szpak, A.; Fiejdasz, S.; Prendota, W. et al. T_1–T_2 Dual-Modal MRI Contrast Agents Based on Superparamagnetic Iron Oxide Nanoparticles with Surface Attached Gadolinium Complexes. *J. Nanopart. Res.* **2014**, *16*, 2678.

Szpak, A.; Kania, G.; Skorka, T. et al. Stable Aqueous Dispersion of Superparamagnetic Iron Oxide Nanoparticles Protected by Charged Chitosan Derivatives. *J. Nanopart. Res.* **2013**, *15*, 1372.

Tang, X. H.; Xie, P.; Ding, Y. et al. Synthesis, Characterization, and in Vitro and in Vivo Evaluation of a Novel Pectin-Adriamycin Conjugate. *Bioorg. Med. Chem.* **2010**, *18*, 1599–1609.

Theiss, A. L.; Laroui, H., Obertone, T. S. et al. Nanoparticle-Based Therapeutic Delivery of Prohibitin to the Colonic Epithelial Cells Ameliorates Acute Murine Colitis. *Inflam. Bowel Dis.* **2011**, *17*, 1163–1176.

Thomas, R.; Park, I. K.; Jeong, Y. Y. Magnetic Iron Oxide Nanoparticles for Multimodal Imaging and Therapy of Cancer. *Int. J. Mol. Sci.* **2013**, *14*, 15910–15930.

Timbie, K.; Nance, E.; Zhang, C. et al. Ultrasound-Mediated Nanoparticle Delivery Across the Bloodbrain Barrier (676.17). *FASEB J* **2014**, *28*.

Tomalia, D. A.; Reyna, L. A.; Svenson, S. Dendrimers as Multipurpose Nanodevices for Oncology Drug Delivery and Diagnostic Imaging. *Biochem. Soc. Trans.* **2007**, *35*, 61–67.

Tummers, Q. R.; Hoogstins, C. E.; Peters, A. A. et al. The Value of Intraoperative Near-Infrared Fluorescence Imaging Based on Enhanced Permeability and Retention of Indocyanine Green: Feasibility and False-Positives in Ovarian Cancer. *PLoS One* **2015**, *10*, e0129766. [PubMed: 26110901].

van de Ven, A. L.; Abdollahi, B.; Martinez, C. J, et al. Modeling of Nanotherapeutics Delivery Based on Tumor Perfusion. *New J. Phys.* **2013**, *15*, 55004.

van Schooneveld, M. M.; Vucic, E.; Koole, R.; Zhou, Y.; Stocks, J.; Cormode, D. P.; Tang, C. Y.; Gordon, R.; Nicolay, K.; Meijerink, A.; Fayad, Z. A.; Mulder, W. J. M. Improved Biocompatibility and Pharmacokinetics of Silica Nanoparticles by Means of a Lipid Coating: A Multimodality Investigation. *Nano Lett.* **2008**, *8*, 2517–2525. [PubMed: 18624389].

van Tilborg, G. A. F.; Mulder, W. J. M.; Deckers, N.; Storm, G.; Reutelingsperger, C. P. M.; Strijkers, G. J.; Nicolay, K. Annexin A5-Functionalized Bimodal Lipid-Based Contrast Agents for the Detection of Apoptosis. *Bioconjate Chem.* **2006**, *17*, 741–749.

Veiseh, O.; Gunn, J.; Zhang, M. Design and Fabrication of Magnetic Nanoparticles for Targeted Drug Delivery and Imaging. *Adv. Drug Deliv. Rev.* **2010**, *62*, 284–304.

Villanueva, F. S.; Jankowski, R. J.; Klibanov, S.; Pina, M. L.; Alber, S. M.; Watkins, S. C.; Brandenburger, G. H.; Wagner, W. R. Microbubbles Targeted to Intercellular Adhesion Molecule-1 Bind to Activated Coronary Artery Endothelial Cells. *Circulation* **1998**, *98*, 1–5. [PubMed: 9665051].

Villanueva, F. S.; Lu, E. X.; Bowry, S.; Kilic, S.; Tom, E.; Wang, J. J.; Gretton, J.; Pacella, J. J.; Wagner, W. R. Myocardial Ischemic Memory Imaging with Molecular Echocardiography. *Circulation*, **2007**, *115*, 345–352. [PubMed: 17210843].

Wan, X.; Song, Y.; Song, N. et al. The Preliminary Study of Immune Superparamagnetic Iron Oxide Nanoparticles for the Detection of Lung Cancer in Magnetic Resonance Imaging. *Carbohydr. Res.* **2016**, *419*, 33–40.

Wang, Y.; Zhou, K.; Huang, G. et al. A Nanoparticle-Based Strategy for the Imaging of a Broad Range of Tumours by Nonlinear Amplification of Microenvironment Signals. *Nat. Mater.* **2014**, *13*, 204–212. [PubMed: 24317187].

Weissleder, R.; Kelly, K.; Sun, E. Y. et al. Cell-Specific Targeting of Nanoparticles by Multivalent Attachment of Small Molecules. *Nat. Biotechnol.* **2005**, *23*, 1418–1423.

Weissleder, R.; Kelly, K.; Sun, E. Y.; Shtatland, T.; Josephson, L. Cell-Specific Targeting of Nanoparticles by Multivalent Attachment of Small Nanoparticles. *Nat. Biotech.* **2005**, *23*, 1418–1423.

Weissleder, R.; Stark, D. D.; Engelstad, B. L. et al. Superparamagnetic Iron Oxide: Pharmacokinetics and Toxicity. *Am. J. Roentgenol.* **1989**, *152* (1), 167–173.

Wen, S.; Li, K.; Cai, H. et al. Multifunctional Dendrimerentrapped Gold Nanoparticles for Dual Mode CT/MR Imaging Applications. *Biomaterials* **2013**, *34* (5), 1570–1580.

Winter, P. M.; Cai, K.; Chen, J.; Adair, C. R.; Kiefer, G. E.; Athey, P. S.; Gaffney, P. J.; Buff, C. E.; Robertson, J. D.; Caruthers, S. D.; Wickline, S. A.; Lanza, G. M. Targeted PARACEST Nanoparticle Contrast Agent for the Detection of Fibrin. *Magn. Reson. Med.* **2006**, *56*, 1384–1388. [PubMed: 17089356].

Winter, P. M.; Caruthers, S. D.; Zhang, H.; Williams, T. A.; Wickline, S. A.; Lanza, G. M. Antiangiogenic Synergism of Integrin-Targeted Fumagillin Nanoparticles and Atorvastatin in Atherosclerosis. *JACC: Cardiovasc. Im.* **2008**, *1*, 624–634.

Wu, D.; Huang, L.; Jiang, M. S.; Jiang, H. Contrast Agents for Photoacoustic and Thermoacoustic Imaging: A Review. *Int. J. Mol. Sci.* **2014**, *15* (12), 23616–23639.

Wu, L.; Cai, X.; Nelson, K. et al. A Green Synthesis of Carbon Nanoparticles from Honey and Their Use in Real-Time Photoacoustic Imaging. *Nano Res.* **2013**, *6*, 312–325. [PubMed: 23824757].

Xie, C.; Upputuri, P. K.; Zhen, X.; Pramanik, M.; Pu, K. Self-Quenched Semiconducting Polymer Nanoparticles for Amplified in Vivo Photoacoustic Imaging. *Biomaterials* **2017,** *119,* 1–8. [PubMed: 27988405].

Yan, L.; Zhang, Y.; Xu, B.; Tian, W. Fluorescent Nanoparticles Based on AIE Fluorogens for Bioimaging. *Nanoscale* **2016,** *8,* 2471–2487. [PubMed: 26478255].

Yan, M.; Zhang, Y.; Xu, K. et al. An in Vitro Study of Vascular Endothelial Toxicity of CdTe Quantum Dots. *Toxicology* **2011,** *282,* 94–103.

Yan, Y.; Yang, Q.; Wang, J. et al. Heteropoly Blue Doped Polymer Nanoparticles: An Efficient Theranostic Agent for Targeted Photoacoustic Imaging and Near-Infrared Photothermal Therapy in Vivo. *J. Mater. Chem. B.* **2017,** *5*(2), 382–387. doi: 10.1039/c6tb02652d.

Yang, C.; Gdor, I.; Amit, Y. et al. *Exciton Dynamics in Cu-doped InAs Colloidal Quantum Dots*; Ultrafast Phenomena XIX: Springer, 2015.

Yang, H. W.; Liu, H. L.; Li, M. L. et al. Magnetic Gold-Nanorod/ PNIPAAmMA Nanoparticles for Dual Magnetic Resonance and Photoacoustic Imaging and Targeted Photothermal Therapy. *Biomaterials* **2013,** *34,* 5651–5660. [PubMed: 23602366].

Yang, R. M.; Fu, C. P.; Li, N. N. et al. Glycosaminoglycan-Targeted Iron Oxide Nanoparticles for Magnetic Resonance Imaging of Liver Carcinoma. *Mater. Sci. Eng. C Mater. Biol. Appl.* **2014,** *45,* 556–563.

Yu, F.; Zhang, L.; Huang, Y. et al. The Magnetophoretic Mobility and Superparamagnetism of Core–Shell Iron Oxide Nanoparticles with Dual Targeting and Imaging Functionality. *Biomaterials* **2010,** *31,* 5842–5848.

Yu, W. W.; Chang, E.; Falkner, J. C.; Zhang, J.; Al-Somali, A. M.; Sayes, C. M.; Johns, J.; Drezek, R.; Colvin, V. L. Forming Biocompatible Nonaggregated and Nanocrystals in Water Using Amphiphilic Polymers. *J. Am. Chem. Soc.* **2007,** *129,* 2871–2879. [PubMed: 17309256].

Zanganeh, S.; Hutter, G.; Spitler, R. et al. Iron Oxide Nanoparticles Inhibit Tumour Growth by Inducing Pro-Inflammatory Macrophage Polarization in Tumour Tissues. *Nat. Nano.* **2016,** *11,* 986–994.

Zhang, F.; Kong, X. Q.; Li, Q. et al. Facile Synthesis of CdTe@GdS Fluorescent-Magnetic Nanoparticles for Tumor-Targeted Dualmodal Imaging. *Talanta.* **2016,** *148,* 108–115.

Zhang, G. J.; Ning, Y. Silicon Nanowire Biosensor and Its Applications in Disease Diagnostics: A Review. *Analyt. Chim. Acta* **2012,** *749,* 1–15.

Zheng, G.; Chen, J.; Li, H.; Glickson, J. D. Rerouting Lipoprotein Nanoparticles to Selected Alternate Receptors for the Targeted Delivery of Cancer Diagnostic and Therapeutic Agents. *Proc. Natl. Acad. Sci. U. S. A.* **2005,** *102,* 17757–17762. [PubMed: 16306263].

Zheng, G.; Patolsky, F.; Cui, Y, et al. Multiplexed Electrical Detection of Cancer Markers with Nanowire Sensor Arrays. *Nat. Biotechnol.* **2005,** *23,* 1294–1301.

Zheng, J.; Dickson, R. M. Individual Water-Soluble Dendrimer-Encapsulated Silver Nanodot Fluorescence. *J. Am. Chem. Soc.* **2002,** *124,* 13982–13983.

Zhou, B.; Yang, J.; Peng, C. et al. PEGylated Polyethylenimineentrapped Gold Nanoparticles Modified with Folic Acid for Targeted Tumor CT Imaging. *Colloids Surf. B Biointerf.* **2016,** *140,* 489–496.

Zhou, N.; Ye, C.; Polavarapu, L. et al. Controlled Preparation of Au/ Ag/SnO 2 Core–Shell Nanoparticles Using a Photochemical Method and Applications in LSPR Based Sensing. *Nanoscale* **2015,** 9025–9032.

Zhu, H.; Tao, J.; Wang, W. et al. Magnetic, Fluorescent, and Thermo-Responsive Fe_3O_4 Rare Earth Incorporated Poly(St NIPAM) Core–Shell Colloidal Nanoparticles in Multimodal Optical/Magnetic Resonance Imaging Probes. *Biomaterials* **2013**, *34*, 2296–2306.

Zhu. J.; Fu, F.; Xiong, Z. et al. Dendrimer-Entrapped Gold Nanoparticles Modified with RGD Peptide and Alpha-Tocopheryl Succinate Enable Targeted Theranostics of Cancer Cells. *Colloids Surf. B Biointerf.* **2015,** *133*, 36–42.

CHAPTER 4

NANOPARTICLES IN MEDICAL IMAGING: A PERSPECTIVE STUDY

MUJTABA AAMIR BHAT[1], ISHFAQ AHMAD WANI[1], NASEER AHMAD HAJAM[1], SAFIKUR RAHMAN[2], and ARIF TASLEEM JAN[1*]

[1]School of Biosciences and Biotechnology, Baba Ghulam Shah Badshah University, Rajouri, India

[2]Department of Botany, M. S College, B. R. Ambedkar University, Muzaffarpur, Bihar, India

*Correspondence author. E-mail: atasleem@bgsbu.ac.in

ABSTRACT

The modern health-care system relies on elucidating molecular architecture's intricacies in terms of variation within or between the individuals. Emphasizing the trade-off that occurs at a larger scale between the cellular components, it becomes imperative to know the cellular heterogeneity perceived for a disease's occurrence. Facing a lot of challenges, elucidation of the mechanisms favoring their survival and expansions inside the host is required to understand the disease's hallmarks for the timely management of the disease. Multidisciplinary approaches offered by the newly driven technology that operates at the nanolevel have opened up new avenues in the disease diagnosis and therapeutics. With minimal side effects, the increased half-life of payloads (drugs and other substances) in systemic circulation and efficacy achieved via target-specific delivery has revolutionized the biomedical field. Application of the nanomaterial-based contrast agents in different imaging modalities such as fluorescence imaging, magnetic resonance imaging, and positron electron tomography has broadened the horizons for early-stage detection and diagnosis of the disease. By offering a promising and more significant clinical benefit, nanomaterials are currently being explored for properties that can potentiate their usage in clinics to detect diseases at early stages and broader employment in the diagnosis of the disease. Having shown a shift therapeutics to prognosis of the disease, this study encompasses the recent developments in the field with prejudice on broadening its application in the imaging system as part of its diagnosis.

4.1 INTRODUCTION

Cancer, a disease of global magnitude, affects millions of people worldwide. Exhibiting significant variation in their molecular architecture (between individuals, within the same individual, and within cells of the same tumor), intratumor heterogeneity and intertumoral diversity emphasized the elucidation of trade-offs that illustrate the hallmarks of cancer. It is perceived that cancer cells perform multiple biological functions as a survival strategy for proliferation, invasion of tissues, and in evading host immune response and cell death. With hosts providing a favorable microenvironment, cancer cells opt for a trade-off mechanism to survive and expand inside the host. In this scenario, it becomes necessary to have a thorough understanding of the regulatory network and interpret tumors' heterogeneity to framework the development of more accurate ways of disease prognosis and therapy (Hausser and Alon, 2020).

Nanomaterials (amalgamated material developing by converging concepts of engineering, biology, chemistry, and others) emerged lately with applications ranging from prognosis to therapeutics of disease (Niemeyer, 2001). Increasing the half-life of drugs in systemic circulation and avoiding their clearance from the body increase the drug's therapeutic potential, besides being used in time-dependent monitoring of the disease. Employing nanotechnology's multidisciplinary approach, their use in the disease diagnosis and therapeutics with minimal side effects and increased efficacy achieved via target-specific delivery to tissues has revolutionized the biomedical field. Nanoparticles (NPs) are currently being explored for properties that potentiate their use in imaging with application aimed at diagnosing the disease. Herein, we present a comprehensive overview of the development of NPs with a focus on application in imaging as part of the diagnosis of the disease.

4.2 NANOPARTICLES IN MEDICAL IMAGING

Imaging (bio- and medical imaging) with visual and intuitional interface enables early detection and diagnosis of a disease. However, a sizeable internal contrast between bone and surrounding tissues helps in differentiation, the difference under normal conditions fails in distinguishing between a normal tissue from a tumor (Cormode et al., 2010; He et al., 2015). With this, it becomes imperative to employ different substances to supplement current devices' ability to achieve timely detection, diagnosis, and treatment of diseases such as cancer. Agents used to enhance contrast (iodine, barium sulfate, gadolinium complexes, etc.) show nonspecific biodistribution, fast clearance, and short half-life in blood (Du et al., 2013; Lee et al., 2014; Razi et al., 2015). They exert renal toxicity and often fail in achieving desired contrast in fat patients to yield accurate results. Having greater challenges associated with them, a shift has been observed toward using nano-based systems to achieve desired results in clinics.

NPs (size 1–100 nm) with high surface-to-volume ratio exhibit unique properties (depending on the components and mode of synthesis) distinct from other molecules. In recent years, exponential usage has led to the development of methods controlling their synthesis precisely in terms of shape and size to tune their uptake inside a particular system. Though NPs target tumors passively via enhanced permeability retention (EPR) effect or undergoes molecular sieving to exert target-specific effect (Hawley et al., 1997; Greish, 2007), their employability to target tumor tissues is enhanced on conjugating them with the tumor-specific molecules such as peptides and antibodies (Davis et al., 2008; Smith et al., 2008). NPs have recently been added to the list of contrast agents owing to their: (1) high surface-to-volume ratio that enables multiple ligand binding for enhancing their sensitivity and specificity in the usage of disease diagnosis, (2) increased half-life in systemic circulation attributed by their physicochemical properties, and (3) enhanced encapsulation or surface ligand marking at predetermined ratios (Massoud and Gambhir, 2003; Wang et al., 2008; Janib et al., 2010). Advances in the approaches that increase surface labeling of NPs have leveraged a surge in their usage for targeting multiple biomolecules and their application in biomedical imaging.

4.2.1 NONRADIOLABELED NPS IN BIOLOGICAL IMAGING

Magnetic NPs such as iron oxide–based NPs that are size-tailored functionalized and exhibit paramagnetic behavior are commonly used as diagnostic tracers in imaging soft tissues, especially brains (Mahmoudi et al., 2011). Their use is limited as they were found inducing cytotoxicity via the elevation of oxidative stress, in turn, associated with DNA damage in human breast cancer (Bhattacharyya et al., 2011). Facilitating surface property (geometry, porosity) utilization, silica (silicon dioxide), NPs are currently being employed for imaging in diagnostics (Hudson et al., 2008; Zhao et al., 2011; Liu et al., 2011; Tang et al., 2012). Encapsulation of silica NPs with organic-doping dyes is utilized in fluorescence resonance energy transfer (FRET) as barcoding tags in understanding the complexity of different signaling modules (Wang and Tan, 2006). Gadolinium (Gd) NPs used as a contrast agent in proton magnetic resonance reduce spin–lattice relaxation time for improving the signal from protons (higher signal-to-noise ratio) for brighter T1-weighted image quality (Bridot et al., 2007; Yang et al., 2008; Ananta et al., 2010; Chen et al., 2010; Karfeld-Sulzer et al., 2011). Impeded by cell membrane translocations, gadolinium-based functionalized polymers, peptides, etc., developed as NPs face difficulty in reproducibility of their fabrications (Silva, 2006). Gd NPs conjugated to bifunctional chelates such as Gd_2SiO_2–DO_2A–BTA exhibited higher uptake and cytotoxicity in HeLa, MDA-MB231, Hep3b tumor cell lines (Kang et al., 2016). Self-assembly of Gd NPs at the intercellular level results in better contrast in magnetic resonance (Cao et al., 2013). Gold NPs, including radiolabeled and those having a coating of silica, Gd, etc., are employed in treating arthritis (as an anti-inflammatory agent) and cancer therapy (Bhattacharyya et al., 2011; Lee et al., 2014). Dual-loaded inorganic NPs such as Gd-loaded dendrimer-entrapped gold NPs (Gd–Au De–NPs) were found beneficial in MRI and CT of the heart, liver, kidneys, etc., in a biological system (Li et al., 2013). AuNPs application with X-ray was found beneficial in inducing cytotoxicity that progresses with the damage of DNA in tumor cells. Quantum dots (QDs) exhibiting high surface-to-volume ratio emit brighter fluorescence (based on valence and conduction band size) on the broader spectrum and undergo minimal degradation as compared to organic dyes used for biomedical applications (Jin et al., 2010). Besides being used in delivering drugs to target sites, graphene QDs (GQDs) exhibit wide usage in cell imaging owing to better biocompatibility, low toxicity, and greater photostability (Shen et al., 2012; Abdullah et al., 2013; Wang et al., 2013, 2014). A latter addition to this involves the use of negatively charged silver nanoscale hexagonal columns (NHCs) and silicone amorphous NPs as sensors to detect positively charged histone tails of cell-free DNA surface-enhanced Raman scattering (SERS) of cancerous tissues (Feng et al., 2010; Lin et al., 2011a,b; Wang et al., 2011; Chen et al., 2012; Ito et al., 2015; Powell et al., 2016). Owing to the high sensitivity of SERS, it finds its usage in early detection and imaging with high accuracy in different biomedical applications.

Nanoparticles in Medical Imaging: A Perspective Study 113

4.2.2 RADIOLABELED NPS IN BIOLOGICAL IMAGING

Radiotracer techniques such as positron emission tomography (PET) or single-photon emission computer tomography (SPECT) employing the use of radiolabeled NPs (greater stability and longer circulation half-life) with greater penetration power are used in imaging of deeper tissues as part of their clinical applications (de Barros et al., 2012). Iron oxide NPs conjugated to RGD (arginine, glycine, and aspartate) peptide-functionalized with DOTA for labeling with ^{64}Cu are considered as suitable contrast agents for imaging in mice with human glioblastoma following usage of unconjugated NPs as blocking agent of RGD receptor, *avb3* integrin (Lee et al., 2008). Similarly, the study employed ^{64}Cu-DOTA and Cy 5.5 cross-linked iron oxide (CLIO) NPs injected subcutaneously xenograft U87MG mice revealed significant delineation of tumor in all three imaging modalities, that is, PET, near-infrared fluorescence (NIRF), and MRI (Xie et al., 2010). Dextranated DTPA–modified fluorescent NPs bearing magnetic properties revealed an increased uptake and their distribution in the ApoE$^{-/-}$ atherosclerotic mouse model (Nahrendorf et al., 2008). Study of the fluorine (^{18}F)-labeled CLIO NPs targeting macrophages was found helpful in detecting inflammation in aortic aneurysm-induced ApoE$^{-/-}$ mice model using a PET–CT imaging system (Nahrendorf et al., 2011). Exploiting the sensitivity of PET and accuracy in localization of PET, gold-based NPs loaded with positron or gamma emitters were employed in aging studies (Shao et al., 2011; Xie et al., 2011; Agarwal et al., 2011; Guerrero et al., 2012). Technetium AuNPs conjugated to RGD were employed in studying metastasis of tumors in C6 human glioma athymic tumor mouse model via imaging based on *avb3* expression (Morales-Avila et al., 2011). Gold nanoshells (NS; spherical NPs with a dielectric core surrounded by the metal shell) are employed in theranostic application owing to their size, shape, and optical properties. Injection of radiolabeled NS designed as ^{64}Cu-labeled NPs with PEG2K-DOTA for chelation in nude mice induced for head and neck squamous carcinoma xenograft was found beneficial for high-resolution PET/CT imaging (Pyayt et al., 2009; Xie et al., 2010). Though PEGylation of gold-based nanorods (NRs) increases the half-life of NPs in circulation, iodine (I^{125})-labeled NRs conjugated to anti-intercellular adhesion molecules-1 were found increasing efficiency in the arthritic mouse model (Shao et al., 2011).

Radiolabeled ^{99m}Tc-PEGylated liposomes were employed for scintigraphic detection of inflammation at infected by exploiting the inherent property of enhanced permeation at inflamed and infected sites (Dams et al., 2000).

Indium (In)-loaded liposomes coated with nonspecific IgG developed for low-density lipoprotein receptor, LOX-1, were beneficial in acquiring images of atherosclerotic plaques in ApoE$^{-/-}$ mice (Li et al., 2010).[111]In-pentetic acid or DTPA-labeled liposomes were useful in scintigraphic imaging following uptake by patients' tumor tissues (Harrington et al., 2001). Following rapid encapsulation of ^{64}Cu-loaded PEGylated liposomes, quantitative imaging was achieved for healthy and tumorous mice with PET (Petersen et al., 2011). Nanomicelles (polymer or lipid-based amphiphilic molecules) having a hydrophobic core surrounded by a hydrophilic shell are also employed in the imaging. Peptide-conjugated polymeric nanomicelles used against the EPHB4 receptor were found to achieve an image in prostate cancer (Zhang et al., 2011a). Annexin A5 (having a strong affinity for phosphatidylserine) conjugated nanomicelles labeled with I^{125} and indocyanine Cy7-like dye were employed for imaging near-infrared spectrum in detecting the cellular apoptotic process (Zhang et al., 2011a,b). In another study, antilymphoma agents conjugated to annexin A5 on injecting to mice bearing EL5 lymphoma helped achieve SPECT and optical imaging (Zhang et al., 2011b). Mice injected with multifunctional nanomicelles conjugated with RGD peptide guiding targeting, ^{64}Cu chelator for labeling, and loaded with anticancer agents, doxorubicin employed in theranostic revealed better imaging of U87MG tumor-bearing mice in PET (Zhang et al., 2011a).

4.3 NPS AS CONTRAST AGENTS IN IMAGING/NANOPARTICLES IN MOLECULAR IMAGING

Molecular imaging refers to the development of molecular probes for noninvasive cellular function visualization and measurement of the molecular processes in living organisms without perturbing their function at the cellular and molecular levels (Weissleder, 2006). NP properties such as size, shape, charge, and hydrophilicity are top contenders in achieving target-specific delivery of NPs that potentiate their imaging with application aimed at diagnosing the disease. Of the different NP systems developed, polyethylene glycol (PEG) providing hydrophilic stealth helps in the adsorption of molecules to serum protein that increases their half-life in the circulation, while positively charged NPs are employed in enhancing endocytosis of molecules for cell labeling (Al-Jamal et al., 2008; Prencipe et al., 2009). Though a wide range of NP systems was designed for increased affinity for a protein expressed on the cell surface of cancerous cells (Choi et al., 2007; Longmire et al., 2008; Popovtzer et al., 2008; McCarthy et al., 2008), they are currently being explored for application as contrast agents in achieving prolonged in vivo stability, increased half-life in circulation, and high specificity for the diseased tissue. Designing of NP-based contrast agents for enhancement in usage in imaging as a multivalent system for intended cells and biomolecules includes the following.

4.3.1 NANOPARTICLE SELECTION

Nanocarrier systems developed over time include inorganic NPs, polymeric NPs, liposomes, dendrimers, etc. As imaging modality requires a specific contrasting module, a large number of NP-based opposite agents, each displaying unique characteristics, were developed for employment in the diagnosis of the diseases. An example is of the gadolinium-labeled NPs developed as microemulsions, liposomes, micelles, nanotubes, etc., which utilizes the paramagnetic properties of gadolinium in providing the desired contrast for magnetic resonance imaging (MRI) (Frias et al., 2004; Sitharaman et al., 2005; Anderson et al., 2006; Lanza et al., 2006; Mulder et al., 2006; Amirbekian et al., 2007). Similarly, iron oxides that exhibit superparamagnetism were utilized in enhancing the contrast in T2-weighted MRI (Weissleder, 2006). Although the sensitivity of gadolinium enriched NPs is lower than iron oxide NPs, their use as a contrasting agent depends on the application for which they are used (van Tilborg et al., 2006). Compared with iron oxide–based NPs that induce negative contrast that fails to ascribe signal loss in MRI, gadolinium-enriched NPs are commonly employed positive contrast agent for accessing accumulation in the target tissues (Briley-Saebo et al., 2007; Lipinski et al., 2008). While deciding the appropriateness in their usage, the iron oxide contrast agent is best suited for targets such as the brain that have to produce homogenous signals for predicted structure in MRI. In contrast, a gadolinium-based contrast agent is used for targets having less predicted structures available such as the abdomen and has chances to signal voids in MRI (Walczak et al., 2008). Fluorescent labeling of the NPs developed as QDs is an excellent contrast agent in fluorescent imaging. With a more comprehensive excitation range and narrow emission window, fluorophores capable of resisting the photobleaching effect increase their molecular imaging (Bruchez et al., 1998; Leatherdale et al., 2002; Sukhanova et al., 2004; Medintz et al., 2005). For substantially thick tissues, fluorescent tomography is advocated with fluorophores of absorption and

116 Diverse Applications of Nanotechnology in the Biological Sciences

emission spectra near-infrared (650–900 nm) (Sosnovik and Weissleder, 2005).

Computed tomography (CT) employs electron-dense elements such as iodine and gold, exhibiting higher atomic number as contrast agents (Hainfeld et al., 2006; Rabin et al., 2006; Hyafil et al., 2007). With the high payload requirement of electron-dense elements to be delivered at the target site, researchers have directed their use as solid NPs, wherever chances of their trapping arise in the internal environment (Mukundan et al., 2006). With this, gold-based NPs have become a choice of element for contrast agents in CT (Lumbroso and Dick, 1987; Kim et al., 2007). Though considered insensitive for imaging under in vivo conditions, solid iodine-based NPs in the atherosclerosis rabbit model were analyzed by CT on being taken up macrophages (Weissleder and Mahmood, 2001; Hyafil et al., 2007). In ultrasound, scanning sound waves is achieved using bubbles (micron sized) of water-insoluble gas, decafluorobutane (Villanueva et al., 1998). Coating bubbles for enhancement instability reduces their flexibility for achieving good contrast, thus compromising a situation of compromise between stability and flexibility (Klibanov, 2006). Additionally, multilayer vesicles and microemulsions have also been put to achieve the desired contrast in ultrasound (Lanza et al., 1996; Huang et al., 2002). Having a wide range of NPs available, a choice based on drug nature is to be made for their efficient delivery. For hydrophobic drugs, incorporation within the hydrophilic core increases their stability, and vice versa is valid for hydrophilic drugs (Winter et al., 2006; Schiffelers et al., 2006).

4.3.2   SIZE AND SYNTHETIC STRATEGIES

Considering the biodistribution of NPs, size plays a critical role in cell types, delivery of payload at the target size, clearance/excretion concerning half-life in the systemic circulation, and the contrast produced. As the size of NPs has a significant impact on infiltration into pathological tissues via leaky vasculature, efforts are being made for the production of particles (both inorganic and lipid-based) of defined size in a controlled manner (van Helden et al., 1980; Murray et al., 1993; Hostetler et al., 1998; Gupta and Gupta, 2005). With the same core size, the diameter of NPs varies with the type of coating used. Having a considerable impact on particles' distribution, thrust has been put on critical analysis for aspects such as excretion from the body via, renal system (Choi et al., 2007). As fine-tuning of NPs in terms of size for desired contrast

Nanoparticles in Medical Imaging: A Perspective Study

is required, it is observed that NPs smaller than 5.5 nm undergo excretion (reduced half-life in circulation) through the renal system. In the case of QDs, an increase in size results in an enhancement in the fluorescence emission (in terms of wavelength); therefore, size-based tuning of wavelength was found to have a considerable effect on the emission of NPs (Michalet et al., 2005). Though iron oxide NPs in size range of <20 nm give excellent contrast in MRI (van Tilborg et al., 2006; McAteer et al., 2007), they were found attributing a better difference in the size range of >20 nm for magnetic particle imaging (MPI) (Gleich and Weizenecker, 2005; Gaumet et al., 2008). As size plays an essential role concerning application of NPs, different techniques such as transmission and scanning electron microscopy, and dynamic light scattering are commonly adopted in determining the size of NPs.

In terms of getting approval for usage, synthesis of NPs of biocompatible material in an aqueous process with coating of water-soluble polymer is advantageous over NPs synthesized in an organic phase (water insoluble) as they require further coating of amphiphilic polymers that render them bioapplicable (Brust et al., 1994; Michalet et al., 2005; Kim et al., 2005; Gupta and Gupta, 2005; Yu et al., 2007; van Schooneveld et al., 2008). The second important thing is addition of desired groups of fluorophores as functionalities during synthesis of NPs. A better example is the dextran coating of iron oxide NPs where dextran is modified for amine group for utilization in conjugating targeted groups or fluorophores for different molecules (McCarthy et al., 2007; Nahrendorf et al., 2008). Another example of this is of the micelle developed on the QDs core; with biocompatible group of PEG or phospholipid with functionalities such as maleimide assembled around the NP (Mulder et al., 2007).

4.3.3 COATING TYPES

Properties of NPs such as biocompatibility, stability in systemic circulation, and toxicity raises concern pertaining to use of materials for the synthesis of NPs. It has progressed with the use of compatible materials such as phospholipids, PEG, and dextran, as a coating in the synthesis of contrast agents for application in the imaging of biological systems (Zhang et al., 2002; Berry et al., 2003; Selvan et al., 2005; van Tilborg et al., 2006; Liu et al., 2007). Addition of functional labels to coating corona of NPs makes them multimodal; increase in hydrophilicity of NPs prevents their aggregation and as such enhancement in biological effects (Mulder et al., 2007). Coating with materials such as PEG that prevents opsonization of NPs within human body increases their half-lives in circulation by decreasing the removal effect of reticuloendothelial system, rendering enhancement in terms of their biological effect at the target site (Allen et al., 1991; Otsuka et al., 2003). Silica employed in delivering the cargo of drugs, genes, proteins, etc. (Chowdhury and Akaike, 2005; Selvan et al., 2005; Vallet-Regi et al., 2007; Slowing et al., 2007) on its application with QDs followed by a coating of PEG enhances the biological effect of the cargo than that loaded on silica alone (van Schooneveld et al., 2008).

Lipid-coated particles displaying enhanced circulatory half-life with low cytotoxicity and aggregation significantly improve the biocompatibility of NPs for their widespread application in biological system. Evading recognition by body's immune system, lipoprotein-based NPs synthesized either (1) material are embedded within the core of lipoprotein or (2) surface of NPs is modified to carry the contrast bringing agents such as dyes and ions that are employed for application in medical imaging (Anderson et al., 2006; Cormode et al., 2007; Huang et al., 2007; Sun et al., 2007). The former approach was employed for encapsulation of iron oxide, etc., in the interior of high-density lipoprotein (HDL). The same approach was used for encapsulation of paramagnetic fluorochromes within NPs for their application as contrast agents in MRI and other imaging techniques (Cormode et al., 2008). Uptake of these particles by macrophages under both in vitro and in vivo conditions was observed in MRI, fluorescence imaging, and CT.

4.3.4 TARGETING STRATEGIES

Targeting NPs to the desired site requires information about the architecture of that site and consideration of the conditions under which it is done. In the case of pathologies such as cancer, increased vasculature (leaky) ensures increased accumulation at the target site using EPR effect (Kwon and Kataoka, 1995; Mulde et al., 2006; Sanz and Fayad, 2008). Having the capability to deliver the payload without putting targeting ligand in use, coating of NPs ensures stability of payload in systemic circulation and its uptake at the target site. Additionally, it has been noted that dextran coating of iron oxide NPs increases its uptake by macrophages (Ruehm et al., 2001). For this, surface labeling of NPs with active molecules such as antibodies, peptides, sugars, and aptamers was found effective in enhancing its uptake at the target site (Kang et al., 2002; Schellenberger et al., 2004; Zheng et al.,

2005; Nahrendorf et al., 2006; van Tilborg et al., 2006; Winter et al., 2006; Amirbekian et al., 2007; Villanueva et al., 2007; Javier et al., 2008). Despite enhancing degree of specificity, factors such as cost, size, and any sort of immunological response, if any, after ligand binding hampers their application in biological systems (Frangioni, 2008). To this, the phage display technique is employed to identify short target peptides that can guide NPs in reaching destination with greater stability and increased circulation half-lives (Nahrendorf et al., 2006). Similarly, screening for appropriate ligand among the library of small molecules against different cell types is employed in choosing appropriate molecules that hold promise for use in achieving targeting of desired cargo (Weissleder et al., 2005). Though biotin–streptavidin linkage is commonly exploited (Paganelli et al., 1997; Serda et al., 2007), its usage in achieving targeting of desired cargo is hampered as it was found eliciting immunogenic response in host that results in shifting to covalent type of linkage for surface attachment of ligands (Torchilin, 2005).

4.3.5 INTERACTION AND CLEARANCE ROUTES

Significant properties that govern the interaction of NPs with molecules in biological system are size- and surface-localized ligands (Rosen et al., 2011). The majority of NP-based contrast agents still lack in their ability to undergo slow renal clearance or avoid nonspecific accumulation in liver, spleen, and lymph nodes (Mohammadian et al., 2016a; Sadeghzadeh et al., 2017). As particles <5.5 nm undergo elimination via renal routs and >200 nm shows accumulation in liver, spleen, etc., it becomes necessary to have NPs with appropriate size that can avoid rapid removal by kidneys or induce toxicity on accumulation in organs such as liver and spleen (Mohammadian et al., 2015, 2016b, 2017; Amirsaadat et al., 2017). As biological interactions are important for contrast agents, surface-localized ligands play active role in avoiding uptake by reticuloendothelial system (RES). With this, NPs (<10 nm) with a neutral surface or having a surface ligand of hydrophilic molecule are stealth for adsorption and opsonization as part of the complement system (Sun et al., 2008; Farajzadeh et al., 2018). In addition to degradation of NPs into toxic entities, nonspecific distribution with delayed clearance results in toxicities ranging from acute to chronic through interactions with the molecules inside the body (Cormode et al., 2009; Surendiran et al., 2009; Lee et al., 2012; Ahn et al., 2013). It became primarily essential to have a predetermination module for determining the type of interaction NPs have and method in determining the breakdown products within the body (Davies, 2008; http://www.fda.gov/nanotechnology/; http://nanoehsalliance.org/). Iron oxide injected in the crystalline form undergoes breakdown into nontoxic forms (released at manageable rates). If the same amount is injected as free ions, it would be toxic (Corot et al., 2006). Gold-based NPs being resistant to breakdown are considered quite inert, while QDs that breakdown into highly toxic cadmium ions need serious consideration before being employed for clinical use (Mancini et al., 2008). With this, it is essential to develop biocompatible and biodegradable NPs that show minimal nonspecific binding and uptake and effective elimination from the body for employment in designing contrast agents for medical imaging.

4.4 MAGNETIC RESONANCE IMAGING

Application of MRI in medical diagnosis and soft tissue imaging is considered an effective and safest tool due to the absence of ionizing radiation (Qiao et al., 2009). In vivo, MRI is used to locate and image the particular cell, besides being effectively used in imaging of malignant cells in conjunction with targeted cancer therapy (Banerjee and Chen, 2008; Jain et al., 2008; Santra et al., 2009). Water protons present in the living tissues interact with the strong magnetic field caused by MRI, which aligns the protons in the field direction. The radiofrequency so generated within the tissues is specific to specific tissues and results in protons' alignment. Once the RF is released, protons return to their ground state, and tissues relax. The contrast in the MRI imaging depends on the relaxation times of two different tissues. Greater is the relaxation times; more is the contrast and vice versa. The relaxation time (time to return to the ground state) is a combination of T1 (longitudinal relaxation time) and T2 (transverse relaxation time). Altering the relaxation times by image contrast agents is used to increase the contrast of MRI-generated images. Such alterations are useful in improving the visibility of the target cells on MRI. Gd-DTPA (DTPA =diethylenetriaminepentaacetic acid) is an essential contrasting agent (Zhu et al., 2006).

4.5 CONTRAST AGENTS FOR MRI

Gadolinium (III) ions are contrasting agents for MRI as they are highly paramagnetic and show high magnetic movements because of their unpaired electrons (Zhu et al., 2006). Though these ions may be toxic to a greater extent, appropriate ligands can create gadolinium chelates that are highly stable and nontoxic (Caravan et al., 1999). Besides enhancing the MRI, gadolinium ions find their use in drug delivery to target cells, for example, Gd-DOTA (DOTA = tetraazacyclododecane tetra acetic acid) is used to target a specific drug to target tumor cells (Park et al., 2008). Gadolinium ions can be incorporated into different types of organics [polymers (Zhang et al., 2011; Park et al., 2008), liposomes (Kamaly and Miller, 2010)] and inorganics [silica NPs and carbon nanotubes (Na et al., 2009; Shao et al., 2011)].

Superparamagnetic iron oxide NP (SPION) is another highly important image contrasting agent that works by decreasing T2 relaxation time and producing dark spots of signal voids (Shapiro et al., 2004). The superparamagnetic property of SPIONs is only applicable to those particles, the size of which is less than 20 mm (Teja and Koh, 2009). Permanently magnetized (ferromagnetic) iron oxides lack this property. Superparamagnetism exhibited by SPIONs and paramagnetism exhibited by gadolinium show similar results as both of these undergo net magnetization when the external magnetic field is applied. The only difference is that superparamagnetic materials show zero net magnetization when external field is removed (Murbe et al., 2008; Teja and Koh, 2009; Qiao et al., 2009). SPIONs are preferred over most pragmatic materials as they exhibit a greater magnetic response (Sun et al., 2008). The saturation magnetization (Ms) value is the point in a sample at which all the individual magnetic moments are uniformly distributed. The Ms value of SPIONs is proportional to the degree of MRI signal interference and contrast enhancement. The Ms value of iron oxide is different in different phases. For imaging applications, the bulk magnetite is the better choice due to its higher Ms value (92 emu/g at 300 K) than that of the maghemite (76 emu/g) (Hong et al., 2007). The bulk strongly depends on the size and morphology of SPIONs and the synthesis due to which Ms of SPIONs rarely reach bulk. The SPIONs have magnetic properties that are affected by the reaction temperature, reactant concentration, and pH (Murbe et al., 2008). It is significant to use SPIONs with high Ms value for in vivo work to reduce iron load. The iron oxide is usually nontoxic and is metabolized into hemoglobin inside the body, but higher doses of iron enhance the chances of toxicity (Gupta and Gupta, 2005; Laurent et al., 2008; Qiao et al., 2009).

4.6 COMPUTED TOMOGRAPHY SCANNING

Imaging of different body organs such as head, chest, pelvis, and abdomen is performed by CT scanning. It is an X-ray-based imaging technique that images an organ by multiple X-rays from different angles, which gives a detailed cross-sectional image. In CT scans, molecules enriched with iodine or gadolinium are used as image contrast agents; however, catheterization requirement, weak contract in larger patients (Hainfeld et al., 2006), their poor circulation half-lives (Rabin et al., 2006), and nonspecific distribution becomes their limiting factors. With the development of different nanomaterial-based contrast agents such as iodinated NPs, gold NPs (Hainfeld et al., 2006; Kim et al., 2007; Kattumuri et al., 2007), bismuth sulfide (Bi_2S_3) NPs (Rabin et al., 2006), limitations of traditional ICAs have been overcome up to a greater extent.

4.6.1 NANOPARTICLES FOR CT

To overcome the loopholes in the use of traditional image contrasting agents, polymer-coated Bi2S3 NPs have emerged as a promising new class of contrast agents (Rabin et al., 2006). These coated nanoparticles' clinical trials were demonstrated on mouse model organs such as vasculature, liver, and lymph nodes. They exhibit fivefold increase in X-ray absorption and displayed greater half-life (Rabin et al., 2006). One among the important group of image contrasting agents is gold NP, which is better CT contrasting agent than Bi_2S_3 NPs due to their increased X-ray absorption (2.7–5.7 times) compared to iodine and their prolonged blood circulation time (4 h in rats compared to 10 min for iodine) (Hainfeld et al., 2006; Kim et al., 2007). Greater blood circulation reduces the speedy renal clearance exhibited by minor contrasting agents (Hong et al., 2007).

Moreover, gold-coated contrasting agents can be best imaged at lower energies (80–100 keV), which are effectively absorbed by soft as well as hard tissues of the body. Hence the patients can be experienced to lesser doses of radiations (Hainfeld et al., 2006). Applying the lesser doses of gold NPs results in their lesser accumulation within the target organs (e.g., liver and spleen) and apparent imaging, which helps clear detection of any kind of malignancy or tumor formation on CT scans (Hainfeld et al., 2006). Different studies have shown the application of different concentrations of gold NPs in CT scan imaging. Commercially available 1.6-nm gold NPs have been used in vivo with a mouse model of mammary carcinoma. For useful imaging of blood pool and hepatoma using rodent as a model organism, 30 nm of gold NPs coated with PEG were used (Kim et al., 2007). This contrasting agent in the both cases (1.6 and 30 nm) does not reveal any toxicity to the tissues under in vivo conditions. The formation of newer and effective contrasting agents depends on modifying or enhancing the available older materials. For example, an image contrasting agent iodine suffers a lot of drawbacks due to

Nanoparticles in Medical Imaging: A Perspective Study 123

a lack of circulation time and tissue specificity. However, when it gets incorporated indifferent nanostructures, its limitations get overcome. Different practices that have been adopted include encapsulation of iodine liposomes (Mukundan et al., 2006; Elrod et al., 2009; Kweon et al., 2010; Zheng et al., 2010), nanoscale coordination polymers of iodine (deKrafft et al., 2009), and incorporating iodine into various other nanostructures (Chrastina and Schnitzer, 2010; Van Herck et al., 2010; Aillon et al., 2010). All these modifications have resulted in a significant increase in the imaging capability of iodine (Liu et al., 2007) due to more extended circulation as compared to traditional CT imaging agents, hence allowing better target imaging besides their molecular imaging (Cai and Chen, 2007).

4.7 POSITRON EMISSION TOMOGRAPHY SCANNING

PET scanning is a diagnostic tool widely utilized in modern-day medicine over MRI, CT scan, and X-rays. This technique provides the body's structural and functional information, for example, glucose metabolism, blood flow, and oxygen use. Unlike MRI, CT scan, and X-rays, it measures the amount of image contrast agent within the target organ (Cai and Chen, 2007). This technique requires radiotracers such as ^{11}C, ^{13}N, ^{15}O, ^{18}F, and ^{64}Cu, incorporated inside the biomolecules to measure their metabolism and distribution. Fluorodeoxyglucose is a radiotracer that is used in PET to determine glucose metabolism. PET techniques have proved to be an important technique for nanomaterial researchers as single-walled nanotubes get radiolabeled to check their biodistribution in vivo (Cai and Chen, 2007; Liu et al., 2007; Hong et al., 2009). Nanomaterials arise as important agents of PET scanning as they mediate radiotracer molecules and help in drug delivery to the target area. However, PET only recognizes a radiolabel rather than a NP in itself; thereby, it is possible to inaccuracy the results following dissociation of the compounds on administration (Cai and Chen, 2007).

A progressive trend in recent times is combining two imaging systems, that is, PET with other techniques to generate the composite image, capable of revealing structural and functional information of the imaging tissue. Large surface area of NPs permits their conjugation to several imaging and therapeutic ligands that attribute them with property for use as contrast agents in multimodal imaging (Cai and Chen, 2007). Though chelated metals (e.g., ^{64}Cu) are capable of labeling macromolecules, they suffer in their usage owing to their charged nature and large size that have great influence on macromolecular structure and as such their pharmacokinetic properties (Herth et al., 2009). This

124 Diverse Applications of Nanotechnology in the Biological Sciences

led to a shift toward the development of the effective radiolabels using other radionuclides (e.g., ^{18}F) with properties in avoiding structural irregularities of the macromolecules under study (Herth et al., 2009). Despite influencing macromolecule structure, ^{64}Cu is still being used in tracking biodistribution of the developed nanotubes under in vivo conditions (Hong et al., 2009). As the PET technique has emerged as a potent diagnostic tool, newer studies need to be carried out to develop more effective NP-based multimodal contrast agents capable of simultaneously retrieving structural and functional data.

4.8 FLUORESCENCE IMAGING

Fluorophores (organic), together with fluorescent proteins, are employed to study molecular and cellular processes in biological systems insensitive, efficient, and noninvasive manner (Nune et al., 2009). On getting incorporated into the macromolecular structure, they are used to study macromolecules' distribution and fate along with their interaction with other cellular entities using techniques such as confocal microscopy and flow cytometry. As conventional fluorophores face the problem of limited penetration, poor signal strength with short imaging time, and high susceptibility to photobleaching, fluorescent labeling of NPs is used in fluorescent-based imaging (Smith et al., 2006; Pan et al., 2008). Fluorescent labeling of NPs enhances the ability to tract their uptake, distribution, and retention, if any, as part of cell labeling and tracking strategy.

Fluorescent tagging extended to different types of nanomaterials, including those employed as contrasting agents, has been used in ascertaining their cellular uptake, biodistribution, and mechanisms of accumulation at tumoral sites; thereby opens up new avenues for their utilization in diagnosing diseases and getting insights of the disease progression at the cellular and molecular levels. Semiconductor QDs (particles having a combination of heavy metals such as CdSe and InP) that possess unique optical properties have successively replaced conventional organic fluorophores for their application in fluorescence-based imaging (Cai et al., 2007). QDs—particles with large molar extinction coefficient and high photostability, with brighter emittance irrespective of tissue depth—are effectively used for application in in vivo imaging (Cai et al., 2007). With absorption in a wide spectral range, emission within a particular wavelength irrespective of input wavelength increases their use as a replacement to organic fluorophores (Gao et al., 2004; Cai et al., 2007). They are often used in tracking the movement of cells and cellular differentiation, besides being used in visualizing tissue distribution of nanomaterial (Ballou et al., 2005; Ballou et al., 2007; Cai et al., 2007; Yang et al., 2009). As fine-tuning for achieving emission of the desired wavelength has been gained through variation in the size and composition of QDs (Gao et al., 2004), optical multiplexing (illumination of probes with the same wavelength resulting in multiple color emission) increased the application of QD fluorescent probes in the imaging system. There are specific concerns related to the use of QD fluorescent tags in the biological system, prominent being possible toxic effect as being a constituent of heavy metals that are themselves toxic and second ability to accumulate in soft tissues over an extended period of time that exert harmful effects in the body (Smith et al., 2006). Despite the possible toxicity inside the biological system, their use is connected to the retention of the fluorescent properties even after two years of injection, if properly coated (Ballou et al., 2007).

In addition to QDs, encapsulation of organic fluorescent molecules in a nanocarrier system that enhances signal strength of fluorophores (owing to a mixture of fluorescent dyes) increases the assay (Wang et al., 2006; Yan et al., 2007). As encapsulation resists photobleaching of dyes, their biodistribution and enhanced clearance make them preferred over conventional organic fluorophores. Being a common entrapment method for organic fluorophores, challenges arise in entrapping hydrophobic dye molecules in the hydrophilic core environment of silica NPs (Wang et al., 2006). Of the different modules adopted for encapsulation, the covalent linking of dye molecules to silica material is commonly employed to synthesize silicon NPs (Stober synthesis method) (Ow et al., 2005; Wang et al., 2006). Another technique involves entrapping hydrophobic dyes using microemulsion method where particles are created in droplets of water in oil emulsion with dye trapped physically inside these particles (Wang et al., 2006). Though, silica-based systems offer intriguing features due to low toxicity and the potential to undergo surface modification. Their use is limited to issues about dye leakage and aggregation inside the biological system (Wang et al., 2006; Yan et al., 2007).

4.9 CONCLUSION AND FUTURE PERSPECTIVE OF NP-BASED CONTRAST AGENTS

Though NPs are recommended for use as diagnostic agents, only a few like iron oxide NPs have so far been put into clinical use (Kiessling et al., 2014). The concerns that limit their usage include difficulty achieving homogeneity in particles with desirable pharmacokinetic properties, besides having the concerns of induction of toxicity for not achieving biodegradation toward elimination from biological systems (Mahapatro and Singh, 2011). Additionally, immunoreactivity for approved contrast agents following their binding to plasma protein that makes their entry to liver, spleen, lymph nodes, bone marrow, etc. hinders their broader applicability. They face the challenges of longer retention of potentially toxic NP components or materials as payload or increased removal from circulation before exerting their effects. Steric stabilization of NPs following the coating of polyethylene glycol is often used as a mitigating strategy to impart stability and reduce their toxic effects within the biological system (Jokerst et al., 2011). It is anticipated that future designing strategies will depend on nanomaterials with optimal characteristics and specific for immune cell subset with enhanced tumor accumulation and clearance from the body.

AUTHOR CONTRIBUTIONS

ATJ conceived the idea. Authors contributed equally in generating a draft of different sections and repeated editing of the contents present in the finalized version of the manuscript.

FUNDING

No funding was available for the study.

ACKNOWLEDGMENT

The authors would like to thank Dr. Majid Kamli and Dr. Irfan Ahmad Rather for their timely suggestion and extending help in getting the manuscript finalized.

KEYWORDS

- nanodelivary
- diagnostics
- nanomaterials
- diseases
- cancer

REFERENCES

Abdullah, Al. N.; Lee, J. E.; In, I.; Lee, H.; Lee, K. D.; Jeong, J. H. et al. Target Delivery and Cell Imaging Using Hyaluronic Acid-Functionalized Graphene Quantum Dots. *Mol. Pharm.* **2013**, *10* (10), 3736–3744.

Agarwal, A.; Shao, X.; Rajian, J. R.; Zhang, H.; Chamberland, D. L.; Kotov, N. A. et al. Dual-Mode Imaging with Radiolabeled Gold Nanorods. *J. Biomed. Opt.* **2011**, *16* (5), 051307.

Ahn, S.; Jung, S. Y.; Lee, S. J. Gold Nanoparticle Contrast Agents in Advanced X-ray Imaging Technologies. *Molecules* **2013**, *18*, 5858–5890.

Aillon, K. L.; El-Gendy, N.; Dennis, C.; Norenberg, J. P.; McDonald, J. et al. Iodinated Nano Clusters as an Inhaled Computed Tomography Contrast Agent for Lung Visualization. *Mol. Pharm.* **2010**, *7*, 1274–1282.

Al-Jamal, W. T.; Al-Jamal, K. T.; Bomans, P. H. et al. Functionalized-Quantum-Dot-Liposome Hybrids as Multimodal Nanoparticles for Cancer. *Small* **2008**, *4* (9), 1406–1415.

Allen, T. M.; Hansen, C.; Martin, F.; Redemann, C.; Yau-Young, A. Liposomes Containing Synthetic Lipid Derivatives of Poly(Ethylene Glycol) Show Prolonged Circulation Half-Lives in Vivo. *Biochim. Biophys. Acta* **1991**, *1066*, 29–36.

Amirbekian, V.; Lipinski, M. J.; Briley-Saebo, K. C.; Amirbekian, S.; Aguinaldo, J. G.; Weinreb, D. B.; Vucic, E.; Frias, J. C.; Hyafil, F.; Mani, V.; Fisher, E. A.; Fayad, Z. A. Detecting and Assessing Macrophages in Vivo to Evaluate Atherosclerosis Noninvasively Using Molecular MRI. *Proc. Natl. Acad. Sci. U. S. A.* **2007**, *104*, 961–966.

Amirsaadat, S.; Pilehvar-Soltanahmadi, Y.; Zarghami, F. et al. Silibinin-Loaded Magnetic Nanoparticles Inhibit hTERT Gene Expression and Proliferation of Lung Cancer Cells. *Artif. Cells Nanomed. Biotechnol.* **2017**, *45* (8), 1649–1656.

Ananta, J. S.; Godin, B.; Sethi, R.; Moriggi, L.; Liu, X.; Serda, R. E. et al. Geometrical Confinement of Gadolinium-Based Contrast Agents in Nanoporous Particles Enhances T1 Contrast. *Nat. Nanotechnol.* **2010**, *5* (11), 815–8121.

Anderson, E. A.; Isaacman, S.; Peadbody, D. S.; Wang, E. Y.; Canary, J. W.; Kirshenbaum, K. Viral Nanoparticles Donning a Paramagnetic Coat, Conjugation of MRI Contrast Agents to the MS2 Capsid. *Nano Lett.* **2006**, *6*, 1160–1164.

Ballou, B.; Ernst, L. A.; Andreko, S.; Harper, T.; Fitzpatrick, J. A. J. et al. Sentinel Lymph Node Imaging Using Quantum Dots in Mouse Tumor Models. *Bioconjug. Chem.* **2007**, *18*, 389–396.

Ballou, B.; Ernst, L. A.; Waggoner, A. S. Fluorescence Imaging of Tumors in Vivo. *Curr. Med. Chem.* **2005**, *12*, 795–805.

Banerjee, S. S.; Chen, D. Multifunctional PH-Sensitive Magnetic Nanoparticles for Simultaneous Imaging, Sensing and Targeted Intracellular Anticancer Drug Delivery. *Nanotechnology* **2008**, *19*, 1–8.

Berry, C. C.; Wells, S.; Charles, S.; Curtis, A. S. G. Dextran and Albumin Derivatised Iron Oxide Nanoparticles, Influence on Fibroblasts in Vitro. *Biomaterials* **2003**, *24*, 4551–4557.

Bhattacharyya, S.; Kudgus, R. A.; Bhattacharya, R.; Mukherjee, P. Inorganic Nanoparticles in Cancer Therapy. *Pharm. Res.* **2011**, *28* (2), 237–259.

Bridot, J. L.; Faure, A. C.; Laurent, S.; Riviere, C.; Billotey, C.; Hiba, B. et al. Hybrid Gadolinium Oxide Nanoparticles, Multimodal Contrast Agents for in Vivo Imaging. *J. Am. Chem. Soc.* **2007**, *129* (16), 5076–5084.

Briley-Saebo, K. C.; Mulder, W. J. M.; Mani, V.; Hyafil, F.; Amirbekian, V.; Aguinaldo, J. G. S.; Fisher, E. A.; Fayad, Z. A. Magnetic Resonance Imaging of Vulnerable Atherosclerotic

Plaques, Current Imaging Strategies and Molecular Imaging Probes. *J. Magn. Reson. Imag.* **2007,** *26,* 460–479.

Bruchez, M. Jr.; Moronne, M.; Gin, P.; Weiss, S.; Alivisatos, A. P. Semiconductor Nanocrystals as Fluorescent Biological Labels. *Science* **1998,** *281,* 2013–2016.

Brust, M.; Walker, M.; Bethell, D.; Schiffrin, D. J.; Whyman, R. Synthesis of Thiol-Derivatised Gold Nanoparticles in a Two-Phase Liquid–Liquid System. *Chem. Commun.* **1994,** 801–802.

Cai, W.; Chen, X. Nanoplatforms for Targeted Molecular Imaging in Living Subjects. *Small* **2007,** *3,* 1840–1854.

Cai, W.; Hsu, A. R.; Li, Z. B.; Chen, X. Are Quantum Dots Ready for in Vivo Imaging in Human Subjects? *Nanoscale Res. Lett.* **2007,** *2,* 265–281.

Cao, C. Y.; Shen, Y. Y.; Wang, J. D.; Li, L.; Liang, G. L. Controlled Intracellular Self-Assembly of Gadolinium Nanoparticles as Smart Molecular MR Contrast Agents. *Sci. Rep.* **2013,** *3,* 1024.

Caravan, P.; Ellison, J. J.; McMurry, T. J.; Lauffer, R. B. Gadolinium(III) Chelates as MRI Contrast Agents, Structure, Dynamics, and Applications. *Chem. Rev.* **1999,** *99,* 2293–2352.

Chen, W. T.; Thirumalai, D.; Shih, T. T.; Chen, R. C.; Tu, S. Y.; Lin, C. I. et al. Dynamic Contrast-Enhanced Folate-Receptor-Targeted MR Imaging Using a Gd-Loaded PEG-Dendrimer-Folate Conjugate in a Mouse Xenograft Tumor Model. *Mol. Imag. Biol.* **2010,** *12* (2), 145–154.

Chen, Y.; Chen, G.; Feng, S.; Pan, J.; Zheng, X.; Su, Y. et al. Label-Free Serum Ribonucleic Acid Analysis for Colorectal Cancer Detection by Surface-Enhanced Raman Spectroscopy and Multivariate Analysis. *J. Biomed. Opt.* **2012,** *17* (6), 067003.

Choi, H. S.; Liu, W.; Misra, P. et al. Renal Clearance of Quantum Dots. *Nat. Biotechnol.* **2007,** *25* (10), 1165–1170.

Chowdhury, E. H.; Akaike, T. Bio-Functional Inorganic Materials, An Attractive Branch of Gene-Based Nano-Medicine Delivery for 21st Century. *Curr. Gene Ther.* **2005,** *5,* 669–676.

Chrastina, A.; Schnitzer, J. E Iodine-125 Radiolabeling of Silver Nanoparticles for in Vivo SPECT Imaging. *Int. J. Nanomed.* **2010,** *5,* 653–659.

Cormode, D. P.; Jarzyna, P. A.; Mulder, W. J. et al. Modified Natural Nanoparticles as Contrast Agents for Medical Imaging. *Adv. Drug Deliv. Rev.* **2010,** *62,* 329–338.

Cormode, D. P.; Mulder, W. J. M.; Fisher, E. A.; Fayad, Z. A. Modified Lipoproteins as Contrast Agents for Molecular Imaging. *Future Lipidol.* **2007,** *2,* 587–590.

Cormode, D. P.; Skajaa, T.; Fayad, Z. A. et al. Nanotechnology in Medical Imaging Probe Design and Applications. *Arterioscler Thromb. Vasc. Biol.* **2009,** *29,* 992–1000.

Cormode, D. P.; Skajaa, T.; van Schooneveld, M. M.; Koole, R.; Jarzyna, P.; Lobatto, M. E.; Calcagno, C.; Barazza, A.; Gordon, R. E.; Zanzonico, P.; Fisher, E. A.; Fayad, Z. A.; Mulder, W. J. M. Nanocrystal Core High-Density Lipoproteins, A Multimodal Molecular Imaging Contrast Agent Platform. *Nano Lett.* **2008,** *8,* 3715–3723.

Corot, C.; Robert, P.; Idee, J-M.; Port, M. Recent Advances in Iron Oxide Nanocrystal Technology for Medical Imaging. *Adv. Drug Deliv. Rev.* **2006,** *58,* 1471–1504.

Dams, E. T.; Oyen, W. J.; Boerman, O. C.; Storm, G.; Laverman, P.; Kok, P. J. et al. 99mTc-PEG Liposomes for the Scintigraphic Detection of Infection and Inflammation, Clinical Evaluation. *J. Nucl. Med.* **2000,** *41* (4), 622–630.

Davies, J. C. Nanotechnology Oversight, an Agenda for the New Administration. Woodrow Wilson International Center for Scholars, Project on Emerging Nanotechnologies. 2008, PEN 13.

Davis, M. E.; Chen, Z.; Shin, D. M. Nanoparticle Therapeutics, an Emerging Treatment Modality for Cancer. *Nat. Rev. Drug Discov.* **2008,** *7* (9), 771–82. [PubMed, 18758474].

de Barros, A. B.; Tsourkas, A.; Saboury, B.; Cardoso, V. N.; Alavi, A. Emerging Role of Radiolabeled Nanoparticles as an Effective Diagnostic Technique. *EJNMMI Res.* **2012,** *2* (1), 39.

deKrafft, K. E.; Xie, Z.; Cao, G.; Tran, S.; Ma, L. et al. Iodinated Nanoscale Coordination Polymers as Potential Contrast Agents for Computed Tomography. *Angew Chem. Int. Edit.* **2009,** *48*, 9901–9904.

Du, Y.; Lai, P. T.; Leung, C. H. et al. Design of Superparamagnetic Nanoparticles for Magnetic Particle Imaging (MPI). *Int. J. Mol. Sci.* **2013,** *14*, 18682–18710.

Elrod, D. B.; Partha, R.; Danila, D.; Casscells, S. W.; Conyers, J. L. An Iodinated Liposomal Computed Tomographic Contrast Agent Prepared from a Diiodophosphatidylcholine Lipid. *Nanomedicine* **2009,** *5*, 42–45.

Farajzadeh, R.; Pilehvar-Soltanahmadi, Y.; Dadashpour, M. et al. Nano-Encapsulated Metformin-Curcumin in PLGA/PEG Inhibits Synergistically Growth and hTERT Gene Expression in Human Breast Cancer Cells. *Artif. Cells Nanomed. Biotechnol.* **2018,** *46* (5), 917–925.

Feng, S.; Chen, R.; Lin, J.; Pan, J.; Chen, G.; Li, Y. et al. Nasopharyngeal Cancer Detection Based on Blood Plasma Surface-Enhanced Raman Spectroscopy and Multivariate Analysis. *Biosens. Bioelectron.* **2010,** *25* (11), 2414–2419.

Frangioni, J. V. New Technologies for Human Cancer Imaging. *J. Clin. Oncol.* **2008,** *26*, 4012–4021.

Frias, J. C.; Williams, K. J.; Fisher, E. A.; Fayad, Z. A. Recombinant HDL-Like Nanoparticles, a Specific Contrast Agent for MRI of Atherosclerotic Plaques. *J. Am. Chem. Soc.* **2004,** *126*, 16316–16317.

Gao, X. H.; Cui, Y. Y.; Levenson, R. M.; Chung, L. W.; Nie, S. In Vivo Cancer Targeting and Imaging with Semiconductor Quantum Dots. *Nat. Biotechnol.* **2004,** *22*, 969–976.

Gaumet, M.; Vargas, A.; Gurny, R.; Delie, F. Nanoparticles for Drug Delivery, the Need for Precision in Reporting Particle Size Parameters. *Eur. J. Pharm. Biopharm.* **2008,** *69*, 1–9.

Gleich, B.; Weizenecker, J. Tomographic Imaging Using the Nonlinear Response of magnetic Particles. *Nat. Biotech.* **2005,** *435*, 1214–1217.

Greish, K. Enhanced Permeability and Retention of Macromolecular Drugs in Solid Tumors, a Royal Gate for Targeted Anticancer Nanomedicines. *J. Drug Target.* **2007,** *15* (7–8), 457–464.

Guerrero, S.; Herance, J. R.; Rojas, S.; Mena, J. F.; Gispert, J. D.; Acosta, G. A. et al. Synthesis and in Vivo Evaluation of the Biodistribution of a 18F-Labeled Conjugate Gold-Nanoparticle-Peptide with Potential Biomedical Application. *Bioconjug. Chem.* **2012,** *23* (3), 399–408.

Gupta, A. K.; Gupta, M. Synthesis and Surface Engineering of Iron Oxide Nanoparticles for Biomedical Applications. *Biomaterials* **2005,** *26*, 3995–4021.

Hainfeld, J. F.; Slatkin, D. N.; Focella, T. M.; Smilowitz, H. M. Gold Nanoparticles, a New X-ray Contrast Agent. *Br. J. Radiol.* **2006,** *79* (939), 248–253. [PubMed, 16498039].

Harrington, K. J.; Mohammadtaghi, S.; Uster, P. S.; Glass, D.; Peters, A. M.; Vile, R. G. et al. Effective Targeting of Solid Tumors in Patients with Locally Advanced Cancers by Radiolabeled Pegylated Liposomes. *Clin. Cancer Res.* **2001,** *7* (2), 243–254.

Hausser, J.; Alon, U. Tumour Heterogeneity and the Evolutionary Trade-Offs of Cancer. *Nat. Rev. Cancer.* **2020,** *20*, 247–257.

130   Diverse Applications of Nanotechnology in the Biological Sciences

Hawley, A. E.; Illum, L.; Davis, S. S. Preparation of Biodegradable, Surface Engineered PLGA Nanospheres with Enhanced Lymphatic Drainage and Lymph Node Uptake. *Pharm. Res.* **1997,** *14* (5), 657–661.

He, W.; Ai, K.; Lu, L. Nanoparticulate X-ray CT Contrast Agents. *Sci. China Chem.* **2015,** *58*, 753–760.

Herth, M. M.; Barz, M.; Moderegger, D.; Allmeroth, M.; Jahn, M. et al. Radioactive Labeling of Defined HPMA-Based Polymeric Structures Using [F-18]FETos for in Vivo Imaging by Positron Emission Tomography. *Biomacromolecules* **2009,** *10*, 1697–1703.

Hong, H.; Zhang, Y.; Sun, J.; Cai, W. Molecular Imaging and Therapy of Cancer with Radiolabeled Nanoparticles. *Nano Today* **2009,** *4*, 399–413.

Hong, J.; Gong, P.; Xu, D.; Sun, H.; Yao, S. Synthesis and Characterization of Carboxyl Functionalized Magnetic Nanogel via "Green" Photochemical Method. *J. Appl. Polym. Sci.* **2007,** *105*, 1882–1887.

Hostetler, M. J.; Wingate, J. E.; Zhong, C. J.; Harris, J. E.; Vachet, R. W.; Clark, M. R.; Londono, J. D.; Green, S. J.; Stokes, J. J.; Wignall, G. D.; Glish, G. L.; Porter, M. D.; Evans, N. D.; Murray, R. W. Alkanethiolate Gold Cluster Molecules with Core Diameters from 1.5 to 5.2 nm, Core and Monolayer Properties as a Function of Core Size. *Langmuir* **1998,** *14*, 17–30.

http://nanoehsalliance.org/.

http://www.fda.gov/nanotechnology/.

Huang, S. L.; Hamilton, A. J.; Pozharski, E.; Nagaraj, A.; Klegerman, M. E.; McPherson, D. D.; MacDonald, R. C. Physical Correlates of the Ultrasonic Reflectivity of Lipid Dispersions Suitable as Diagnostic Contrast Agents. *Ultrasound Med. Biol.* **2002,** *28*, 339–348.

Huang, X.; Bronstein, L. M.; Retrum, J.; Dufort, C.; Tsvetkova, I.; Aniagyei, S.; Stein, B.; Stucky, G.; McKenna, B.; Rennes, N.; Baxter, D.; Kao, C. C.; Dragnea, B. Self-Assembled Virus-Like Particles with Magnetic Cores. *Nano Lett.* **2007,** *7*, 2407–2416.

Hudson, S. P.; Padera, R. F.; Langer, R.; Kohane, D. S. The Biocompatibility of Mesoporous Silicates. *Biomaterials* **2008,** *29* (30), 4045–4055.

Hyafil, F.; Cornily, J. C.; Feig, J. E.; Gordon, R.; Vucic, E.; Amirbekian, V.; Fisher, E. A.; Fuster, V.; Feldman, L. J.; Fayad, Z. A. Noninvasive Detection of Macrophages Using a Nanoparticulate Contrast Agent for Computed Tomography. *Nat. Med.* **2007,** *13*, 636–641.

Ito, H.; Hasegawa, K.; Hasegawa, Y.; Nishimaki, T.; Hosomichi, K.; Kimura, S. et al. Silver Nanoscale Hexagonal Column Chips for Detecting Cell-Free DNA and Circulating Nucleosomes in Cancer Patients. *Sci. Rep.* **2015,** *5*, 10455.

Jain, T. K.; Richey, J.; Strand, M.; Leslie-Pelecky, D. L.; Flask, C. A. et al. Magnetic Nanoparticles with Dual Functional Properties, Drug Delivery and Magnetic Resonance Imaging. *Biomaterials* **2008,** *29*, 4012–4021.

Janib, S. M.; Moses, A. S.; MacKay, J. A. Imaging and Drug Delivery Using Theranostic Nanoparticles. *Adv. Drug. Deliv. Rev.* **2010,** *62*, 1052–1063.

Javier, D. J.; Nitin, N.; Levy, M.; Ellington, A.; Richards-Kortum, R. Aptamertargeted Gold Nanoparticles as Molecular-Specific Contrast Agents for Reflectance Imaging. *Bioconjug. Chem.* **2008,** *19*, 1309–1312.

Jin, Y.; Jia, C.; Huang, S. W.; O'Donnell, M.; Gao, X. Multifunctional Nanoparticles as Coupled Contrast Agents. *Nat. Commun.* **2010,** *1*, 41.

Jokerst, J. V.; Lobovkina, T.; Zare, R. N. et al. Nanoparticle PEGylation for Imaging and Therapy. *Nanomedicine* **2011,** *6*, 715–728.

Kamaly, N.; Miller, A. D. Paramagnetic Liposome Nanoparticles for Cellular and Tumour Imaging. *Int. J. Mol. Sci.* **2010,** *11,* 1759–1776.

Kang, H. W.; Weissleder, R.; Bogdanov, A. Targeting of MPEG-Protected Polyamino Acid Carrier to Human E-Selectin in Vitro. *Amino Acids* **2002,** *23,* 301–308.

Kang, M. K.; Lee, G. H.; Jung, K. H.; Jung, J. C.; Kim, H. K.; Kim, Y. H. et al. Gadolinium Nanoparticles Conjugated with Therapeutic Bifunctional Chelate as a Potential T1 Theranostic Magnetic Resonance Imaging Agent. *J. Biomed. Nanotechnol.* **2016,** *12* (5), 894–908.

Karfeld-Sulzer, L. S.; Waters, E. A.; Kohlmeir, E. K.; Kissler, H.; Zhang, X.; Kaufman, D. B. et al. Protein Polymer MRI Contrast Agents, Longitudinal Analysis of Biomaterials in Vivo. *Magn. Reson. Med.* **2011,** *65* (1), 220–228.

Kattumuri, V.; Katti, K.; Bhaskaran, S.; Boote, E.; Casteel, S. et al. Gum Arabic as a Phytochemical Construct for the Stabilization of Gold Nanoparticles, in Vivo Pharmacokinetics and X-ray-Contrast-Imaging Studies. *Small* **2007,** *3,* 333–341.

Kiessling, F.; Mertens, M. E.; Grimm, J. et al. Nanoparticles for Imaging, Top or Flop? *Radiology* **2014,** *273,* 10–28.

Kim, B. S.; Qiu, J. M.; Wang, J. P.; Taton, T. A. Magnetomicelles, Composite Nanostructures from Magnetic Nanoparticles and Cross-Linked Amphiphilic Block Copolymers. *Nano Lett.* **2005,** *5,* 1987–1991.

Kim, D.; Park, S.; Lee, J. H.; Jeong, Y. Y.; Jon, S. Antibiofouling Polymer-Coated Gold Nanoparticles as a Contrast Agent for in Vivo X-ray Computed Tomography Imaging. *J. Am. Chem. Soc.* **2007,** *129,* 7661–7665.

Klibanov, A. L. Microbubble Contrast Agents Targeted Ultrasound Imaging and Ultrasound-Assisted Drug-Delivery Applications. *Invest. Radiol.* **2006,** *41,* 354–362.

Kweon, S.; Lee, H.; Hyung, W. J.; Suh, J.; Lim, J. S. et al. Liposomes Coloaded with Iopamidol/ Lipiodol as a RES-Targeted Contrast Agent for Computed Tomography Imaging. *Pharm. Res.* **2010,** *27,* 1408–1415.

Kwon, G. S.; Kataoka, K. Block Copolymer Micelles as Long-Circulating Drug Vehicles. *Adv. Drug Deliv. Rev.* **1995,** *16,* 295–309.

Lanza, G. M.; Wallace, K. D.; Scott, M. J.; Cacheris, W. P.; Abendschein, D. R.; Christy, D. H.; Sharkey, A. M.; Miller, J. G.; Gaffney, P. J.; Wickline, S. A. A Novel Site-Targeted Ultrasonic Contrast Agent with Broad Biomedical Application. *Circulation* **1996,** *94,* 3334–3340.

Lanza, G. M.; Winter, P. M.; Caruthers, S. D.; Hughes, M. S.; Cyrus, T.; Marsh, J. N.; Neubauer, A. M.; Partlow, K. C.; Wickline, S. A. Nanomedicine Opportunities for Cardiovascular Disease with Perfluorocarbon Nanoparticles. *Nanomedicine* **2006,** *1,* 321–329.

Laurent, S.; Forge, D.; Port, M.; Roch, A.; Robic, C. et al. Magnetic Iron Oxide Nanoparticles, Synthesis, Stabilization, Vectorization, Physicochemical Characterizations, and Biological Applications. *Chem. Rev.* **2008,** *108,* 2064–2110.

Leatherdale, C. A.; Woo, W-K.; Mikulec, F. V.; Bawendi, M. G. On the Absorption Cross Section of CdSe Nanocrystal Quantum Dots. *J. Phys. Chem. B.* **2002,** *116,* 7619–7622.

Lee, H. Y.; Li, Z.; Chen, K.; Hsu, A. R.; Xu, C.; Xie, J. et al. PET/MRI Dual-Modality Tumor Imaging Using Arginine-Glycine-Aspartic (RGD)-Conjugated Radiolabeled Iron Oxide Nanoparticles. *J. Nucl. Med.* **2008,** *49* (8), 1371–1379.

Lee, J. H.; Park, G.; Hong, G. H. et al. Design Considerations for Targeted Optical Contrast Agents. *Quant. Imag. Med. Surg.* **2012,** *2,* 266–273.

Lee, J.; Chatterjee, D. K.; Lee, M. H.; Krishnan, S. Gold Nanoparticles in Breast Cancer Treatment, Promise and Potential Pitfalls. *Cancer Lett.* **2014,** *347* (1), 46–53.

Lee, S. H.; Kim, B. H.; Na, H. B. et al. Paramagnetic Inorganic Nanoparticles as T1 MRI Contrast Agents. *Wires Nanomed. Nanobiotechnol.* **2014,** *6*, 196–209.

Li, D.; Patel, A. R.; Klibanov, A. L.; Kramer, C. M.; Ruiz, M.; Kang, B. Y. et al. Molecular Imaging of Atherosclerotic Plaques Targeted to Oxidized LDL Receptor LOX-1 by SPECT/CT and Magnetic Resonance. *Circ. Cardiovasc. Imag.* **2010,** *3* (4), 464–472.

Li, K.; Wen, S.; Larson, A. C.; Shen, M.; Zhang, Z.; Chen, Q. et al. Multifunctional Dendrimer-Based Nanoparticles for in Vivo MR/CT Dual-Modal Molecular Imaging of Breast Cancer. *Int. J. Nanomed.* **2013,** *8*, 2589–2600.

Lin, D.; Feng, S.; Pan, J.; Chen, Y.; Lin, J.; Chen, G. et al. Colorectal Cancer Detection by Gold Nanoparticle Based Surface-Enhanced Raman Spectroscopy of Blood Serum and Statistical Analysis. *Opt. Express* **2011a,** *19* (14), 13565–13577.

Lin, J.; Chen, R.; Feng, S.; Pan, J.; Li, Y.; Chen, G. et al. A Novel Blood Plasma Analysis Technique Combining Membrane Electrophoresis with Silver Nanoparticle-Based SERS Spectroscopy for Potential Applications in Noninvasive Cancer Detection. *Nanomed.* **2011b,** *7* (5),655–663.

Lipinski, M. J.; Briley-Saebo, K. C.; Mani, V.; Fayad, Z. A. "Positive Contrast" Inversion-Recovery with Oxide Nanoparticles-Resonant Water Suppression Magnetic Resonance Imaging, a Change for the Better? *J. Am. Coll. Cardiol.* **2008,** *52*, 492–494.

Liu, H. L.; Ko, S. P.; Wu, J. H.; Jung, M. H.; Min, J. H.; Lee, J. H.; An, B. H.; Kim, Y. K. One-Pot Polyol Synthesis of Monosize PVP-Coated Sub-5nm Fe3O4 Nanoparticles for Biomedical Applications. *J. Magnet. Magnet. Mater*. **2007,** *310*, E815–E817.

Liu, Y.; Mi, Y.; Zhao, J.; Feng, S. S. Multifunctional Silica Nanoparticles for Targeted Delivery of Hydrophobic Imaging and Therapeutic Agents. *Int. J. Pharm.* **2011,** *421* (2), 370–378.

Liu, Z.; Cai, W.; He, L.; Nakayama, N.; Chen, K. et al. In Vivo Biodistribution and Highly Efficient Tumour Targeting of Carbon Nanotubes in Mice. *Nat. Nanotechnol.* **2007,** *2*, 47–52.

Longmire, M.; Choyke, P. L.; Kobayashi, H. Clearance Properties of Nano-Sized Particles and Molecules as Imaging Agents, Considerations and Caveats. *Nanomed.* -UK **2008,** *3* (5), 703–717.

Lumbroso, P.; Dick, C. E. X-ray Attenuation Properties of Radiographic Contrast Media. *Med. Phys.* **1987,** *14*, 752–758.

Mahapatro, A.; Singh, D. K. Biodegradable Nanoparticles Are Excellent Vehicle for Site Directed in-Vivo Delivery of Drugs and Vaccines. *J. Nanobiotechnol.* **2011,** *9*, 55.

Mahmoudi, M.; Sahraian, M. A.; Shokrgozar, M. A.; Laurent, S. Superparamagnetic Iron Oxide Nanoparticles, Promises for Diagnosis and Treatment of Multiple Sclerosis. *ACS Chem. Neurosci.* **2011,** *2* (3), 118–140.

Mancini, M. C.; Kairdolf, B. A.; Smith, A. M.; Nie, S. Oxidative Quenching and Degradation of Polymer-Encapsulated Quantum Dots, New Insights into the Long-Term Fate and Toxicity of Nanocrystals in Vivo. *J. Am. Chem. Soc.* **2008,** *130*, 10836–10837.

Massoud, T. F.; Gambhir, S. S. Molecular Imaging in Living Subjects, Seeing Fundamental Biological Processes in a New Light. *Genes Dev.* **2003,** *17*, 545–580.

McAteer, M. A.; Sibson, N. R.; von zur Muhlen, C.; Schneider, J. E.; Lowe, A. S.; Warrick, N.; Channon, K. M.; Anthony, D. C.; Choudhury, R. P. In Vivo Magnetic Resonance Imaging of Acute Brain Inflammation Using Microparticles of Iron Oxide. *Nat. Med.* **2007,** *13*, 1253–1258.

McCarthy, J. R.; Kelly, K. A.; Sun, E. Y.; Weissleder, R. Targeted Delivery of Multifunctional Magnetic Nanoparticles. *Nanomedicine* **2007**, *2*, 153–167.

McCarthy, J. R.; Weissleder, R. Multifunctional Magnetic Nanoparticles for Targeted Imaging and Therapy. *Adv. Drug Deliv. Rev.* **2008**, *60* (11), 1241–1251. [PubMed, 18508157].

Medintz, I. L.; Uyeda, H. T.; Goldman, E. R.; Mattoussi, H. Quantum Dot Bioconjugates for Imaging, Labelling and Sensing. *Nat. Mater.* **2005**, *4*, 435–446.

Michalet, X.; Pinaud, F. F.; Bentolila, L. A.; Tsay, J. M.; Doose, S.; Li, J. J.; Sundaresan, G.; Wu, A. M.; Gambhir, S. S.; Weiss, S. Quantum Dots for Live Cells, in Vivo Imaging, and Diagnostics. *Science* **2005**, *307*, 538–544.

Mohammadian, F.; Abhari, A.; Dariushnejad, H. et al. Effects of Chrysin-PLGA-PEG Nanoparticles on Proliferation and Gene Expression of miRNAs in Gastric Cancer Cell Line. *Iran J. Cancer Prev.* **2016b**, *9*, e4190.

Mohammadian, F.; Abhari, A.; Dariushnejad, H. et al. Upregulation of Mir-34a in AGS Gastric Cancer Cells by a PLGA-PEG-PLGA Chrysin Nano Formulation. *Asian Pac. J. Cancer Prev.* **2015**, *16*, 8259–8263.

Mohammadian, F.; Pilehvar-Soltanahmadi, Y.; Mofarrah, M. et al. Down Regulation of miR-18a, miR-21 and miR-221 Genes in Gastric Cancer Cell Line by Chrysin-Loaded PLGA-PEG Nanoparticles. *Artif. Cells Nanomed. Biotechnol.* **2016a**, *44*, 1972–1978.

Mohammadian, F.; Pilehvar-Soltanahmadi, Y.; Zarghami, F. et al. Upregulation of miR-9 and Let-7a by Nanoencapsulated Chrysin in Gastric Cancer Cells. *Artif. Cells Nanomed. Biotechnol.* **2017**, *45*, 1201–1206.

Morales-Avila, E.; Ferro-Flores, G.; Ocampo-Garcia, B. E.; De Leon-Rodriguez, L. M.; Santos-Cuevas, C. L.; Garcia-Becerra, R. et al. Multimeric System of 99mTc-Labeled Gold Nanoparticles Conjugated to c [RGDfK(C)] for Molecular Imaging of Tumor avb3 Expression. *Bioconjug. Chem.* **2011**, *22* (5), 913–922.

Mukundan, S.; Ghaghada, K. B.; Badea, C. T.; Kao, C.-Y.; Hedlund, L. W.; Provanzale, J. M.; Johnson, G. A.; Chen, E.; Bellamkonda, R. V.; Annapragada, A. A Liposomal Nanoscale Contrast Agent for Preclinical CT in Mice. *AJR* **2006**, *186*, 300–307.

Mulder, W. J. M.; Douma, K.; Koning, G. A.; van Zandvoort, M. A.; Lutgens, E.; Daemen, M. J.; Nicolay, K.; Strijkers, G. J. Liposome-Enhanced MRI of Neointimal Lesions in the apoE-KO Mouse. *Magn. Res. Med.* **2006**, *55*, 1170–1174.

Mulder, W. J. M.; Griffioen, A. W.; Strijkers, G. J.; Cormode, D. P.; Nicolay, K.; Fayad, Z. A. Magnetic and Fluorescent Nanoparticles for Multimodality Imaging. *Nanomedicine* **2007**, *2*, 307–324.

Mulder, W. J. M.; Strijkers, G. J.; Briley-Saeboe, K. C.; Frias, J. C.; Aguinaldo, J. G. S.; Vucic, E.; Amirbekian, V.; Tang, C.; Chin, P. T. K.; Nicolay, K.; Fayad, Z. A. Molecular Imaging of Macrophages in Atherosclerotic Plaques Using Bimodal PEG-Micelles. *Magn. Res. Med.* **2007**, *58*, 1164–1170.

Murbe, J.; Rechtenbach, A.; Topfer, J. Synthesis and Physical Characterization of Magnetite Nanoparticles for Biomedical Applications. *Mater. Chem. Phys.* **2008**, *110*, 426–433.

Murray, C. B.; Norris, D. J.; Bawendi, M. G. Synthesis and Characterization of Nearly Monodisperse CdE (E_S, Se, Te) Semiconductor Nanocrystallites. *J. Am. Chem. Soc.* **1993**, *115*, 8706–8715.

Na, H. B.; Song, I. C.; Hyeon, T. Inorganic Nanoparticles for MRI Contrast Agents. *Adv. Mater.* **2009**, *21*, 2133–2148.

Nahrendorf, M.; Jaffer, F. A.; Kelly, K. A.; Sosnovik, D. E.; Aikawa, E.; Libby, P.; Weissleder, R. Noninvasive Vascular Cell Adhesion Molecule-1 Imaging Identifies Inflammatory Activation of Cells in Atherosclerosis. *Circulation* **2006,** *114,* 1504–1511.

Nahrendorf, M.; Keliher, E.; Marinelli, B.; Leuschner, F.; Robbins, C. S.; Gerszten, R. E. et al. Detection of Macrophages in Aortic Aneurysms by Nanoparticle Positron Emission Tomography-Computed Tomography. *Arterioscler Thromb. Vasc. Biol.* **2011,** *31* (4), 750–757.

Nahrendorf, M.; Zhang, H.; Hembrador, S.; Panizzi, P.; Sosnovik, D. E.; Aikawa, E.; Libby, P.; Swirski, F. K.; Weissleder, R. Nanoparticle PET-CT Imaging of Macrophages in Inflammatory Atherosclerosis. *Circulation* **2008,** *117,* 379–387.

Niemeyer, C. M. Nanoparticles, Proteins, and Nucleic Acids, Biotechnology Meets Materials Science. *Angew Chem. Int. Ed.* **2001,** *40* (22), 4128–4158.

Nune, S. K.; Gunda, P.; Thallapally, P. K.; Lin, Y.; Forrest, M. L. et al. Nanoparticles for Biomedical Imaging. *Expert Opin. Drug Deliv.* **2009,** *6,* 1175–1194.

Otsuka, H.; Nagasaki, Y.; Kataoka, K. PEGylated Nanoparticles for Biological and Pharmaceutical Applications. *Adv. Drug Deliv.* **2003,** *55,* 403–419.

Ow, H.; Larson, D. R.; Srivastava, M.; Baird, B. A.; Webb, W. W. et al. Bright and Stable Core-Shell Fluorescent Silica Nanoparticles. *Nano Lett.* **2005,** *5,* 113–117.

Paganelli, G.; Chinol, M.; Maggiolo, M.; Sidoli, A.; Corti, A.; Baroni, S.; Siccardi, A. G. The Three-Step Pretargeting Approach Reduces the Human Anti-Mouse Antibody Response in Patients Submitted to Radioimmunoscintigraphy and Radioimmunotherapy. *Eur. J. Nucl. Med.* **1997,** *24,* 350–351.

Pan, J.; Wang, Y.; Feng. S. Formulation, Characterization, and in Vitro Evaluation of Quantum Dots Loaded in Poly(lactide)-Vitamin E TPGS Nanoparticles for Cellular and Molecular Imaging. *Biotechnol. Bioeng.* **2008,** *101,* 622–633.

Park, J.; Lee, J.; Jung, J.; Yu, D.; Oh, C. et al. Gd-DOTA Conjugate of RGD as a Potential Tumor-Targeting MRI Contrast Agent. *ChemBioChem* **2008,** *9,* 2811–2813.

Petersen, A. L.; Binderup, T.; Rasmussen, P.; Henriksen, J. R.; Elema, D. R.; Kjaer, A. et al. ^{64}Cu Loaded Liposomes as Positron Emission Tomography Imaging Agents. *Biomaterials* **2011,** *32* (9), 2334–2341.

Popovtzer, R.; Agrawal, A.; Kotov, N. A. et al. Targeted Gold Nanoparticles Enable Molecular CT Imaging of Cancer. *Nano Lett.* **2008,** *8* (12), 4593–4596. [PubMed, 19367807].

Powell, J. A.; Venkatakrishnan, K.; Tan, B. Programmable SERS Active Substrates for Chemical and Biosensing Applications Using Amorphous/Crystalline Hybrid Silicon Nanomaterial. *Sci. Rep.* **2016,** *6,* 19663.

Prencipe, G.; Tabakman, S. M.; Welsher, K.; Liu, Z. et al. PEG Branched Polymer for Functionalization of Nanomaterials with Ultralong Blood Circulation. *J. Am. Chem. Soc.* **2009,** *131* (13), 4783–4787. [PubMed, 19173646].

Pyayt, A. L.; Fattal, D. A.; Li, Z.; Beausoleil, R. G. Nanoengineered Optical Resonance Sensor for Composite Material Refractive-Index Measurements. *Appl. Opt.* **2009,** *48* (14), 2613–2618.

Qiao, R.; Yang, C.; Gao, M. Superparamagnetic Iron Oxide Nanoparticles, from Preparations to in Vivo MRI Applications. *J. Mater. Chem.* **2009,** *19,* 6274–6293.

Rabin, O.; Perez, J. M.; Grimm, J.; Wojtkiewicz, G.; Weissleder, R. An X-ray Computed Tomography Imaging Agent Based on Long-Circulating Bismuth Sulphide Nanoparticles. *Nat. Mater.* **2006,** *5,* 118–122.

Razi, M.; Dehghani. A.; Beigi, F. et al. The Peep of Nanotechnology in Reproductive Medicine: Amini-Review. *Int. J. Med. Lab.* **2015,** *2*, 1–15.

Rosen, J. E.; Yoffe, S.; Meerasa, A. et al. Nanotechnology and Diagnostic Imaging, New Advances in Contrast Agent Technology. *J. Nanomed. Nanotechnol.* **2011,** *2*, 115.

Ruehm, S. G.; Corot, C.; Vogt, P.; Kolb, S.; Debatin, J. F. Magnetic Resonance Imaging of Atherosclerotic Plaque with Ultrasmall Super Paramagnetic Particles of Iron Oxide in Hyperlipidemic Rabbits. *Circulation* **2001,** *103*, 415–422.

Sadeghzadeh, H.; Pilehvar-Soltanahmadi, Y.; Akbarzadeh, A. et al. The Effects of Nanoencapsulated Curcumin-Fe3O4 on Proliferation and hTERT Gene Expression in Lung Cancer Cells. *Anticancer Agents Med. Chem.* **2017,** 17.

Santra, S.; Kaittanis, C.; Grimm, J.; Perez, J. M. Drug/Dye-Loaded, Multifunctional Iron Oxide Nanoparticles for Combined Targeted Cancer Therapy and Dual Optical/Magnetic Resonance Imaging. *Small* **2009,** *5*, 1862–1868.

Sanz, J.; Fayad, Z. A. Imaging of Atherosclerotic Cardiovascular Disease. *Nature* **2008,** *451*, 953–957.

Schellenberger, E. A.; Sosnovik, D.; Weissleder, R.; Josephson, L. Magneto/Optical Annexin V, a Multimodal Protein. *Bioconjug. Chem.* **2004,** *15*, 1062–1067.

Schiffelers, R. M.; Banciu, M.; Metselaar, J. M.; Storm, G. Therapeutic Application of Long Circulation Liposomal Glucocorticoids in Auto-Immune Diseases and Cancer. *J. Liposome Res.* **2006,** *16*, 185–194.

Selvan, S. T.; Tan, T. T.; Ying, J. Y. Robust, Non-Cytotoxic, Silica-Coated CdSe Quantum Dots with Efficient Photoluminescence. *Adv. Mater.* **2005,** *17*, 1620–1625.

Serda, R. E.; Adolphi, N. L.; Bisoffi, M.; Sillerud, L. O. Targeting and Cellular Trafficking of Magnetic Nanoparticles for Prostate Cancer Imaging. *Mol. Imag.* **2007,** *6*, 277–288.

Shao, X.; Agarwal, A.; Rajian, J. R.; Kotov, N. A.; Wang, X. Synthesis and Bioevaluation of 125I-Labeled Gold Nanorods. *Nanotechnology* **2011,** *22* (13), 135102.

Shao, X.; Zhang, H.; Rajian, J. R.; Chamberland, D. L.; Sherman, P. S.; Quesada, C. A. et al. ^{125}I-Labeled Gold Nanorods for Targeted Imaging of Inflammation. *ACS Nano.* **2011,** *5* (11), 8967–8973.

Shao, Y.; Liu, L.; Song, S.; Cao, R.; Liu, H. et al. A Novel One-Step Synthesis of Gd3+-Incorporated Mesoporous SiO2 Nanoparticles for Use as an Efficient MRI Contrast Agent. *Contrast Media Mol. Imag.* **2010,** *6*, 110–118.

Shapiro, E. M.; Skrtic, S.; Sharer, K.; Hill, J. M.; Dunbar, C. E. et al. MRI Detection of Single Particles for Cellular Imaging. *Proc. Natl. Acad. Sci. U. S. A.* **2004,** *101*, 10901–10906.

Shen, J.; Zhu, Y.; Yang, X.; Li, C. Graphene Quantum Dots, Emergent Nanolights for Bioimaging, Sensors, Catalysis and Photovoltaic Devices. *Chem. Commun.* **2012,** *48* (31), 3686–3699.

Silva, G. A. Neuroscience Nanotechnology, Progress, Opportunities and Challenges. *Nat. Rev. Neurosci.* **2006,** *7* (1), 65–74.

Sitharaman, B.; Kissell, K. R.; Hartman, K. B.; Tran, L. A.; Baikalov, A.; Rusakova, I.; Sun, Y.; Khant, H. A.; Ludtke, S. J.; Chiu, W.; Laus, S.; Toth, W.; Helm, L.; Merbach, A. E. Superparamagnetic Gadonanotubes Are High-Performance MRI Contrast Agents. *Chem. Commun.* **2005,** 3915–3917.

Slowing, I. I.; Trewyn, B. G.; Lin, V. S. Y. Mesoporous Silica Nanoparticles for Intracellular Delivery of Membrane-Impermeable Proteins. *J. Am. Chem. Soc.* **2007,** *129*, 8845–8849.

Smith, A. M.; Dave, S.; Nie, S. M.; True, L.; Gao, X. H. Multicolor Quantum Dots for Molecular Diagnostics of Cancer. *Expert Rev. Mol. Diagn.* **2006,** *6*, 231–244.

136   Diverse Applications of Nanotechnology in the Biological Sciences

Smith, A. M.; Duan, H. W.; Mohs, A. M.; Nie, S. M. Bioconjugated Quantum Dots for in Vivo Molecular and Cellular Imaging. *Adv. Drug Deliv. Rev.* **2008,** *60* (11), 1226–1240.

Sosnovik, D.; Weissleder, R. Magnetic Resonance and Fluorescence Based Molecular Imaging Technologies. *Progress Drug Res.* **2005,** *62,* 86–114.

Sukhanova, A.; Devy, M.; Venteo, L.; Kaplan, H.; Artemyev, M.; Oleinikov, V.; Klinov, D.; Pluot, M.; Cohen, J. H. M.; Nabiev, I. Biocompatible Fluorescent Nanocrystals for Immunolabeling of Membrane Proteins and Cells. *Anal. Biochem.* **2004,** *324,* 60–67.

Sun, C.; Lee, J. S.; Zhang, M. Magnetic Nanoparticles in MR Imaging and Drug Delivery. *Adv. Drug Deliv. Rev.* **2008,** *60,* 1252–1265.

Sun, J.; DuFort, C.; Daniel, M-C.; Murali, A.; Chen, C.; Gopinath, K.; Stein, B.; De, M.; Rotello, V. M.; Holzenburg, A.; Kao, C. C.; Dragnea, B. Core-Controlled Polymorphism in Virus-Like Particles. *Proc. Natl. Acad. Sci. U. S. A.* **2007,** *104,* 1354–1359.

Surendiran, A.; Sandhiya, S.; Pradhan, S. et al. Novel Applications of Nanotechnology in Medicine. *Indian J. Med. Res.* **2009,** *130,* 689–701.

Tang, F.; Li, L.; Chen, D. Mesoporous Silica Nanoparticles, Synthesis, Biocompatibility and Drug Delivery. *Adv. Mater.* **2012,** *24* (12), 1504–1534.

Teja, A. S.; Koh, P. Synthesis, Properties, and Applications of Magnetic Iron Oxide Nanoparticles. *Prog. Cryst. Growth Charact. Mater.* **2009,** *55,* 22–45.

Torchilin, V. P. Recent Advances with Liposomes as Pharmaceutical Carriers. *Nat. Rev. Drug Discov.* **2005,** *4,* 145–160.

Vallet-Regi, M.; Balas, F.; Arcos, D. Mesoporous Materials for Drug Delivery. *Angew. Chem. Int. Ed.* **2007,** *46,* 7548–7558.

van Helden, A. K.; Jansen, J. W.; Vrij, A. Preparation and Characterization of Spherical Monodisperse Silica Dispersions in Nonaqueous Solvents. *J. Colloid Interf. Sci.* **1980,** *78,* 354–368.

Van Herck, J. L.; De Meyer, G. R. Y.; Martinet, W.; Salgado, R. A.; Shivalkar, B. et al. Multi-Slice Computed Tomography with N1177 Identifies Ruptured Atherosclerotic Plaques in Rabbits. *Basic Res. Cardiol.* **2010,** *105,* 51–59.

van Schooneveld, M. M.; Vucic, E.; Koole, R.; Zhou, Y.; Stocks, J.; Cormode, D. P.; Tang, C. Y.; Gordon, R.; Nicolay, K.; Meijerink, A.; Fayad, Z. A.; Mulder, W. J. M. Improved Biocompatibility and Pharmacokinetics of Silica Nanoparticles by Means of a Lipid Coating, a Multimodality Investigation. *Nano Lett.* **2008,** *8,* 2517–2525.

van Tilborg, G. A. F.; Mulder, W. J. M.; Deckers, N.; Storm, G.; Reutelingsperger, C. P. M.; Strijkers, G. J.; Nicolay, K. Annexin A5-Functionalized Bimodal Lipid-Based Contrast Agents for the Detection of Apoptosis. *Bioconjate Chem.* **2006,** *17,* 741–749.

Villanueva, F. S.; Jankowski, R. J.; Klibanov, S.; Pina, M. L.; Alber, S. M.; Watkins, S. C.; Brandenburger, G. H.; Wagner, W. R. Microbubbles Targeted to Intercellular Adhesion Molecule-1 Bind to Activated Coronary Artery Endothelial Cells. *Circulation* **1998,** *98,* 1–5.

Villanueva, F. S.; Lu, E. X.; Bowry, S.; Kilic, S.; Tom, E.; Wang, J. J.; Gretton, J.; Pacella, J. J.; Wagner, W. R. Myocardial Ischemic Memory Imaging with Molecular Echocardiography. *Circulation* **2007,** *115* (3), 345–352.

Walczak, P.; Zhang, J.; Gilad, A. A.; Kedziorek, D. A.; Ruiz-Cabello, J.; Young, R. G.; Pittenger, M. F.; Van Zijl, P. C. M.; Huang, J.; Bulte, J. W. M. Dual-Modality Monitoring of Targeted Intraarterial Delivery of Mesenchymal Stem Cells after Transient Ischemia. *Stroke* **2008,** *39,* 1569–1574.

Nanoparticles in Medical Imaging: A Perspective Study

Wang, L.; Tan, W. Multicolor FRET Silica Nanoparticles by Single Wavelength Excitation. *Nano Lett.* **2006**, *6* (1), 84–88.

Wang, L.; Wang, K.; Santra, S.; Zhao, X.; Hilliard, L. R. et al. Watching Silica Nanoparticles Glow in the Biological World. *Anal. Chem.* **2006**, *78*, 646–654.

Wang, X.; Qian, X.; Beitler, J. J.; Chen, Z. G.; Khuri, F. R.; Lewis, M. M. et al. Detection of Circulating Tumor Cells in Human Peripheral Blood Using Surface-Enhanced Raman Scattering Nanoparticles. *Cancer Res.* **2011**, *71* (5), 1526–1532.

Wang, X.; Sun, X.; Lao, J.; He, H.; Cheng, T.; Wang, M. et al. Multifunctional Graphene Quantum Dots for Simultaneous Targeted Cellular Imaging and Drug Delivery. *Colloids Surf. B Biointerf.* **2014**, *122*, 638–644.

Wang, X.; Yang, L.; Chen, Z. et al. Application of Nanotechnology in Cancer Therapy and Imaging. *Ca. Cancer J. Clin.* **2008**, *58*, 97–110.

Wang, Z.; Xia, J.; Zhou, C.; Via, B.; Xia, Y.; Zhang, F. et al. Synthesis of Strongly Green-Photoluminescent Graphene Quantum Dots for Drug Carrier. *Colloids Surf B Biointerf.* **2013**, *112*, 192–196.

Weissleder, R. Molecular Imaging in Cancer. *Science* **2006**, *312* (5777), 1168–1171.

Weissleder, R.; Kelly, K.; Sun, E. Y. et al. Cell-Specific Targeting of Nanoparticles by Multivalent Attachment of Small Molecules. *Nat. Biotechnol.* **2005**, *23* (11), 1418–23.

Weissleder, R.; Mahmood, U. Molecular Imaging. *Radiology* **2001**, *219*, 316–333.

Winter, P. M.; Neubauer, A. M.; Caruthers, S. D.; Harris, T. D.; Robertson, J. D.; Williams, T. A.; Schmieder, A. H.; Hu, G.; Allen, J. S.; Lacy, E. K.; Zhang, H. Y.; Wickline, S. A.; Lanza, G. M. Endothelial Alpha (v) beta(3) Integrin-Targeted Fumagillin Nanoparticles Inhibit Angiogenesis in Atherosclerosis. *Arterioscler Thromb Vasc Biol.* **2006**, *26*, 2103–2109.

Xie, H.; Diagaradjane, P.; Deorukhkar, A. A.; Goins, B.; Bao, A.; Phillips, W. T. et al. Integrin avb3-Targeted Gold Nanoshells Augment Tumor Vasculature-Specific Imaging and Therapy. *Int. J. Nanomed.* **2011**, *6*, 259–269.

Xie, H.; Wang, Z. J.; Bao, A.; Goins, B.; Phillips, W. T. In Vivo PET Imaging and Biodistribution of Radiolabeled Gold Nanoshells in Rats with Tumor Xenografts. *Int. J. Pharm.* **2010**, *395* (1–2), 324–330.

Xie, J.; Chen, K.; Huang, J.; Lee, S.; Wang, J.; Gao, J. et al. PET/NIRF/MRI Triple Functional Iron Oxide Nanoparticles. *Biomaterials* **2010**, *31* (11), 3016–3022.

Yan, J.; Estevez, M. C.; Smith, J. E.; Wang, K.; He, X. et al. Dye-Doped Nanoparticles for Bioanalysis. *Nano Today* **2007**, *2*, 44–50.

Yang, H.; Lu, C.; Liu, Z.; Jin, H.; Che, Y.; Olmstead, M. M. et al. Detection of a Family of Gadoliniumcontaining Endohedral Fullerenes and the Isolation and Crystallographic Characterization of One Member as a Metal-Carbide Encapsulated Inside a Large Fullerene Cage. *J. Am. Chem. Soc.* **2008**, *130* (51), 17296–17300.

Yang, L.; Mao, H.; Wang, Y. A.; Cao, Z.; Peng, X. et al. Single Chain Epidermal Growth Factor Receptor Antibody Conjugated Nanoparticles for in Vivo Tumor Targeting and Imaging. *Small* **2009**, *5*, 235–243.

Yu, W. W.; Chang, E.; Falkner, J. C.; Zhang, J.; Al-Somali, A. M.; Sayes, C. M.; Johns, J.; Drezek, R.; Colvin, V. L. Forming Biocompatible and Nonaggregated Nanocrystals in Water Using Amphiphilic Polymers. *J. Am. Chem. Soc.* **2007**, *129*, 2871–2879.

Zhang, R.; Lu, W.; Wen, X.; Huang, M.; Zhou, M.; Liang, D. et al. Annexin A5-Conjugated Polymeric Micelles for Dual SPECT and Optical Detection of Apoptosis. *J. Nucl. Med.* **2011**, *52* (6), 958–964.

Zhang, R.; Xiong, C.; Huang, M.; Zhou, M.; Huang, Q.; Wen, X. et al. Peptide-Conjugated Polymeric Micellar Nanoparticles for Dual SPECT and Optical Imaging of EphB4 Receptors in Prostate Cancer Xenografts. *Biomaterials* **2011a,** *32* (25),5872–5879.

Zhang, W.; Yong, D.; Huang, J.; Yu, J.; Liu, S. et al. Fabrication of Polymer-Gadolinium (III) Complex Nanomicelle from Poly(ethylene glycol)- Polysuccinimide Conjugate and Diethylenetriaminetetraacetic Acid-Gadolinium as Magnetic Resonance Imaging Contrast Agents. *J. Appl. Polym. Sci.* **2011b,** *120*, 2596–2605.

Zhang, Y.; Kohler, N.; Zhang, M. Surface Modification of Superparamagnetic Magnetite Nanoparticles and Their Intracellular Uptake. *Biomaterials* **2002,** *23*, 1553–1561.

Zhao, Y.; Sun, X.; Zhang, G.; Trewyn, B. G.; Slowing, I. I.; Lin, V. S. Interaction of Mesoporous Silica Nanoparticles with Human Red Blood Cell Membranes, Size and Surface Effects. *ACS Nano.* **2011,** *5* (2), 1366–1375.

Zheng, G.; Chen, J.; Li, H.; Glickson, J. D. Rerouting Lipoprotein Nanoparticles to Selected Alternate Receptors for the Targeted Delivery of Cancer Diagnostic and Therapeutic Agents. *Proc. Natl. Acad. Sci. U. S. A.* **2005,** *102*, 17757–17762.

Zheng, J.; Allen, C.; Serra, S.; Vines, D.; Charron, M. et al. Liposome Contrast Agent for CT-Based Detection and Localization of Neoplastic and Inflammatory Lesions in Rabbits, Validation with FDG-PET and Histology. *Contrast Media Mol. Imag.* **2010,** *5*, 147–154.

Zhu, D.; White, R. D.; Hardy, P. A.; Weerapreeyakul, N.; Sutthanut, K. et al. Biocompatible Nanotemplate-Engineered Nanoparticles Containing Gadolinium, Stability and Relaxivity of a Potential MRI Contrast Agent. *J. Nanosci. Nanotechnol.* **2006,** *6*, 996–1003.

CHAPTER 5

NANOTECHNOLOGY IN HEALTHCARE MANAGEMENT

IFRAH MANZOOR[1,2], MUZAFAR AHMAD RATHER[1], SAIMA SAJOOD[3], SHOWKEEN MUZAMIL BASHIR[1*], SOHAIL HASSAN[4], MANZOOR-U-REHMAN[1], and RABIA HAMID[5*]

[1]*Division of Veterinary Biochemistry, Faculty of Veterinary Sciences and Animal Husbandry, SKUAST-Kashmir, Shuhama, Srinagar, Jammu and Kashmir, India*

[2]*Department of Biochemistry, University of Kashmir, Hazratbal, Srinagarar, Jammu and Kashmir, India*

[3]*Department of Biotechnology, University of Kashmir, Hazratbal, Srinagarar-190006, Jammu and Kashmir, India.*

[4]*Department of Microbiology, University of Veterinary and Animal Sciences, Lahore, Pakistan*

[5]*Department of Nanotechnology, University of Kashmir, Hazratbal, Srinagarar, Jammu and Kashmir, India*

Corresponding authors. E-mail: showkeen@skuastkashmir.ac.in/ showkeen.muzamil82@gmail.com; rabeyams@gmail.com.

ABSTRACT

Nanotechnology developments have revolutionized the health-care system in this age of nanoscience by developing nanostructures that greatly enhance the diagnosis and therapeutic aspects of many chronic diseases such as cancer, diabetes, infectious diseases, neurodegenerative diseases, blood disorders, and orthopedic issues. Nanotherapeutic technologies of multifunctionalities have tremendous potential to fill the gaps that exist in the existing therapeutic domain. In the field of cancer management, nanomedicines are already a reality. It has improved the permeability and successful targeting of drugs with less to no toxicity. Nanorobotics, in the field of surgery, has ameliorated procedures needed for complex surgeries, from cutting to the healing of scars. In this chapter, we will discuss the importance of nanotechnology in health-care management.

5.1 INTRODUCTION

Nanotechnology, nanomaterials science provides apt potential in different engineering materials and is currently the growing and evolving scientific technology. It is defined as the study to control, manipulate, and create systems according to their atomic or molecular specifications (Morrison et al., 2008). According to the US National Science and Technology Council, this approach's power presents miniaturized manipulation of matters at atomic, molecular, and supramolecular levels (sized between 1 and 100 nm) to create newer structures and devices (Rocco et al., 2000). The great visionary late Nobel physicist Richard P Feynman first conceived the concept of molecular manufacturing in 1959 and proposed that devices and materials might one day have the potential to revolutionize every atomic specification (Feynman, 1993). A number of "bottom-up" and "top-down" approaches for nanostructure synthesis were developed irrespective of their field or discipline. Bottom-up technique underlies the formation of larger, more structured nanostructures such as nanotubes and quantum dots (QDs) accomplished by increasing or assembling the building blocks of atoms or molecules (Mohamed et al., 1998). However, the "top-down" approach subsequently uses more refined tools for breaking, cutting, or etching techniques (Guo, 2005) like machining of bulk or film, surface finishing, and molding using lithography (Majumder et al., 2007) for creating correspondingly smaller structures.

During past decades, nanomaterials have received significant attention in medicine, diagnostics, therapeutics, drug delivery, etc. by overwhelming the limitations of those fields. Owing to nanomaterials' distinct physical and chemical properties, nanotechnology has become a focused issue in global public health. However, health issues about using nanoparticles that can be oxidized, reduced, or dissolved include cytotoxicity and even genotoxicity. The small-sized nanoparticles are considered to confer toxicity by crossing the human body's physiological barrier to cause genotoxicity, carcinogenic potential, and oxidative stress in organelles, mitochondrial stress, and cell wall disturbances. In addition to this, the cost factor becomes a further barrier to its use (Liu et al., 2014; Guo et al., 2012a,b; Wu et al., 2013). Overall, unprecedented resilience and efficiency, enhanced durability and versatility, unique physicochemical properties of nonmaterial overcome such weaknesses and are therefore commonly used to maintain good health.

5.2 NANOTECHNOLOGY AND HEALTH-CARE MANAGEMENT

Nanoscience offers the ability to build a host of incredibly powerful products, including nanomedicines and nanodevices. In biomedicine, nanomedicines and nanodevices are interrelated with biotechnology and information technology and can bring various benefits. Products based on nanotechnology capable of overcoming traditional methods' limitations were used for medical diagnosis with early detection of diseases, targeted clinical treatment, and regenerative medicine to regenerate damaged tissues. Significant applications of nanotechnology for the management of healthcare are represented in the figure (Fig. 5.1) and include:

a. Prevention of diseases.
b. Disease diagnosis.
c. Treatment of diseases.

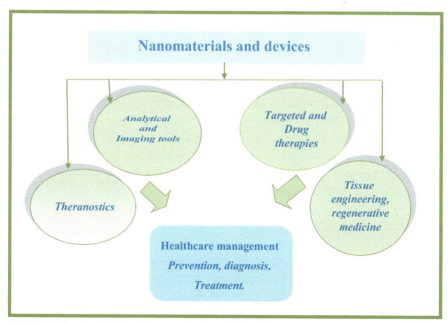

FIGURE 5.1 Application of nanotechnology in health-care management.

5.2.1 ROLE OF NANOTECHNOLOGY IN PREVENTION

The disease management preventive approach is always better than the reactive one in terms of health-care expenses. Prevention and early intervention help to avert the upstream disease; however, a reactive approach offers treatments only after the health issues arise. Nanotechnology seems to be an upcoming need for the future as it provides an effective way to avert diseases beforehand. Various technical tools involved in disease prevention have been employed over time and are mentioned next.

5.2.1.1 NANOMASK

Facemasks help in the control and prevention of infectious respiratory disease transmission. Nanomasks with nanosized pores help to mitigate airborne transmission or exposure of pathogenic bacteria/viruses in health-care settings. Various viable viruses such as the influenza A virus, MERS-CoV, and SARS-CoV virus 80–140 nm in diameter have been detected in aerosols in the air (Shiu et al., 2019; Sim et al., 2014). Compared to conventional masks made from microfibers, a nanomask can filter nanosized particles because they comprise multiple layers of different types of nanofibres. Multilayering of the mask helps in the free movement of air to facilitate comfortable breathing. The use of various kinds of nanofibres in different layers of these masks serves several purposes, like converting air pollutants to harmless substances by incorporating titanium dioxide and other semiconductor composite nanofibres (Kong Polytechnic University, 2016). Li and Gong in 2015 developed a novel electrospun nanofiber mask that efficiently filters out the PM2.5 particles and, at the same time, preserves a good level of breathability (Xingzhou and Gong, 2015). Besides, a KAIST (Korea Advanced Institute of Science and Technology) research has developed a mask with nanofibres with a diameter of 100~500 nm aligned in orthogonal or unidirectional directions reusability. It helps to relieve changes arising from a shortage of facemasks.

5.2.1.2 NANOFILTERS

Minimizing the impact of pathogenic bacteria and viruses can help in abating most of the health issues. The nanofilters, with nanosized pores, can capture and remove the virus particles from the air passing through them. Various components of personal protective equipment are made of nanomaterial that minimize the pore size of the material, decrease or limit nanosized organisms'

Nanotechnology in Healthcare Management 143

entry like bacteria and viruses. For example, robust gloves and footwear are produced from the ceramic nanocomposite armor system.

5.2.1.3 ANTIBACTERIAL COATINGS

The functional coating of equipment involved in the health and food industry improves the health-care system with several diseases' minimum risk. Medical devices such as knee and hip implants become contaminated in a proper clinical setting due to bacterial invasion during handling and implantation. To overcome this problem, TiO_2 or silver gold nanoparticle-based coatings have been used to inhibit microbial growth and eliminate pathogens (Krasimir, 2019). Coating tooth surfaces with antibacterial nanocoating (zinc oxide, silver, and polyethylenimine) inhibits bacterial growth and adhesion via various mechanisms, thereby eliminating bacteria while maintaining tooth surfaces' integrity, the presence of biological fluids like saliva (Abou Neel et al., 2015). Nanotechnology offers myriad of applications in the food industry by preventing bacteria from adhering to food processing equipment after their active coating. Coating of food contact sources with nanoparticles enables sustained release of antimicrobial compounds to check bacteria and bacterial/fungal spores' growth, thereby increasing the shelf life of packaged foods (Elingarami et al., 2014).

5.2.1.4 NANOSENSORS

A nanobiosensor produces an electrical signal from a physiochemical signal that enables the detection of bioagents such as nucleic acids, interleukins, antibodies, pathogens, and their associated toxins. They exhibit ultrasensitivity in biomarker detection to rapidly identify the pathogens' source to contain their spread and epidemics. With smaller dimensions and larger surface area, thousands of nanoparticles can be placed on a single nanosensor to be actively used to detect the presence of various pathogens. Nanobiosensors can be classified into various types of biosensors; nanoparticle-based biosensors make use of metallic nanoparticles as a biochemical signal enhancer; nanotube-based sensors involve carbon nanotubes as enhancers of the sensitivity and efficiency while nanowire biosensors use nanowires as charge carriers. Nanobiosensors that utilize QDs as the contrast agents are known as QDs-based sensors. Biosensors' emergence has undoubtedly improved the food quality estimation, diagnosis of numerous diseases and related malfunctions, and environmental monitoring. The recent outbreak of COVID-19 is believed to originate from the Huanan Wholesale Seafood

Market and other area markets, and these infections have either transmitted directly or indirectly to humans from market civets. Reliable and sensitive testing by biosensors could have helped on-site virus detection to check potential contamination and proven beneficial in its prevention (Biswas, 2019; Goswami et al., 2019, Ravindran et al., 2019; Malik et al., 2013).

5.2.2 ROLE OF NANOTECHNOLOGY IN DISEASE DIAGNOSIS

The standards for molecular diagnostics have been redefined by nanodiagnostics and represent integrating diagnostics with therapeutics employing new and essential approaches to recognize biomolecules, including proteins, nucleic acids, and antibodies (Table 5.1). These promising approaches increase sensitivity when individual nanoscale particles like nanoparticles, nanowires, nanopores, and nanotubes are tacked together with them (Manikandan et al., 2018).

5.2.2.1 NANOBIOASSAY

Advancements in nanotechnology have helped the development of more sensitive and reproducible diagnostic tools like bioassays. They have allowed timely detection of diseases and identifying potential "biomarkers" of disease even at low concentrations. The selection of nanomaterials considering their biotoxicity and stability is crucial for assay performance. Optimization of the physiological conditions like buffering system, temperature, presence of surfactants, surface modification of the nanoparticle should be done to retain stability and functionality of the detection targets, recognition elements, and dispersed nanoparticles. Apart from the use of nanoscale systems for biomolecular screening approach in general (viz. electrochemical and electromechanical) (Jain, 2007), strategies based on nanoparticles, including gold, silica, silver, and QDs, are long-established for disease diagnostics because of their ultrasensitivity as compared to conventional reporter molecules. Carbon dots and QDs are nanoparticles of carbon and tiny semiconductor particles, respectively, with high quantum yields. Studies have reported that QDs conjugated with certain materials make them useful in monitoring a specific molecule after getting attached. "Multiple disease–detecting bioassay" is also a result of a bioassay plate designed by arranging QDs made from diverse materials specific for a biomolecule.

Similarly, the microphysiometer composed of multiple carbon nanotubes operates at normal cellular physiological pH levels. These sensors find applications in monitoring the insulin levels in real time as current flow through the sensor increases with an increase in insulin molecules (Eckert et al., 2013). In contrast to this, significant efforts have been made to achieve noninvasive glucose analysis using glucose sensor technology (Oliver et al., 2009). A microfluidic device with multiple carbon nanotube/dihydropyran walls in the composite sensor has been used for the electrochemical detection of insulin (Snider et al., 2008; DiSanto et al., 2015). A novel gene quantification platform named QD electrophoretic mobility shift assay (QEMSA) demonstrates an incredible resolution, capable of detecting DNA copy number variation and quantification of DNA methylation level in human cancer (Zhang et al., 2010).

Moreover, colorimetric sensor is already in commercial development with several successful tests against various pathogens and environmental samples (Elghanian et al., 1997). Over the years, gold nanoparticle– and upconversion nanoparticle–assisted colorimetry has come up with promising biolabeling applications displaying quick and on-site detection with unaided eyes (Wang et al., 2012; Xue et al., 2008; Wu et al., 2014; Su et al., 2013; Deng et al., 2015). This kind of sensor offers advantages like low cost and less cumbersome for the detection within less period of time (Xu et al., 2014).

5.2.2.2 NANOCHIP

Nanochips are micrometer-sized analytical devices developed by Nanogen company from individual atoms and molecules that employ the power of an electronic current to separate DNA oligos based on charge and size on an array. Earlier data considered nanochips as a distant prospect. However, in a recent proof-of-concept study published in Nature Biomedical Engineering, a uniquely designed microfluidic nanochip could detect low levels of ovarian cancer–associated markers in the blood, which paves the way for early detection of cancer.

In another study, nanochip integrated into the microfluidic chip was inserted into the point-of-care system, which enables processing of blood sample and its subsequent detection and automatic data analysis within 30 min, thereby helping in the clinical diagnosis of sepsis through quantitative analysis of procalcitonin biomarker (Abbasi, 2019; Larry, 2001; Sun et al., 2019). The Nanochip-based system finds commercial application in the field of genomic diagnostics that facilitates the identification of specific DNA sequences within minutes and offers maximum accuracy in detecting SNPs (Jain, 2014). Based on the complementary base pairing of the target DNA sequences with the probes on the nanochip's specific sites, fluorescence is detected at the sites where DNA molecules hybridize perfectly with target DNA sequences on the chip (Richard, 2005).

5.2.2.3 NANOFLUIDIC

Nano/microfluidic technologies have potential advantages in diagnostics and monitoring infectious diseases. Meanwhile, nanofluidics offers a robust platform for the requisite liquid biopsies in precision medicine, some of which are fully commercialized by now (Ion Torrent), some almost (Oxford Nanopore technologies), and some have not yet cleared the academic research phase (Segerink and Eijkel, 2014). Conventionally existing diagnostic instruments are usually not cost-effective and time-consuming; however, miniaturized nanofluidic devices ensure quick and more accurate medical diagnosis even using small fluid volumes. In particular, these nano/microfluidic diagnostic technologies are inexpensive and portable and can be effectively used to detect bloodborne- or body fluid–associated diseases on a large scale (Lo, 2017; Lee et al., 2010). Chou et al. (2008) put forward a nanofluidic biosensor that uses surface-enhanced Raman spectroscopy (SERS) to detect the conformational states of the beta-amyloid peptide considered as a marker for Alzheimer's disease (AD). Moreover, nanofluidics may allow separation of biomolecules based on molecular differences, such as phosphorylation status or difference in oxidative stress states (Napoli et al., 2010).

Strategies involving "Lab-On-A-Chip" devices have been exploited by doctors to monitor different health parameters constantly on a "real-time" basis. Doctors can do a complete assessment of blood lipid profile, blood oxygen level, and blood pressure by just reading the signals from a chip that a patient is wearing. These tests are less time-consuming and reduced misdiagnosing and improper handling. Nanofluidic devices are characterized by microfabricated fluidic channels capable of analyzing nanoliter size DNA samples. Other necessary accessories include heaters, temperature sensors, electrophoretic chambers, and fluorescence detectors.

5.2.2.4 IMAGING TECHNOLOGY

Invasive diagnostic tools are useful in imaging physiological and pathophysiological changes within the patient body after administering contrast agents or medicinal radiocompounds. Molecular imaging has evolved as an essential imaging technique concerned with the visualization of molecular biomarkers during disease diagnoses, such as gene expression and protein synthesis or degradation. Molecular imaging focuses on disease-associated molecular signatures that allow early detection of diseases and paves the way for appropriate therapies. Imaging contributes to continuous monitoring and evaluation of disease along with cost-effective optimization of treatment. Nanotechnology focuses on improving the diagnosis process by introducing new imaging techniques (Kim et al., 2018).

Nanotechnology has improved the imaging technique by enhancing the sensitivity of specific imaging agents created by tailoring the surface properties of nanoparticles by coupling them to numerous atoms of radioisotopes, lanthanides, or fluorophores. These provide imaging signals that display higher sensitivity and increased specificity (Sullivan and Ferrari, 2004; Wickline and Lanza, 2003).

For the first time, liquid perfluorocarbon nanoparticles incorporated with antibody ligands directed against cross-linked fibrin were used as molecular targeting agents (Cormode et al., 2009; He et al., 2013; Dayton and Ferrara, 2002; Baghbani et al., 2016). Radioactive substances or gadolinium compounds conjugated with the lipid layer of the nanoparticles have been used in imaging modalities, including MRI or scintigraphic imaging (Morawski et al., 2004). Nanoparticles are used as contrast agents to characterize lesions and detailed tissue imaging with longer circulation time and find applications in diagnosing thrombosis, atherosclerosis, and angiogenesis during tumor development. Superparamagnetic iron oxides (50–500 nm), also known as monocrystalline iron oxide nanoparticles, are extensively used in magnetic resonance imaging with or without magnetic particle labels (Bhowmik et al., 2009; Magro et al., 2018). Superparamagnetic iron oxide nanoparticles (SPIONS), 5–50 nm, have found clinical applications in magnetic resonance imaging, magnetic particle imaging, and biosensing. Iron oxide nanoparticles viz. ferucarbotran and ferumoxide show a high liver uptake, which augments their use on a large scale for primary lesion analysis in the liver as in the case of liver cancer (Fig. 5.2). However, the FDA has approved the oral administration of ferumoxsil as a contrasting agent for gastrointestinal imaging. Other iron oxide nanoparticles like ferumoxtran and ferumoxytol have not cleared clinical trials for imaging lymph node metastasis. Ferumoxtran-based USPIO has been utilized for insulitis imaging (Dadfar et al., 2019).

Given the right targeting molecule, the nanoparticles eventually enable visualization of pathogenic changes in blood vessels. Dendrimer-coated particles can be used for labeling living cells and visualization and quantification of tumor vascularization. Magnetodendrimers of this kind has been successfully employed in laboratory animals to observe the migration of transplanted cells in the body to evaluate the overall functionality of the transplant (Bulte et al., 2001). Gadolinium dendrimers have become agents for contrast enhancement (Mekuria et al., 2017). However, in multiple sclerosis patients, the use of ultrasmall SPIONS conjugated with gadolinium chelates enhances active lesions' visibility compared to gadolinium alone (Dadfar et al., 2019). QDs are semiconducting nanoparticles that act as fluorophores and emit several desired colors when stimulated by light. QDs can exhibit a size quantization effect that can prove helpful for diagnostic purposes as emission can range from visible light to IR or UV radiations of the spectrum (Xu et al., 2016; Bilan et al., 2015; Hildebrandt et al., 2017). Imaging using QDs is a preferred technique compared to modern fluorescent dyes due to the longer lifetime of light emission.

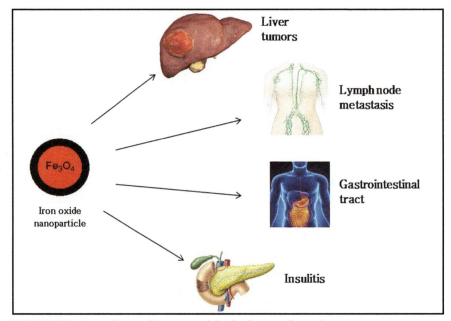

FIGURE 5.2 Use of iron oxide nanoparticles in diseases diagnosis.

5.2.2.5 IMPLANTED SENSORS

Implantable nanosensors (<1 µm) could be inserted under the skin to facilitate a continuous monitoring system and led to significant advances in in vivo diagnostics. Ex vivo biosensing has been extensively used like that in glucose level monitoring; however, they fail to provide continuous physiologic monitoring. Implantable nanosensors could be either at a fixed place or circulating in the blood, allowing the noninvasive detection of biomarkers for disease diagnosis. These sensors may produce magnetic, electrical, optical, or acoustic signals for analyte detection and may interface with wearable devices. Implantable glucose monitoring enables measurement of glucose levels 24 h with a user-friendly approach. One such explainable nanosensor used an injectable polyacrylamide hydrogel to immobilize a fluorescent boronic acid derivative and worked for even 140 days. An injectable glucose sensor has also been made in the form of a hydrogel microparticle and has achieved a linear response within the physiologic range (Balaconis and Clark, 2013; Balaconis et al., 2011; Singh and McShane, 2011).

Engineered glucose nanosensors can convert variations in glucose concentrations into changes in fluorescence by binding to the analyte of interest like lectins and boronic acid derivatives (Pickup et al., 2005; McCartney et al., 2001; Ballerstadt and Schulz, 2000). Some implantable nanosensors have an advantage over implanted electrochemical electrodes, as they can be excited using near-infrared light that can penetrate several centimeters of tissue, facilitating noninvasive detection. Likewise, this type of implantable glucose sensor consists of encapsulated, fluorescent-tagged glucose-receptor molecules (e.g., glucose-binding protein). After excitation using near-infrared light, the fluorescence can be detected near the skin (Pickup et al., 2008). Implantable biosensors may help monitor variability in metabolic parameters, including blood glucose level, blood pressure, and markers for inflammation, which allows early diagnosis of infectious and noninfectious diseases. Even in cancers, implantable biosensors prove to provide real-time and user-friendly health monitoring that undoubtedly may help in early disease diagnosis bypassing the inconclusive imaging or lengthy biopsies (Siontorou et al., 2017; Narayan and Verma, 2017).

5.2.2.6 BARCODE AMPLIFICATION

The barcode assay based on nanotechnology is an ultrasensitive technique that uses short oligonucleotides as surrogate targets for small molecule detection, including proteins and nucleic acids. Mirkin *et al.* developed a fluorophore-based biobarcode amplification assay for high-sensitivity proteins and low complexity (Fig. 5.3). Most sandwich-type biobarcode amplification assays have been applied for rapid and efficient detection of biological macromolecules such as virus particles, food toxins, biomarkers of tumor, neurotransmitters, and cytokines.

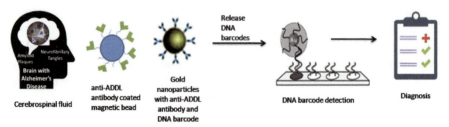

FIGURE 5.3 Alzheimer's disease diagnosis by Barcode amplification.

Thaxton et al. developed a new biobarcode PSA (prostate-specific antigen) assay with increased sensitivity in contrast to conventional immunoassays. PSA-specific antibodies were conjugated to the surface of gold nanoparticles through terminal barcode DNA tosyl modification. High levels of amyloid beta-derived diffusible ligand (ADDL) have been implicated in the neurotoxicity in AD (Lambert et al., 1998; Gong et al., 2003). Biobarcode amplification helps in AD detection (Lambert et al., 2003; Nam et al., 2003) by exposing the cerebrospinal fluid to magnetic beads coated with anti-ADDL monoclonal antibodies. ADDLs present in the CSF bind to these anti-ADDL antibodies, separated using a magnetic field and washed. After the addition of DNA: gold nanoparticle conjugates linked to secondary antibodies, hybridization of complementary DNA sequences occurs; however, unreacted particles are dissociated by applying a magnetic field. The elution release of reacted barcode DNA is done by high temperature and low-salt conditions and further analyzed (Georganopoulou et al., 2005). Due to high sensitivity and specificity, biobarcode detection technology is of significant importance in the clinical field. It can be widely used for the detection of DNA and viruses with pandemic potential with a higher sensitivity than conventional ELISA (Wang et al., 2019).

5.2.2.7 PREGNANCY TEST

A pregnancy test measures the human chorionic gonadotropin (HcG) hormone in urine. This kit uses gold nanoparticles (less than 50 nm in diameter). If HcG hormone-detected nanoparticles reflect red or pink color indicates pregnancy positive, otherwise blue or gray indicates pregnancy negative (Rojanathanes et al., 2008). This results in the reduction of time of detection.

Nanotechnology in Healthcare Management

TABLE 5.1 Represents Usage of Nanomaterials for Early Diagnosis of Specific Diseases.

Nanomaterial	Principle	Application	Reference
Silicon nanowires coated with antibody and covered with polydimethylsiloxane	Laser microdissection (LMD)	Detection and isolation of tumor cells in circulation	Lu et al. (2013)
Gold nanoparticle–coated antibody of influenza A	Dynamic light scattering (DLS)	Detects influenza viruses	Driskell et al. (2011)
Nano-MRI agent	By magnetic resonance molecular imaging (MRI)	Binding to avβ3-integrin are detected	Liu et al. (2013)
Iron oxide magnetic nanoparticles surfaced with polydopamine	By MRI and photothermal cancer therapy	Locating cancerous cells clusters	Wu et al. (2015)
Molecular sentinel (MS) plasmonic nanoprobe	By surface-enhanced Raman scattering (SERS)	Detects an RNA target related to viral infection	Wang et al. (2013)
Nanoparticles of iron oxide covered with proteases (matrix metalloproteinases, cathepsins)	Mass spectrometry	For detection of cancer and liver fibrosis	Kwong et al. (2013)
Gold nanoparticle	Colorimetric assay	For detection of viable emetic *Bacillus cereus*.	Li et al. (2018)
Plasmonic nanomaterial (Au-NP)	Colorimetric DNA sensor	Detection of MERS-CoV, MTB, and HIV oligonucleotides	Teengam et al. (2017)
Nanoparticles of perfluorohydrocarbons	Ultrasonic imaging	Targeted imagining	Dayton and Ferrara (2002)
Nanoplasmonic biosensor	By localized surface plasmon resonance (LSPR)	For detection of Alzheimer's disease (AD)	Kim et al. (2018)
Nanoflares	By fluorescence	For detecting cancer cells from blood	Halo et al. (2014)

5.2.3 ROLE OF NANOTECHNOLOGY IN TREATMENT

Nanotechnology has transfigured health-care strategies in recent years and is intended to have a remarkable impact on offering better health facilities. Nanotherapeutics is a current nanotechnology application that has gained increasing interest in research in the modern world's present-day medical sector. Chronic, severe diseases such as cancer, diabetes, gastrointestinal issues, bacterial/viral infections, cardiac disease, and orthopedic problems are a concern due to the lack of target specificity and drug resistance, which are major therapeutic hurdles. Nanotechnology plays a major role in solving these problems by improving pharmacokinetic properties, bioavailability, and targeting medicines. The development of multifunctional nanotherapeutics has revolutionized the entire drug therapy strategy with significant potential to fill the existing lacunae in the current therapeutic domain. Drug-conjugated nanoparticles allowed drug delivery more effectively and precisely with few side effects of traditional therapeutic agents (Wagner et al., 2006; Freitas, 2005). Over the past two decades, around 20 nanomedicines have been approved by the US FDA, and hundreds of nanocarrier-based products are currently available at different preclinical and clinical development stages. We will address the therapeutic role of nanotechnology in this section of the chapter.

5.2.3.1 NANOTECHNOLOGY IN THE TREATMENT OF CANCERS

The role of nanotechnology in cancer treatments is already a reality by providing a wide range of nanotools and possibilities, from earlier diagnostics and imaging to more efficient and more targeted therapies, that can even circumvent MDR (Sun et al., 2019; van der Meel et al., 2019; Majidinia et al., 2020). In the abovementioned sections, we mentioned cancer diagnosis and imaging for tumor staging and planning of therapy. It is at the forefront and enabled to deliver active drug in a complex disease environment (Chen et al., 2019; Yue et al., 2018). Up to now, a range of therapeutic approaches have been applied extensively to treat cancers in clinics such as chemotherapy, targeted therapy, gene therapy, radiotherapy, immunotherapy, and phototherapy (Wang et al., 2013, Prasad, 2012; McCarthy and Weissleder, 2008) and nanoparticles have been extensively investigated for cancer therapies such as chemotherapy, gene therapy, radiotherapy, immunotherapy, and phototherapy (Guisasola et al., 2018). Extrinsic stimuli such as light, ultrasound, temperature, and magnetic field are typically used locally to cause shift at the tumor site, while intrinsic stimuli such as acid pH overexpressed receptor protein may release payload in response to the tumor environment. At present, these nanoparticles are divided mainly into two main groups: inorganic nanomedicines (QDs, nanogold, nanomesoporous silicone, nanocarbons, nanocalcium materials, etc.) and organic nanomedicines (natural and synthetic nanopolymers, copolymers, self-assembled dendrimers, etc.). Those nanomedicines have their characteristic in cancer care. The tumor-targeting approaches include the effect of passive EPR-based targeting (enhanced permeability and retention), active targeting, and stimulus-responsive targeting. The basic concepts of first-generation anticancer nanomedicines such as Doxil (for Kaposi's sarcoma) have a passive ERP targeting strategy in which simple lipid vesicles shielded with PEG (polyethylene glycol) prolong circulation time and prevent immunogenic response. However, anticancer nanomedicines of the next generation have tissue-specific ligand coatings, as in BIND-014 (prostate cancer), which results in proper localization of nanomedicines at targeted tumors, similarly stimuli-controlled release like ThermoDox (hepatocellular carcinoma) that increases the accumulation of drug at the tumor site. To date, only 16 anticancer nanomedicines are approved and 75 are in clinical trials, out of these recently approved anticancer nanomedicines (He et al., 2019) like APEALEA (micellar paclitaxel), which was approved in 2018 by EMA for ovarian cancer and VYXEOS (liposomal daunorubicin and cytarabine), which was approved by the FDA in 2017 for treating leukemia. In Phase III clinical studies, VYXEOS showed a 31% reduction of deaths and ONIVYDE (liposomal irinotecan), which was approved by the FDA in 2015 for treating pancreatic cancer. With a better understanding of cancer's molecular mechanism, efforts are being made to come with better nanoparticle distribution and penetration into the tumor, and best-benefited nanomedicines according to a patient (Gupta et al., 2006). As a result, several novels, next-generation nanomedicine strategies have come into view. For instance, nanomedicines are being developed to target newly identified biomarkers, similarly targeted nanoparticles are being developed to carry new anticancer therapeutic agents, such as siRNA (Gomes-da-Silva et al., 2012; Xu et al., 2017; Lin et al., 2020), and genome editing systems like CRISPR/Cas (Lu et al., 2018; Miller et al., 2017; Lin et al., 2020) and TALEN.

5.2.3.2 *NANOTECHNOLOGY IN THE TREATMENT OF DIABETES*

In treating diabetes, glucose concentration management near the average level depends upon the controlled release of hormones, especially insulin. In managing diabetes, nanotechnology plays a vital role by the following sections.

5.2.3.2.1 Modification of Insulin

Nanotechnology helps in the prevention of its degradation, inactivation, digestion, and increases its permeability by providing carrier systems, like liposomes, solid lipid nanoparticles, nanoparticles, nanoemulsions, pH-responsive nanovesicles. These systems improve the bioavailability and efficacy of hormone (Pan et al., 2002; Sarmento et al., 2007; Li et al., 2013; Tai et al., 2014: Yu et al., 2018). Thus, with nanotechnology insulin delivered orally, through respiratory tracts, various nanomedicines are in different phases of clinical trials like dry powder formulation of insulin Exubera an inhaled product.

5.2.3.2.2 Artificial β Cells

Synthetic artificial β cells can be utilized for regulated insulin delivery using membrane fusion (Chen et al., 2018). These mimic the natural β cells. These artificial β cells have multicompartmental vesicle-in-vesicle superstructures, inner lipid nanovesicles are loaded with insulin, and other acts like outer membrane with a glucose metabolism system like GLUT 2, GOx/CAT, gramicidin A, and membrane fusion machinery. Changing body glucose levels results in the change in pH inside, which ultimately triggers the fusion and release of insulin. With the assistance of F127 thermogel, artificial β cells could be directly transplanted subcutaneously for maintaining glucose level without requiring immunosuppressant drugs (Chen et al., 2019).

5.2.3.2.3 Artificial Pancreas

It is a device that can be implanted in the body for delivering insulin when needed. Continuously monitoring glucose levels by sensor causes insulin release in the bloodstream by an infusion pump (Kropff and DeVries, 2016). This could be an everlasting solution for type 1 diabetic patients (Breton et al., 2012).

5.2.3.3 NANOTECHNOLOGY IN THE TREATMENT OF RESPIRATORY DISORDERS

Nanotechnology is a boon and bane for the respiratory system by treating the respiratory disorder, fatal respiratory infections like tuberculosis, and by

Nanotechnology in Healthcare Management

affecting lung anatomy. Nanotherapeutics improves the efficacy of treatment mainly due to therapies based on inhalation that provide noninvasive means for the delivery of drugs with limited side effects (Anderson et al., 2020). Nanomedicines function in respiratory disease treatment discussed in the following sections.

5.2.3.3.1 Pulmonary Tuberculosis

In pulmonary tuberculosis (TB) cases, a major problem in treatment involves MDR, poor patient compliance due to the prolonged administration of drugs, often with severe side effects. Nanomedicines have the potential to overcome these complications. Nanocarrier-based drug delivery system provides ample potential in the field of respiratory medicines in TB (Pham et al., 2015). Ahmad and coworkers have reported the effect of the direct delivery of antituberculosis drugs by inhaled alginate nanoparticles encapsulation (isoniazid, rifampicin, and pyrazinamide). Encapsulation of active pharmaceutical agents showed better bioavalibity and higher efficiency (Ahmad et al., 2005; Pandey et al., 2003).

5.2.3.3.2 Lung Cancers

The role of nanotechnology in treating cancers is already mentioned previously. Still, here we discuss nanomedicines related to lung cancers, for example, Genexol PM (in which polymeric paclitaxel is loaded in micelle); it's in a Phase II trial of non–small cell lung cancer patients. Another drug is Aurimmune Cyt-6091; it's in Phase I trial in patients with adenocarcinoma of the lungs.

5.2.3.3.3 Chronic Pulmonary Diseases

These include asthma and chronic obstructive pulmonary disease, treated with adrenergic stimulants, and inhaled corticosteroids that can only alleviate symptoms and do not entirely palliate the loss of the aerobic function caused by these diseases. Long time usage of these drugs results in severe side effects. To overcome these side effects, nanomedicines provide a sight that results in the target release and high pharmacological potency (Lopes Da Silva et al., 2017; Sadikot, 2018). Inhaled nanotherapeutics provides an in-site for targeted, reduced drug dosage (Blank et al., 2017). Gene therapies also provide sight for treating lung diseases; it improves patient compliance by lowering dose frequency (Kaczmarek et al., 2016; Patel et al., 2019). Similarly, thymulin-analog gene methionine serum thymus factor (MSTF) inhibits inflammation, collagen deposition, and smooth muscle hypertrophy in the allergic asthma murine model (da Silva et al., 2014) to minimize inflammation in asthma patients.

5.2.3.3.4 Cystic Fibrosis

Nanotherapeutics provides sight for the treatment of diseases such as cystic fibrosis, the development of drug-loaded NPs to penetrate the thickened layer of mucus showed efficacy of antibacterial drugs in the fight against bacteria (Suk et al., 2011; Deacon et al., 2015; Ong et al., 2019). Gene- and protein-centric therapies are also promising methods for the treatment of cystic fibrosis, such as CRISPR/Cas9 gene-editing technology provided through NPs alter CFTR genes (McNeer et al., 2015) and the prevention of CFTR protein degradation by inhibiting proteasomal degradation with encapsulated PS-341 within PLGA-PEG NPs (Vij et al., 2010).

5.2.3.4 NANOTECHNOLOGY IN THE TREATMENT OF NEURODEGENERATIVE DISORDERS

In the case of neurodegenerative disorders, nanotechnology provides various nanocarrier systems that play an important role in delivering active pharmaceutical agents by overcoming major difficulties like passing the blood–brain barrier (BBB) (Teleanu et al., 2018, 2019). The role of nanotechnology in the treatment of these disorders is as follows.

5.2.3.4.1 Alzheimer's Disease

In various investigations, polymeric nanocarriers loaded with therapeutic agents like ROCK II siRNA, curcumin, Tet-1, B6-peptide, memantine have shown increased stability, reduced drug toxicity, improved memory and learning abilities, reduced oxidative stress and reduced inflammatory stress, and ability to eliminate Aβ aggregation (Mathew et al., 2012; Liu et al., 2013; Wen et al., 2014; Sánchez-López et al., 2018; Pacheco et al., 2020; Salehi et al., 2020); similarly, in vitro study brought out that magnetic NPs loaded with antitransferrin receptor antibodies (OX26) results in extracellular accretion of Aβ aggregates could be promising strategy for treating AD.(Cui et al., 2018).

5.2.3.4.2   Parkinson's Disease

In treating Parkinson's disease (PD), nano-based delivery system provides realistic approach in targeted release of therapeutic agent (Garbayo et al., 2014; Torres-Ortega et al., 2019; Salehi et al., 2020). In vivo experiment with animal model showed NP-loaded therapeutic agents like chemical compounds, genes, and proteins improve locomotion, reduced dopaminergic neuronal loss, enhanced level of dopamine (Lu et al., 2015; Huang et al., 2013; Saraiva et al., 2016; Yurek et al., 2017; Hu et al., 2018; Niu et al., 2017).

5.2.3.5   NANOTECHNOLOGY AND TISSUE ENGINEERING

It is one of the significant applications of nanotechnology in tissue engineering. It has completely changed the strategies of organ transplant, artificial implants, grafting, and healing. Nanocomposites provide a base for the fabrication of tissue. In this portion, we will discuss bone and wound healing strategies.

5.2.3.5.1   Role in Bone Engineering

Bone implants mechanical strength increase due to surface modification with nanostructures; it prevents loosening, corrosion, bacterial infections, subsequent failure, and with extended life (Pleshko et al., 2012). Several nanocomposites used in tissue engineering have demonstrated increased mesenchymal differentiation in the osteogenic stem cells (Christenson et al., 2007; Hwang et al., 2019; Siddiqui et al., 2020), increased osteoblast proliferation (Hussain, 2019), promoted osteointegration, supported healing of bone diseases, and promoted specific protein interactions.

5.2.3.5.2   Role in Tissue Repair (Wound Healing)

One of nanotechnology's major advantages is tissue recovery, regeneration, and repair up to original structural characteristics and functions. The most promising advancement in nanotechnology offers the facile, biodegradable, harmless nanomaterial that opens new approach to wound healing. Nanotechnology provides efficient delivering structures like polycaprolactone, poly(L-lactic acid)-*co*-poly(ε-caprolactone), chitosan, gelatin, polyethylene glycol nanofiber, nanoparticles, hydrogels, etc. to deliver wound healing therapeutic agents like growth factors [fibroblast growth factor (FGF), platelet-derived growth factor (PDGF), vascular endothelial growth factor (VEGF), etc.]; various phytodrugs with characteristic antimicrobial and anti-inflammatory property; various therapeutic nucleic acids (DNA, siRNA, etc.); and other compounds like nitric oxide, etc. (Barrientos et al., 2008; Krausz et al., 2015; Chereddy et al., 2015; Randeria et al., 2015; Joshi et al., 2019; Chakrabarti et al., 2019; Sharma et al., 2020; Bhattacharya et al., 2019; Blanco-Fernandez et al., 2020).

5.3 NANOMEDICAL TECHNICAL TOOLS/DEVICES

5.3.1 NANOROBOTS

Nanodevices can be injected into the patient's body and can pass through the smallest vessels and can, thereby, approach the individual cells to perform diagnosis or treatment by altering structure and composition in the molecular or submolecular level. These nanotools improve biomedical instruments for surgery, pharmacokinetics, targeted drug delivery, disease monitoring, and thus provide efficient health-care management (Hogg and frreitas, 2010; Khawaja, 2011). These nanodevices can be controlled remotely by surgeons to perform the desired task.

5.3.2 DENTIFROBOTS

Medical nanobots capable of regeneration and repairing gums and teeth can even prevent the calculus accumulation. Dentifrobots mouthwash or toothpaste removes the organic residues left on the occlusal surfaces of teeth (Kumar and Vijayalakshmi, 2006; Reddy, 2014). Thus, it provides a new dimension to dental health

5.3.3 RESPIROCYTE

It is artificial red cell; its main application includes transfusable blood substitution, treatment, and prevention of various blood disorders like anemia, maintenance of artificial breathing, perinatal, neonatal disorders, etc.

Nanotechnology in Healthcare Management

5.3.4 NANOTWEEZERS

Devices developed for manipulation and interrogation of nanostructures. Help nanodevices to move and positioning of these devices in the body

5.3.5 NANOCOMPUTERS

Nanocomputer deals with the structures at a molecular level and provides increasingly smaller and faster computers. These can be biochemical or organic such as DNA computers.

5.4 TOXICITY AND SAFETY CONCERNS OF NANOMATERIAL

The most critical feature of nanomaterials' use, in particular nanotherapeutic agents, is the effective handling, least accumulation, and nontoxic effect of these compounds. However, if the nanomaterial is inert, it can pose no problem, but when it reaches the body, it may pose serious effects if it is active. Nanoparticles pose a major health threat as they are capable of breaching physiological barriers inside the body that cause serious damage to cells and even can kill membranes, organelles, modify DNA-coded messages and contribute to reactive oxygen species production, etc. (Xia et al., 2006; Madkour, 2020). Generating reactive oxygen species results in the activation of different signal cascades, which ultimately affects the proper functioning of cells such as proliferation, adhesion, and differentiation. The presence of nanoparticles in the blood will give rise to thrombosis and platelet aggregation, which eventually increases the risk of strokes. Studies have shown that high doses of nanotubes cause chronic lung inflammation, granuloma formation, and fibrosis (De Jong and Borm, 2008). Other types of nanoparticles, such as carbon-based nanoparticles, gold nanoparticles, and silver nanoparticles, have frequently been documented to have cytotoxic effects (Zhao et al., 2008; Donaldson et al., 2010; Granet et al., 2019; Ha et al., 2020). As far as toxicity is concerned, nanoparticles pose a significant parameter for concluding any nanoproduct efficacy. Specific parameters are recognized by researchers (such as scale, zeta potential, and nanoparticle solubility), which should be taken into consideration when assessing the toxicity of the nanoformulations.

5.5 CONCLUSION

Nanotechnology provides ample opportunities for its preventive, diagnostic as well as therapeutic applications. Besides these nanotechnology applications, its role can be seen in various health-care management systems, such as in surgeries, dressing materials, and medical devices. However, during these processes starting from manufacturing to disposal, they are likely to get introduced to the environment, and thus, it becomes necessary to determine the safety measures. Also, they diversify ecotoxicological effects of nano-content from placebo drugs (e.g., sugar molecules) to euthanasia drugs (death-inducing molecules). Most nanodrugs licensed by the FDA, such as Abraxane and Doxil, also help reduce the toxicity of the encapsulated drugs.

On the contrary, a small subset of nanoparticles currently under preclinical study, for example, carbon and nanoparticles centered on metals, usually demonstrate cytotoxic properties. Besides nanoparticles' ability to aid in the diagnosis of disease much earlier than ever possible, they may also exhibit harmful effects based mainly on ROS production, cellular compartment destruction, and immunomodulation. Nonetheless, as long as healthy organs are protected by selective targeting, the inherent toxic properties will serve as a "double-edged sword" of these nanoparticles that may gain to ablate diseased tissue. Besides, toxicity monitoring can be conducted in vivo following many techniques (e.g., surface modification) to reduce nanoparticle toxicity.

KEYWORDS

- **nanotechnology**
- **nanomedicine**
- **nanotubes**
- **chemotherapeutic technologies**
- **health care system**

REFERENCES

Abbasi, J. Novel Nanochip Paves Way for Early Ovarian Cancer Detection. *JAMA* **2019,** *321* (18), 1759. doi:10.1001/jama.2019.5460

Abou Neel, E. A.; Laurent Bozec, L.; Perez, R. A.; Hae-Won, K.; Knowles Jonathan, C. Nanotechnology in Dentistry: Prevention, Diagnosis, and Therapy. *Int. J. Nanomed.* **2015,** *10,* 6371–6394.

Ahmad, Z.; Sharma, S.; Khuller, G. K. Inhalable Alginate Nanoparticles as Antitubercular Drug Carriers against Experimental Tuberculosis. *Int. J. Antimicrob. Agents* **2005,** *26,* 298–303.

Anderson, C. F.; Grimmett, M. E.; Domalewski, C. J.; Cui, H. Inhalable Nanotherapeutics to Improve Treatment Efficacy for Common Lung Diseases. *Wiley Interdiscipl. Rev.: Nanomed. Nanobiotechnol.* **2020,** *12* (1), e1586.

Awasthi, R.; Roseblade, A.; Hansbro, P. M.; Rathbone, M. J.; Dua, K.; Bebawy, M. Nanoparticles in Cancer Treatment: Opportunities and Obstacles. *Curr. Drug Targets* **2018,** *19* (14), 1696–1709.

Baghbani, F.; Moztarzadeh, F.; Mohandesi, J. A.; Yazdian, F.; Mokhtari-Dizaji, M. Novel Alginate-Stabilized Doxorubicin-Loaded Nanodroplets for Ultrasounic Theranosis of Breast Cancer. *Int. J. Biol. Macromol.* **2016,** *93,* 512–519.

Balaconis, M. K.; Billingsley, K.; Dubach, J. M.; Cash, K. J.; Clark, H. A. The Design and Development of Fluorescent Nano-Optodes for in Vivo Glucose Monitoring. *J. Diab. Sci. Technol.* **2011,** *5* (1), 68–75.

Balaconis, M. K.; Clark, H. A. Gel Encapsulation of Glucose Nanosensors for Prolonged in Vivo Lifetime. *J. Diab. Sci. Technol.* **2013,** *7* (1), 53–61.

Ballerstadt, R.; Schultz, J. S. A Fluorescence Affinity Hollow Fiber Sensor for Continuous Transdermal Glucose Monitoring. *Analyt. Chem.* **2000,** *72* (17), 4185–4192.

Barrientos, S.; Stojadinovic, O.; Golinko, M. S.; Brem, H.; Tomic-Canic, M. Growth Factors and Cytokines in Wound Healing. *Wound Repair Regen.* **2008,** *16* (5), 585–601.

Bhattacharya, D.; Ghosh, B.; Mukhopadhyay, M. Development of Nanotechnology for Advancement and Application in Wound Healing: A Review. *IET Nanobiotechnol.* **2019,** *13* (8), 778–785.

Bhowmik, D.; Chiranjib, C. R.; Jayakar, B. Role of Nanotechnology in Novel Drug Delivery System. *J. Pharm. Sci. Technol.* **2009,** *1* (1), 20–35.

Bilan, R.; Fleury, F.; Nabiev, I.; Sukhanova, A. Quantum Dot Surface Chemistry and Functionalization for Cell Targeting and Imaging. *Bioconjug. Chem.* **2015,** *26* (4), 609–624.

Biswas, B.; Warr, L. N.; Hilder, E. F.; Goswami, N.; Rahman, M. M.; Churchman, J. G.; Vasilev, K.; Pan, G.; Naidu, R. Biocompatible Functionalisation of Nanoclays for Improved Environmental Remediation. *Chem. Soc. Rev.* **2019,** *48,* 3740–3770.

Blanco-Fernandez, B.; Castaño, O.; Mateos-Timoneda, M. A.; Engel, E.; Perez Amodio, S. Nanotechnology Approaches in Chronic Wound Healing. *Adv. Wound Care* **2020.**

Blank, F.; Fytianos, K.; Seydoux, E.; Rodriguez-Lorenzo, L.; Petri-Fink, A.; Von Garnier, C.; Rothen-Rutishauser, B. Interaction of Biomedical Nanoparticles with the Pulmonary Immune System. *J. Nanobiotechnol.* **2017,** *15* (1), 6.

Breton, M.; Farret, A.; Bruttomesso, D.; Anderson, S.; Magni, L.; Patek, S.; ... & Toffanin, C. Fully Integrated Artificial Pancreas in Type 1 Diabetes: Modular Closed-Loop Glucose Control Maintains Near Normoglycemia. *Diabetes* **2012,** *61* (9), 2230–2237.

Bulbake, U.; Doppalapudi, S.; Kommineni, N.; Khan, W. Liposomal Formulations in Clinical Use: An Updated Review. *Pharmaceutics* **2017,** *9* (2), 12.

Bulte, J. W.; Douglas, T.; Witwer, B.; Zhang, S. C.; Strable, E.; Lewis, B. K.; Duncan, I. D. Magnetodendrimers Allow Endosomal Magnetic Labeling and in Vivo Tracking of Stem Cells. *Nat. Biotechnol.* **2001,** *19* (12), 1141–1147.

Chakrabarti, S.; Chattopadhyay, P.; Islam, J.; Ray, S.; Raju, P. S.; Mazumder, B. Aspects of Nanomaterials in Wound Healing. *Curr. Drug Deliv.* **2019,** *16* (1), 26–41.

Chen, Q.; Wang, C.; Zhang, X.; Chen, G.; Hu, Q.; Li, H.; Yang, G. In Situ Sprayed Bioresponsive Immunotherapeutic Gel for Post-Surgical Cancer Treatment. *Nat. Nanotechnol.* **2019,** *14* (1), 89–97.

Chen, Z.; Wang, J.; Sun, W.; Archibong, E.; Kahkoska, A. R.; Zhang, X.; Gu, Z. Synthetic Beta Cells for Fusion-Mediated Dynamic Insulin Secretion. *Nat. Chemi. Biol.* **2018,** *14* (1), 86.

Chen, Z.; Wang, Z.; Gu, Z. Bioinspired and Biomimetic Nanomedicines. *Acc. Chem. Res.* **2019,** *52* (5), 1255–1264.

Chereddy, K. K.; Lopes, A.; Koussoroplis, S.; Payen, V.; Moia, C.; Zhu, H.; Préat, V. Combined Effects of PLGA and Vascular Endothelial Growth Factor Promote the Healing of Non-Diabetic and Diabetic Wounds. *Nanomed.: Nanotechnol. Biol. Med.* **2015,** *11*(8), 1975–1984.

Chiappini, C.; De Rosa, E.; Martinez, J. O.; Liu, X.; Steele, J.; Stevens, M. M.; Tasciotti, E. Biodegradable Silicon Nanoneedles Delivering Nucleic Acids Intracellularly Induce Localized In Vivo Neovascularization. *Nat. Mater.* **2015a,** *14* (5), 532–539.

Chiappini, C.; Martinez, J. O.; De Rosa, E.; Almeida, C. S.; Tasciotti, E.; Stevens, M. M. *ACS Nano* **2015b,** *9* (5), 5500–5509.

Chou, I. H. et al. Nanofluidic Biosensing for Beta-Amyloid Detection Using Surface Enhanced Raman Spectroscopy. *Nano Lett.* 2008, *8* (6), 1729–1735.

Christenson, E. M.; Anseth, K. S.; van den Beucken, J. J.; Chan, C. K.; Ercan, B.; Jansen, J. A.; Ramakrishna, S. Nanobiomaterial Applications in Orthopedics. *J. Orthopaed. Res.* **2007,** *25* (1), 11–22.

Clogston, J. D.; Hackley, V. A.; Prina-Mello, A.; Puri, S.; Sonzini, S.; Soo, P. L. Sizing up the Next Generation of Nanomedicines. *Pharm. Res* **2020,** *37* (1), 6.

Cormode, D. P.; Skajaa, T.; Fayad, Z. A.; Mulder, W. J. Nanotechnology in Medical Imaging: Probe Design and Applications. *Arterioscler Thromb. Vasc. Biol.* **2009,** *29* (7), 992–1000. doi:10.1161/ATVBAHA.108.165506

Cui, N.; Lu, H.; Li, M. Magnetic Nanoparticles Associated PEG/PLGA Block Copolymer Targeted with Anti-Transferrin Receptor Antibodies for Alzheimer's Disease. *J. Biomed. Nanotechnol.* **2018,** *14* (5), 1017–1024.

Da Silva, A. L.; Martini, S. V.; Abreu, S. C.; Samary, C. D. S.; Diaz, B. L.; Fernezlian, S.; Suk, J. S. DNA Nanoparticle-Mediated Thymulin Gene Therapy Prevents Airway Remodeling in Experimental Allergic Asthma. *J. Control. Release* **2014,** *180*, 125–133.

Dadfar, S. M.; Roemhild, K.; Drude, N. I.; von Stillfried, S.; Knüchel, R.; Kiessling, F.; Lammers, T. Iron Oxide Nanoparticles: Diagnostic, Therapeutic and Theranostic Applications. *Adv. Drug Deliv. Rev.* **2019,***138*, 302–325.

Dayton, P. A.; Ferrara, K. W. Targeted Imaging Using Ultrasound. *J. Magnet. Resonan. Imag.* **2002,** *16* (4), 362–377.

De Jong, W. H.; Borm, P. J. Drug Delivery and Nanoparticles: Applications and Hazards. *Int. J. Nanomed.* **2008,** *3* (2), 133.

Deacon, J.; Abdelghany, S. M.; Quinn, D. J. et al. Antimicrobial Efficacy of Tobramycin Polymeric Nanoparticles for *Pseudomonas aeruginosa* Infections in Cystic Fibrosis: Formulation, Characterisation and Functionalisation with Dornase Alfa (DNase). *J. Control. Release* **2015,** *198*, 55–61.

Nanotechnology in Healthcare Management   163

Deng, H.-H.; Li, G.-W.; Hong, L.; Liu, A.-L.; Chen, W.; Lin, X.-H.; Xia, X.-H. Colorimetric Sensor Based on Dual-Functional Gold Nanoparticles: Analyte-Recognition and Peroxidase-Like Activity. *Chemical* **2015**, *147*, 257–261.

DiSanto, R. M.; Subramanian, V.; Gu, Z. Recent advances in Nanotechnology for Diabetes Treatment. *Wiley Interdiscipl. Rev.: Nanomed. Nanobiotechnol.* **2015**, *7* (4), 548–564.

Donaldson, K.; Murphy, F. A.; Duffin, R.; Poland, C. A. Asbestos, Carbon Nanotubes and the Pleural Mesothelium: A Review of the Hypothesis Regarding the Role of Long Fibre Retention in the Parietal Pleura, Inflammation and Mesothelioma. *Particle Fibre Toxicol.* **2010**, *7* (1), 5.

Driskell, J. D.; Jones, C. A.; Tompkins, S. M.; Tripp, R. A. One-Step Assay for Detecting Influenza Virus Using Dynamic Light Scattering and Gold Nanoparticles. *Analyst* **2011**, *136* (15), 3083–3090.

Eckert, M. A.; Vu, P. Q.; Zhang, K.; Kang, D.; Ali, M. M.; Chenjie, X.; Weian, Z. Novel Molecular and Nanosensors for in Vivo Sensing. *Theranostics* **2013**, *3* (8), 583–594.

Elingarami, S.; Liu, M.; Fan, J.; He, N. Applications of Nanotechnology in Gastric Cancer. Detection and Prevention by Nutrition. *J. Nanosci. Nanotechnol.* **2014**, *14*, 1.

Feynman, R. Infinitesimal Machinery. Lecture Reprinted in the J. Microelectromech. Syst., **1993**.

Freitas Jr, R. A. What Is Nanomedicine? *Nanomed.: Nanotechnol., Biol. Med.* **2005**, *1* (1), 2–9.

Garbayo, E.; Estella-Hermoso de Mendoza, A.; J Blanco-Prieto, M. Diagnostic and Therapeutic Uses of Nanomaterials in the Brain. *Curr. Med. Chem.* **2014**, *21* (36), 4100–4131.

Georganopoulou, D. G.; Chang, L.; Nam, J. M.; Thaxton, C. S.; Mufson, E. J.; Klein, W. L.; Mirkin, C. A. Nanoparticle-Based Detection in Cerebral Spinal Fluid of a Soluble Pathogenic Biomarker for Alzheimer's Disease. *Proc. Natl. Acad. Sci. U. S. A.* **2005**, *102* (7), 2273–2276.

Gomes-da-Silva, L. C.; Fonseca, N. A.; Moura, V.; de Lima, M. C. P.; Simoes, S.; Moreira, J. N. Lipid-Based Nanoparticles for siRNA Delivery in Cancer Therapy: Paradigms and Challenges. *Acc. Chem. Res.* **2012**, *45*, 1163–1171.

Gong, Y.; Chang, L.; Viola, K. L.; Lacor, P. N.; Lambert, M. P.; Finch, C. E.; Klein, W. L. Alzheimer's Disease-Affected Brain: Presence of Oligomeric Aβ Ligands (ADDLs) Suggests a Molecular Basis for Reversible Memory Loss. *Proc. Natl. Acad. Sci. U. S. A.* **2003**, *100* (18), 10417–10422.

Goswami, N.; Bright, R.; Visalakshan, R. M.; Biswas, B.; Zilm, P.; Vasilev, K. Core-in-Cage Structure Regulated Properties of Ultra-Small Gold Nanoparticles. *Nanoscale Adv.* **2019**, *1*, 2356–2364.

Granet, R.; Faure, R.; Ntoutoume, G. M. A. N.; Mbakidi, J. P.; Leger, D. Y.; Liagre, B.; Sol, V. Enhanced Cytotoxicity of Gold Porphyrin Complexes after Inclusion in Cyclodextrin Scaffolds Adsorbed on Polyethyleneimine-Coated Gold Nanoparticles. *Bioorganic Med. Chem. Lett.* **2019**, *29* (9), 1065–1068.

Guisasola, E.; Asin, L.; Beola, L.; La Fuente, J. M.; Baeza, A.; Valletregi, M. Beyond Traditional Hyperthermia: In Vivo Cancer Treatment with Magnetic-Responsive Mesoporous Silica Nanocarriers. *ACS Appl. Mater. Interf.* **2018**, *10* (15), 12518–12525.

Guo, N. L.; Wan, Y. W.; Denvir, J.; Porter, D. W.; Pacurari, M. et al. Multiwalled Carbon Nanotube-Induced Gene Signatures in the Mouse Lung: Potential Predictive Value for Human Lung Cancer Risk and Prognosis. *J. Toxicol. Environ. Health* **2012a**, *A75*, 1129–1153.

Guo, N. L.; Wan, Y. W.; Denvir, J.; Porter, D. W.; Pacurari, M.; Wolfarth, M. G.; Qian, Y. Multiwalled Carbon Nanotube-Induced Gene Signatures in the Mouse Lung: Potential Predictive Value for Human Lung Cancer Risk and Prognosis. *J. Toxicol. Environ. Health, Part A* **2012b**, *75* (18), 1129–1153.

Guo, P. A Special Issue on Bionanotechnology-Preface. *J. Nanosci. Nanotechnol.* **2005**, *5*, 12.

Gupta, U.; Agashe, H. B.; Asthana, A.; Jain, N. K. Dendrimers: Novel Polymeric Nanoarchitectures for Solubility Enhancement. *Biomacromolecules* **2006**, *7* (3), 649–658.

Ha, M. K.; Chung, K. H.; Yoon, T. H. Heterogeneity in Biodistribution and Cytotoxicity of Silver Nanoparticles in Pulmonary Adenocarcinoma Human Cells. *Nanomaterials* **2020**, *10* (1), 36.

Halo, T. L.; McMahon, K. M.; Angeloni, N. L.; Xu, Y.; Wang, W.; Chinen, A. B.; ... & Mirkin, C. A. NanoFlares for the Detection, Isolation, and Culture of Live Tumor Cells from Human Blood. *Proc. Natl. Acad. Sci. U. S. A.* **2014**, *111* (48), 17104–17109.

Hanazaki, K.; Nosé, Y.; Brunicardi, F. C. Artificial Endocrine Pancreas. *J. Am. Coll. Surg.* **2001**, *193* (3): 310–322.

He, H.; Liu, L.; Morin, E. E.; Liu, M.; Schwendeman, A. Survey of Clinical Translation of Cancer Nanomedicines—Lessons Learned from Successes and Failures. *Acc. Chem. Res.* **2019**, *52* (9), 2445–2461.

He, Wen-Jie, Hosseinkhani, H.; Hong, Po-Da; Chiang, Chiao-Hsi; Yu, Dah-Shyong. Magnetic Nanoparticles for Imaging Technology. *Int. J. Nanotechnol.* **2013**, *10*, 930–944. doi: 10.1504/IJNT.2013.058120.

Hee An, et al. Gold Nanoparticles-Based Barcode Analysis for Detection of Norepinephrine. *J. Biomed. Nanotechnol.* **2016**, *12*, 357–365, doi: 10.1166/jbn.2016.2185

High, K. A.; Roncarolo, M. G. Gene therapy. *New Engl. J. Med.* **2019**, *381* (5), 455–464.

Hogg, T.; Freitas Jr, R. A. Chemical Power for Microscopic Robots in Capillaries. *Nanomed. Nanotechnol., Biol. Med.* **2010**, 6 (2), 298–317.

Hu, K.; Chen, X.; Chen, W.; Zhang, L.; Li, J.; Ye, J.; ... & Guan, Y. Q. Neuroprotective Effect of Gold Nanoparticles Composites in Parkinson's Disease Model. *Nanomed. Nanotechnol., Biol. Med.* **2018**, *14* (4), 1123–1136.

Huang, R.; Ma, H.; Guo, Y.; Liu, S.; Kuang, Y.; Shao, K.; An, S. Angiopep-Conjugated Nanoparticles for Targeted Long-Term Gene Therapy of Parkinson's Disease. *Pharm. Res.* **2013**, *30* (10), 2549–2559.

Hussain, Z. Nanotechnology Guided Newer Intervention for Treatment of Osteoporosis: Efficient Bone Regeneration by Up-Regulation of Proliferation, Differentiation and Mineralization of Osteoblasts. *Int. J. Polym. Mater. Polym. Biomater.* **2019**, 1–13.

Hwang, J. H.; Han, U.; Yang, M.; Choi, Y.; Choi, J.; Lee, J. M.; Hong, J. H. Artificial Cellular Nano-Environment Composed of Collagen-Based Nanofilm Promotes Osteogenic Differentiation of Mesenchymal Stem Cells. *Acta Biomater.* **2019**, *86*, 247–256.

Jagjeevanrao, L.; Prasanna, R.; Sowjanya, A. Manasasaisita: Drastic Changes in Medical Field by the Invention of Nanobots. *Int. J. Image Process. Vision Sci.* **2012**, *1*, 2278–1110.

Jain, K. K. Nanodiagnostics: Application of Nanotechnology in Molecular Diagnostics. *Expert Rev. Mol. Diagnos.* **2014**, *3* (2), 153–161.

Jain, K. K. Use of Nanoparticles for Drug Delivery in Glioblastoma Multiforme. *Expert Rev. Neurother.* **2007**, *7*, 363–372.

Joshi, K.; Chattopadhyay, P.; Mazumdar, B. Nanotechnology for Tissue Regeneration. In *Nanotechnology: Therapeutic, Nutraceutical, and Cosmetic Advances*; 2019, p 197.

Kaczmarek, J. C.; Patel, A. K.; Kauffman, K. J.; Fenton, O. S.; Webber, M. J.; Heartlein, M. W.; Anderson, D. G. Polymer–Lipid Nanoparticles for Systemic Delivery of mRNA to the Lungs. *Angewandte Chem. Int. Ed.* **2016,** *55* (44), 13808–13812.

Kaneti, L.; Bronshtein, T.; Malkah Dayan, N.; Kovregina, I.; Letko Khait, N.; Lupu-Haber, Y.; Fliman, M.; Schoen, B. W.; Kaneti, G.; Machluf, M. Nanoghosts as a Novel Natural Nonviral Gene Delivery Platform Safely Targeting Multiple Cancers. *Nano Lett.* **2016,** *16,* 1574–1582.

Kang, J. K.; Kim, J. C.; Shin, Y.; Han, S. M.; Won, W. R.; Her, J.; Park, J. Y.; Oh, K. T. Principles and Applications of Nanomaterial-Based Hyperthermia in Cancer Therapy. *Arch. Pharm. Res.* **2020,** *43* (1), 46–57.

Karve, S.; Werner, M. E.; Sukumar, R.; Cummings, N. D.; Copp, J. A.; Wang, E. C.; ... & Wang, A. Z. Revival of the Abandoned Therapeutic Wortmannin by Nanoparticle Drug Delivery. *Proc. Natl. Acad. Sci. U. S. A.* **2012,** *109* (21), 8230–8235.

Kaushik, A.; Jayant, R. D.; Bhardwaj, V.; and Nair, M. Personalized Nanomedicine for CNS Diseases. *Drug Discov. Today* **2018,** *23*, 1007–1015.

Khawaja, A. M. The Legacy of Nanotechnology: Revolution and Prospects in Neurosurgery. *Int. J. Surg.* **2011,** *9* (8), 608–614.

Kim, H.; Lee, J. U.; Song, S.; Kim, S.; Sim, S. J. A Shape-Code Nanoplasmonic Biosensor for Multiplex Detection of Alzheimer's Disease Biomarkers. *Biosens. Bioelect.* **2018,** *101*, 96–102.

Kim, Y.; Park, E. J.; Na, D. H. Recent Progress in Dendrimer-Based Nanomedicine Development. *Arch. Pharm. Res.* **2018,** *41* (6), 571–582.

Koushki, N.; Katbab, A. A.; Tavassoli, H.; Jahanbakhsh, A.; Majidi, M.; Bonakdar, S. A New Injectable Biphasic Hydrogel Based on Partially Hydrolyzed Polyacrylamide and Nanohydroxyapatite as Scaffold for Osteochondral Regeneration. *RSC Adv.* **2015,** *5* (12), 9089–9096.

Kranz, L. M. et al. Systemic RNA Delivery to Dendritic Cells Exploits Antiviral Defence for Cancer Immunotherapy. *Nature* **2016,** 534, 396–401.

Krasimir, V. Nanoengineered Antibacterial Coatingsand Materials: A Perspective. *Coatings* **2019,** *9*, 654.

Krausz, A. E.; Adler, B. L.; Cabral, V.; Navati, M.; Doerner, J.; Charafeddine, R. A.; Harper, S. Curcumin-Encapsulated Nanoparticles as Innovative Antimicrobial and Wound Healing Agent. *Nanomed.: Nanotechnol., Biol. Med.* **2015,** *11* (1), 195–206.

Kreiter, S. et al. Mutant MHC Class II Epitopes Drive Therapeutic Immune Responses to Cancer. *Nature* **2015,** 520, 692–696.

Kropff, J.; DeVries, J. H. Continuous Glucose Monitoring, Future Products, and Update on Worldwide Artificial Pancreas Projects. *Diab. Technol. Therap.* **2016,** *18* (S2), S2–S53.

Kwong, G. A.; Von Maltzahn, G.; Murugappan, G.; Abudayyeh, O.; Mo, S.; Papayannopoulos, I. A.; Schuppan, D. Mass-Encoded Synthetic Biomarkers for Multiplexed Urinary Monitoring of Disease. *Nat. Biotechnol.* **2013,** *31* (1), 63.

Lambert, M. P.; Barlow, A. K.; Chromy, B. A.; Edwards, C.; Freed, R.; Liosatos, M.; Wals, P. Diffusible, Nonfibrillar Ligands Derived from Aβ1–42 are Potent Central Nervous System Neurotoxins. *Proc. Natl. Acad. Sci. U. S. A.* **1998,** *95* (11), 6448–6453.

Larry, J. K. Microchips, Microarrays, Biochips and Nanochips: Personal Laboratories for the 21st Century. Clin. Chim. Acta **2001,** *307, 1–2*, 219–223.

Lee, W. G.; Kim, Y. G.; Chung, B. G.; Demirci, U.; Khademhosseini, A. Nano/Microfluidics for Diagnosis of Infectious Diseases in Developing Countries. *Adv. Drug Deliv. Rev.* Mar 18 **2010**, *62* (4–5), 449–457.

Li, F.; Li, F.; Yang, G.; Aguilar, Z. P.; Lai, W.; Xu, H. Asymmetric Polymerase Chain Assay Combined with Propidium Monoazide Treatment and Unmodified Gold Nanoparticles for Colorimetric Detection of Viable Emetic *Bacillus cereus* in Milk. *Sens. Actuat. B: Chemical* **2018**, *255*, 1455–1461.

Li, X.; Qi, J.; Xie, Y.; Zhang, X.; Hu, S.; Xu, Y.; Yi, Lu; Wu, W. Nanoemulsions Coated with Alginate/Chitosan as Oral Insulin Delivery Systems: Preparation, Characterization, and Hypoglycemic Effect in Rats. *Int. J. Nanomed.* **2013**, *8*, 23–32.

Lin, Y. X.; Wang, Y.; Blake, S.; Yu, M.; Mei, L.; Wang, H.; Shi, J. RNA Nanotechnology-Mediated Cancer Immunotherapy. *Theranostics* **2020**, *10* (1), 281.

Liu, L.; Sun, M.; Li, Q.; Zhang, H.; Alvarez, P. J. J. et al. Genotoxicity and Cytotoxicity of Cadmium Sulfide Nanomaterials to Mice: Comparison Between Nanorods and Nanodots. *Environ. Eng. Sci.* **2014**, *31*, 373–380.

Liu, Y.; Yang, Y.; Zhang, C. A Concise Review of Magnetic Resonance Molecular Imaging of Tumor Angiogenesis by Targeting Integrin $\alpha v\beta 3$ with Magnetic Probes. *Int. J. Nanomed.* **2013**, *8*, 1083.

Liu, Z.; Gao, X.; Kang, T.; Jiang, M.; Miao, D.; Gu, G.; Chen, H. B6 Peptide-Modified PEG-PLA Nanoparticles for Enhanced Brain Delivery of Neuroprotective Peptide. *Bioconjug. Chem.* **2013**, *24* (6), 997–1007.

Lo, R. C. Microfluidics Technology: Future Prospects for Molecular Diagnostics. *Adv. Health Care Technol.* **2017**, *3*, 3–17.

Lopes Da Silva, A.; Ferreira Cruz, F.; Rieken, P.; Rocco, M.; Morales, M. M. New Perspectives in Nanotherapeutics for Chronic Respiratory Diseases. *Biophys. Rev.* **2017**, *9*, 793–803.

Lu, C. T.; Jin, R. R.; Jiang, Y. N.; Lin, Q.; Yu, W. Z.; Mao, K. L.; Zhao, Y. Z. Gelatin Nanoparticle-Mediated Intranasal Delivery of Substance P Protects against 6-Hydroxydopamine-Induced Apoptosis: An in Vitro and in Vivo Study. *Drug Design Dev. Therap.* **2015**, *9*, 1955.

Lu, Y. T.; Zhao, L.; Shen, Q.; Garcia, M. A.; Wu, D.; Hou, S.; Lichterman, J. NanoVelcro Chip for CTC enumeration in prostate cancer patients. *Methods* **2013**, *64* (2), 144–152.

Lu, Y.; Xue, J. X.; Deng, T.; Zhou, X. J.; Yu, K.; Huang, M. J. A Phase I Trial of PD-1 Deficient Engineered T Cells with CRISPR/Cas9 in Patients with Advanced Non-Small Cell Lung Cancer. *J. Clin. Oncol.* **2018**, *36*, 3050.

Madkour, L. H. Pathways for Nanoparticle (NP)-Induced Oxidative Stress. In *Nanoparticles Induce Oxidative and Endoplasmic Reticulum Stresses*; Springer: Cham, 2020; pp 285–328.

Magro, M.; Baratella, D.; Bonaiuto, E.; de A Roger, J.; Vianello, F. New Perspectives on Biomedical Applications of Iron Oxide Nanoparticles. *Curr. Med. Chem.* **2018**, *25* (4), 540–555.

Majidinia, M.; Mirza-Aghazadeh-Attari, M.; Rahimi, M.; Mihanfar, A.; Karimian, A.; Safa, A.; Yousefi, B. Overcoming Multidrug Resistance in Cancer: Recent Progress in Nanotechnology and New Horizons. *IUBMB Life* **2020**, *72* (5), 855–871. https://doi.org/10.1002/iub.2215

Majumder, D. D.; Banerjee, R.; Ulrichs, C. H.; Mewis, I.; Goswami, A. Nanomaterials: Science of Bottom-Up and Topdown. *IETE Tech. Rev.* **2007**, *24* (1), 9–25.

Malik, P.; Katyal, V.; Malik, V.; Asatkar, A.; Inwati, G.; Mukherje, T. K. Nanobiosensors: Concepts and Variations. *Int. Scholar. Res. Notices* **2013**, 1–9.

Nanotechnology in Healthcare Management 167

Manikandan, D.; Jeyachandra, V.; .Manikandan, A. Biobarcode Nanoassay for Rapid Identification and Detection of Rifampicin-Resistanct Mycobacterium Tuberculosis in Sputum Samples. *Int. J. Pure Appl. Maths.* **2018**, *119* (12), 2185–2203.

Mathew, A.; Fukuda, T.; Nagaoka, Y.; Hasumura, T.; Morimoto, H.; Yoshida, Y.; Kumar, D. S. Curcumin Loaded-PLGA Nanoparticles Conjugated with Tet-1 Peptide for Potential Use in Alzheimer's Disease. *PLoS One* **2012**, *7* (3).

McCarthy, J. R.; Weissleder, R. Multifunctional Magneti Nanoparticles for Targeted Imaging and Therapy. *Adv. Drug Deliv. Rev.* **2008**, *60* (11), 1241–1251.

McCartney, L. J.; Pickup, J. C.; Rolinski, O. J.; Birch, D. J. Near-Infrared Fluorescence Lifetime Assay for Serum Glucose Based on Allophycocyanin-Labeled Concanavalin A. *Analyt. Biochem.* **2001**, *292* (2), 216–221.

McNeer, N.; Anandalingam, K.; Fields, R.; et al. Nanoparticles That Deliver Triplex-Forming Peptide Nucleic Acid Molecules Correct F508del CFTR in airway epithelium. *Nat. Commun.* **2015**, *6*, 6952.

Mekuria, S. L.; Debele, T. A.; Tsai, H. C. Encapsulation of Gadolinium Oxide Nanoparticle (Gd_2O_3) Contrasting Agents in PAMAM Dendrimer Templates for Enhanced Magnetic Resonance Imaging in Vivo. *ACS Appl. Mater. Interf.* **2017**, *9* (8), 6782–6795.

Miller, J. B.; Zhang, S.; Kos, P.; Xiong, H.; Zhou, K.; Perelman, S. S. et al. Non-Viral CRISPR/Cas Gene Editing in Vitro and in Vivo Enabled by Synthetic Nanoparticle Co-delivery of Cas9 mRNA and sgRNA. *Angew. Chem. Int. Ed.* **2017**. *56*, 1059–1063.

Miller, M. A.; Chandra, R.; Cuccarese, M. F.; Pfirschke, C.; Engblom, C.; Stapleton, S.; Weissleder, R. Radiation Therapy Primes Tumors for Nanotherapeutic Delivery via Macrophage-Mediated Vascular Bursts. *Sci. Transl. Med.* **2017**, *9* (392).

Mohamed, M. B.; Ismail, K. Z.; Link, S.; El-Sayed, M. A. Thermal Reshaping of Gold Nanorods in Micelles. *J. Phys. Chem. B* **1998**, *102*, 9370–9374.

Morawski, A. M.; Winter, P. M.; Crowder, K. C.; Caruthers, S. D.; Fuhrhop, R. W.; Scott, M. J.; Wickline, S. A. Targeted Nanoparticles for Quantitative Imaging of Sparse Molecular Epitopes with MRI. *Magnet. Resonan. Med.: Off. J. Int. Soc. Magnet. Resonan. Med.* **2004**, *51* (3), 480–486.

Morrison, D.; Dokmeci, M.; Demirci, U.; Khademhosseini, A. Biomedical Nanostructures; Edited by Kenneth, G.; Craig, H.; Cato, T. L.; Lakshmi, N.; John Wiley & Sons, Inc., 2008.

Nam, J. M. et al. Nanoparticle-Based Bio-Bar Codes for the Ultrasensitive Detection of Proteins. *Science* **2003**, *301*, 1884–1886. doi: 10.1126/science.1088755

Napoli, M.; Eijkel, J. C. T.; Pennathur, S. Nanofluidic Technology for Biomolecule Applications: A Critical Review. *Lab Chip* **2010**, *10*, 957–985.

Narayan, R. J.; Verma, N. Nanomaterials as Implantable Sensors. In *Materials for Chemical Sensing*; Cesar Paixão, T., Reddy, S., Eds.; Springer: Cham, 2017; pp 123–139.

Niu, S.; Zhang, L. K.; Zhang, L.; Zhuang, S.; Zhan, X.; Chen, W. Y.; Guan, Y. Q. Inhibition by Multifunctional Magnetic Nanoparticles Loaded with Alpha-Synuclein RNAi Plasmid in a Parkinson's Disease Model. *Theranostics* **2017**, *7* (2), 344.

Oh, B. K. et al. A Fluorophore-Based Bio-Barcode Amplification Assay for Proteins. *Small* **2006**, *2*, 103–108. doi: 10.1002/smll.200500260.

Oliver, N. S.; Toumazou, C.; Cass, A. E.; Johnston, D. G. Glucose Sensors: A Review of Current and Emerging Technology. *Diabet. Med.* **2009**, *26*, 197–210.

Ong, V.; Mei, V.; Cao, L.; Lee, K.; Chung, E. J. Nanomedicine for Cystic Fibrosis. *SLAS Technol. Transl. Life Sci. Innov.* **2019**, *24* (2), 169–180.

Pacheco, C.; Sousa, F.; Sarmento, B. Chitosan-Based Nanomedicine for Brain Delivery: Where Are We Heading? *React. Funct. Polym.* **2020,** *146,* 104430.

Pan, Y.; Li, Y-J.; Zhao, H-Y.; Zheng, J-M.; Xu, H.; Wei, G.; Hao, J-S.; Cui, F-D. Bioadhesive Polysaccharide in Protein Delivery System: Chitosan Nanoparticles Improve the Intestinal Absorption of Insulin in Vivo. *Int. J. Pharm.* **2002,** *249,* 139–147.

Pandey, R.; Sharma, A.; Zahoor, A.; Sharma, S.; Khuller, G. K.; Prasad, B. Poly (dl-Lactide-*co*-Glycolide) Nanoparticle-Based Inhalable Sustained Drug Delivery System for Experimental Tuberculosis. *J. Antimicrob. Chemother.* **2003,** *52,* 981–986.

Parhizkar, M.; Mahalingam, S.; Homer-Vanniasinkam, S.; Edirisinghe, M. *Latest Developments in Innovative Manufacturing to Combine Nanotechnology with Healthcare,* 2018.

Patel, A. K.; Kaczmarek, J. C.; Bose, S.; Kauffman, K. J.; Mir, F.; Heartlein, M. W.; ... & Anderson, D. G. Inhaled Nanoformulated mRNA Polyplexes for Protein Production in Lung Epithelium. *Adv. Mater.* **2019,** *31* (8), 1805116.

Patel, G. M.; Patel, G. C.; Patel, R. B.; Patel, J. K.; Patel, M. Nanorobot: A Versatile Tool in Nanomedicine. *J. Drug Target.* **2006,** *14* (2), 63–67.

Pham, D. D.; Fattal, E.; Tsapis, N. Pulmonary Drug Delivery Systems for Tuberculosis Treatment. *Int. J. Pharm.* **2015,** *478* (2), 517–529.

Pickup, J. C.; Hussain, F.; Evans, N. D.; Rolinski, O. J.; Birch, D. J. Fluorescence-Based Glucose Sensors. *Biosens. Bioelectr.* **2005,** *20* (12), 2555–2565.

Pickup, J. C.; Zhi, Z. L.; Khan, F.; Saxl, T.; Birch, D. J. Nanomedicine and Its Potential in Diabetes Research and Practice. *Diab. /Metabol. Res. Rev.* **2008,** *24* (8), 604–610.

Pleshko, N.; Grande, D. A.; Myers, K. R. Nanotechnology in Orthopaedics. *JAAOS-J. Am. Acad. Orthopaed. Surg.* **2012,** *20* (1), 60–62.

Prasad, P. N. *Introduction to Nanomedicine and Nanobioengineering,* Vol. 7. John Wiley & Sons, 2012.

Randeria, P. S.; Seeger, M. A.; Wang, X. Q.; Wilson, H.; Shipp, D.; Mirkin, C. A.; Paller, A. S. siRNA-Based Spherical Nucleic Acids Reverse Impaired Wound Healing in Diabetic Mice by Ganglioside GM3 Synthase Knockdown. *Proc. Natl. Acad. Sci. U. S. A.* **2015,** *112* (18), 5573–5578.

Ravindran Girija, A.; Balasubramanian, S.; Bright, R.; Cowin, A. J.; Goswami, N.; Vasilev, K. Ultrasmall Gold Nanocluster Based Antibacterial Nanoaggregates for Infectious Wound Healing. *ChemNanoMat.* **2019.**

Richard, E. Semiconductor Packaging Technologies Advance DNA Analysis Systems. *IVD Technology Magazine* Apr 4, 2005.

Rocco, M.; Williams, S.; Alivisato, P. Nanotechnology Research Directions: IWGN, Loyola College in Maryland, "National Nanotechnology Initiative: Leading to the Next Industrial Revolution," A Report by the Interagency Working Group on Nanoscience, Engineering and Technology Committee on Technology, National Science and Technology Council, Washington, DC, 2000.

Rojanathanes, R.; Sereemaspun, A.; Pimpha, N.; Buasorn, V.; Ekawong, P.; Wiwanitkit, V. Gold Nanoparticle as an Alternative Tool for a Urine Pregnancy Test. *Taiwanese J. Obstetr. Gynecol.* **2008,** *47* (3), 296–299.

Rosenblum, D.; Joshi, N.; Tao, W.; Karp, J. M.; Peer, D. Progress and Challenges towards Targeted Delivery of Cancer Therapeutics. *Nat. Commun.* **2018,** *9* (1), 1–12.

Sadikot, R. T. The Potential Role of Nanomedicine in Lung Diseases. *Med. Res. Arch.* **2018,** *6* (5), 1–9.

Sahin, U.; Türeci, Ö. Personalized Vaccines for Cancer Immunotherapy. *Science* **2018,** *359,* 1355–1360.

Saini, R.; Saini, S. Nanotechnology and Surgical Neurology. *Surg. Neurol. Int.* **2010,** *1.*

Salani, M.; Roy, S.; Fissell IV, W. H. Innovations in Wearable and Implantable Artificial Kidneys. *Am. J. Kidney Dis.* **2018,** *72* (5), 745–751.

Salehi, B.; Calina, D.; Docea, A. O.; Koirala, N.; Aryal, S.; Lombardo, D.; Martins, N. Curcumin's Nanomedicine Formulations for Therapeutic Application in Neurological Diseases. *J. Clin. Med.* **2020,** *9* (2), 430.

Sánchez-López, E.; Ettcheto, M.; Egea, M. A.; Espina, M.; Cano, A.; Calpena, A. C.; García, M. L. Memantine Loaded PLGA PEGylated Nanoparticles for Alzheimer's Disease: In Vitro and in Vivo Characterization. *J. Nanobiotechnol.* **2018,** *16* (1), 32.

Sapalidis, K.; Kosmidis, C.; Laskou, S.; Katsaounis, A.; Mantalobas, S.; Passos, I.; Zarogoulidis, P. Targeted Nanotechnology from Bench to Bedside. *Curr. Cancer Drug Targets* **2019,** *19* (1), 3–4.

Saraiva, C.; Paiva, J.; Santos, T.; Ferreira, L.; Bernardino, L. MicroRNA-124 Loaded Nanoparticles Enhance Brain Repair in Parkinson's Disease. *J. Control. Release* **2016,** *235,* 291–305.

Sarmento, B.; Ribeiro, A. J.; Veiga, F.; Ferreira, D. C.; Neufeld, R. J. Insulin-Loaded Nanoparticles Are Prepared by Alginate Ionotropic Pre-Gelation Followed by Chitosan Polyelectrolyte Complexation. *J. Nanosci. Nanotechnol.* **2007,** *7* (8), 2833–2841.

Segerink, L. I.; Eijkel, J. C. T. Nanofluidics in Point of Care Applications. (Frontier) *Lab Chip* **2014,** *14,* 3201–3205.

Serini, S.; Cassano, R.; Trombino, S.; Calviello, G. Nanomedicine-Based Formulations Containing ω-3 Polyunsaturated Fatty Acids: Potential Application in Cardiovascular and Neoplastic Diseases. *Int. J. Nanomed.* **2019,** *14,* 2809.

Sharma, P.; Guleria, P.; Kumar, V. Green Nanotechnology for Bioactive Compounds Delivery. In *Biotechnological Production of Bioactive Compounds*; Elsevier, 2020; pp 391–407.

Shiu, E.; Leung, N.; Cowling, B. Controversy Around Airborne Versus Droplet Transmission of Respiratory Viruses: Implication for Infection Prevention, 2019.

Shukla, A.; Dasgupta, N.; Ranjan, S.; Singh, S.; Chidambram, R. Nanotechnology towards Prevention of Anaemia and Osteoporosis: From Concept to Market. *Biotechnol. Biotechnol. Equip.* **2017,** *31* (5), 863–879. doi: 10.1080/13102818.2017.1335615

Siddiqui, N.; Madala, S.; Parcha, S. R.; Mallick, S. P. Osteogenic Differentiation Ability of Human Mesenchymal Stem Cells on Chitosan/Poly (Caprolactone)/Nano Beta Tricalcium Phosphate Composite Scaffolds. *Biomed. Phys. Eng. Express* **2020,** *6* (1), 015018.

Sim, S. W.; Moey, K. S.; Tan, N. C. The Use of Facemasks to Prevent Respiratory Infection: A Literature Review in the Context of the Health Belief Model. *Singapore Med. J.* **2014,** *55* (3), 160–167.

Singh, S.; McShane, M. Role of Porosity in Tuning the Response Range of Microsphere-Based Glucose Sensors. *Biosens Bioelectron.* Jan 15, **2011,** *26* (5), 2478–2483. doi: 10.1016/j. bios.2010.10.036. Epub Oct 30, 2010.

Siontorou, C. G.; Nikoleli, G. P. D.; Nikolelis, D. P.; Karapetis, S.; Tzamtzis, N.; Bratakou, S. Point-of-Care and Implantable Biosensors in Cancer Research and Diagnosis. In *Next Generation Point-of-care Biomedical Sensors Technologies for Cancer Diagnosis*; Chandra, P., Tan, Y., Singh, S., Eds.; Springer: Singapore, 2017; pp 115–132.

Snider, R. M.; Ciobanu, M.; Rue, A. E.; Cliffe, D. E. A Multiwalled Carbon Nanotube/ Dihydropyran Composite Film Electrode for Insulin Detection in A Microphysiometer Chamber. *Analyt. Chim. Acta* **2008,** *609* (1), 44–52.

Stoeva, S. I. et al. Multiplexed Detection of Protein Cancer Markers with Biobarcoded Nanoparticle Probes. *J. Am. Chem. Soc.***2006,** *128*, 8378–8379.

Su, H.; Zhao, H.; Qiao, F.; Chen, L.; Duan, R.; Ai, S. Colorimetric Detection of *Escherichia coli* O157:H7 Using Functionalized Au@Pt Nanoparticles as Peroxidase Mimetics. *Analyst* **2013,** *138*, 3026–3031.

Suk, J. S.; Lai, S. K.; Boylan, N. J. et al. Rapid Transport of Muco-Inert Nanoparticles in Cystic Fibrosis Sputum Treated with N-Acetyl Cysteine. *Nanomed. (London, England)* **2011,** *6*, 365–375.

Sullivan, D. C.; Ferrari, M. Nanotechnology and Tumor Imaging: Seizing an Opportunity. *Mol. Imag.* **2004.**

Sun, L. L.; Leo, Y. S.; Zhou, X.; Willie, Ng; Wong, T. I.; Deng, J. Localized Surface Plasmon Resonance Based Point-of-Care System for Sepsis Diagnosis. *Mater. Sci. Energy Technol.* **2019,** *3*. 10.1016/j.mset.2019.10.007.

Tai, W.; Mo, R.; Di, J.; Subramanian, V.; Gu, X.; Buse, J. B.; Gu, Z. Bio-Inspired Synthetic Nanovesicles for Glucose-Responsive Release of Insulin. *Biomacromolecules* **2014,** *15*, 3495–3502.

Teengam, P.; Siangproh, W.; Tuantranont, A.; Vilaivan, T.; Chailapakul, O.; Henry, C. S. Multiplex Paper-Based Colorimetric DNA Sensor Using Pyrrolidinyl Peptide Nucleic Acid-Induced AgNPs Aggregation for Detecting MERS-CoV, MTB, and HPV oligonucleotides. *Anal. Chem.* **2017,** *89*, 5428–5435.

Teleanu, D. M.; Chircov, C.; Grumezescu, A. M.; Teleanu, R. I. Neuronanomedicine: An Up-To-Date Overview. *Pharmaceutics* **2019,** *11* (3), 101.

Teleanu, D. M.; Chircov, C.; Grumezescu, A. M.; Volceanov, A.; Teleanu, R. I. Blood-Brain Delivery Methods Using Nanotechnology. *Pharmaceutics* **2018,** *10* (4), 269.

Thaxton, C. S.; Elghanian, R.; Thomas, A. D.; Stoeva, S. I.; Lee, J-S.; Smith, N. D.; Schaeffer, A. J.; Klocker, H.; Horninger, W.; Bartsch, G.; Mirkin, C. A. Nanoparticle-Based Bio-Barcode Assay Redefines "Undetectable" PSA and Biochemical Recurrence after Radical Prostatectomy. *Proc. Natl. Acad. Sci. U. S. A.* Nov **2009,** *106* (44) 18437–18442

The Hong Kong Polytechnic University. Superfilter Nanomask Protects from Invisible Killers. *Science Daily*, 6 June 2016. <www.sciencedaily.com/releases/2016/06/160606101244. htm>.

Toledano Furman, N. E.; Lupu-Haber, Y.; Bronshtein, T.; Kaneti, L.; Letko, N.; Weinstein, E.; Baruch, L.; Machluf, M. Reconstructed Stem Cell Nanoghosts: A Natural Tumor Targeting Platform. *Nano Lett.* **2013,** *13*, 3248–3255.

Torres-Ortega, P. V.; Saludas, L.; Hanafy, A. S.; Garbayo, E.; Blanco-Prieto, M. J. Micro-and Nanotechnology Approaches to Improve Parkinson's Disease Therapy. *J. Control. Release* **2019,** *295*, 201–213.

Vaiserman, A.; Koliada, A.; Zayachkivska, A.; Lushchak, O. Nanodelivery of Natural Antioxidants: An Anti-Aging Perspective. *Front. Bioeng. Biotechnol.* **2019,** *7*.

van der Meel, R.; Sulheim, E.; Shi, Y.; Kiessling, F.; Mulder, W.; Lammers, T. Smart Cancer Nanomedicine. *Nat. Nanotechnol.* **2019,** *14* (11), 1007–1017.

VanOsdol, J.; Ektate, K.; Ramasamy, S.; Maples, D.; Collins, W.; Malayer, J. et al. Sequential HIFU Heating and Nanobubble Encapsulation Provide Efficient Drug Penetration from

Stealth and Temperature Sensitive Liposomes in Colon Cancer. *J. Control. Release* **2017,** *247*, 55–63.

Venishetty, V. K.; Chede, R.; Komuravelli, R.; Adepu, L.; Sistla, R.; Diwan, P. V. Design and Evaluation of Polymer Coated Carvedilol Loaded Solid Lipid Nanoparticles to Improve the Oral Bioavailability: A Novel Strategy to Avoid Intraduodenal Administration. *Colloids Surf. B: Biointerf.* **2012,** *95*, 1–9.

Vij, N.; Min, T.; Marasigan, R.; et al. Development of PEGylated PLGA Nanoparticle for Controlled and Sustained Drug Delivery in Cystic Fibrosis. *J. Nanobiotechnol.* **2010,** *8*, 22.

Wagner, V.; Dullaart, A.; Bock, A. K. et al. The Emerging Nanomedicine Landscape. *Nat. Biotechnol.* **2006,** *24*, 1211–1217.

Wagner, V.; Dullaart, A.; Bock, A. K.; Zweck, A. The Emerging Nanomedicine Landscape. *Nat. Biotechnol.* **2006,** *24* (10), 1211–1217.

Wang, C.; Cheng, L.; Liu, Z. Upconversion Nanoparticles for Photodynamic Therapy and Other Cancer Therapeutics. *Theranostics* **2013,** *3* (5), 317.

Wang, H. N.; Fales, A. M.; Zaas, A. K.; Woods, C. W.; Burke, T.; Ginsburg, G. S.; Vo-Dinh, T. Surface-Enhanced Raman Scattering Molecular Sentinel Nanoprobes for Viral Infection Diagnostics. *Analyt. Chim. Acta* **2013,** *786*, 153–158.

Wang, H.; Mooney, D. J. Biomaterial-Assisted Targeted Modulation of Immune Cells in Cancer Treatment. *Nat. Mater.* **2018,** *17* (9), 761–772.

Wang, X.; Li, Y.; Wang, J.; Wang, Q.; Xu, L.; Du, J.; Yan, S.; Zhou, Y.; Fu, Q.; Wang, Y.; Zhan, L. A Broad-Range Method to Detect Genomic DNA of Multiple Pathogenic Bacteria Based on the Aggregation Strategy of Gold Nanorods. *Analyst* **2012,** *137*, 4267–4273.

Wang, Y.; Jin, M.; Chen, G.; Cui, X.; Zhang, Y.; Li, M.; .. Wang, J. Bio-Barcode Detection Technology and Its Research Applications: A Review. *J. Adv. Res.* **2019,** *20*, 23–32.

Wen, X.; Wang, L.; Liu, Z.; Liu, Y.; Hu, J. Intracranial Injection of PEG-PEI/ROCK II-siRNA Improves Cognitive Impairment in a Mouse Model of Alzheimer's Disease. *Int. J. Neurosci.* **2014,** *124* (9), 697–703.

Wickline, S. A.; Lanza, G. M. Nanotechnology for Molecular Imaging and Targeted Therapy. *Circulation* **2003,** *107*, 1092–1095. doi: 10.1161/01.CIR.0000059651.17045.77.

Wijesena, R. Top Six "Game Changing" Nanotech Developments for Diabetes Treatment, 2015.

Wu, J.; Ding, T.; Sun, J. Neurotoxic Potential of Iron Oxide Nanoparticles in the Rat Brain Striatum and Hippocampus. *Neurotoxicology* **2013,** *34*, 243–253.

Wu, M.; Zhang, D.; Zeng, Y.; Wu, L.; Liu, X.; Liu, J. Nanocluster of superparamagnetic Iron Oxide Nanoparticles Coated with Poly (Dopamine) for Magnetic Field-Targeting, Highly Sensitive MRI and Photothermal Cancer Therapy. *Nanotechnology* **2015,** *26* (11), 115102.

Wu, S.; Duan, N.; Shi, Z.; Fang, C.; Wang, Z. Simultaneous Aptasensor for Multiplex Pathogenic Bacteria Detection Based on Multicolor Upconversion Nanoparticles Labels. *Anal. Chem.* **2014,** *86*, 3100–3107.

Xia, T.; Kovochich, M.; Brant, J.; Hotze, M.; Sempf, J.; Oberley, T.; Nel, A. E. Comparison of the Abilities of Ambient and Manufactured Nanoparticles to Induce Cellular Toxicity According to an Oxidative Stress Paradigm. *Nano Lett.* **2006,** *6* (8), 1794–1807.

Xing, L. X.; Shi, Q. S.; Zheng, K. L.; Shen, M.; Ma, J.; Li, F.; et al. Ultrasound-Mediated Microbubble Destruction (UMMD) Facilitates the Delivery of CA19-9 Targeted and Paclitaxel Loaded mPEG-PLGA-PLL Nanoparticles in Pancreatic Cancer. *Theranostics* **2016,** *6*, 1573–1587.

Xingzhou, L.; Gong, Y. Design of Polymeric Nanofiber Gauze Mask to Prevent Inhaling PM2.5 Particles from Haze Pollution. *J. Chem.* **2015**, *60392*, 1–5.

Xu, G.; Zeng, S.; Zhang, B.; Swihart, M. T.; Yong, K. T.; Prasad, P. N. New Generation Cadmium-Free Quantum Dots for Biophotonics and Nanomedicine. *Chem. Rev.* **2016**, *116* (19), 12234–12327.

Xu, R. et al. An Injectable Nanoparticle Generator Enhances Delivery of Cancer Therapeutics. *Nat. Biotechnol.* **2016**, *34*, 414–418.

Xu, X. D.; Wu, J.; Liu, Y. L.; Saw, P. E.; Tao, W.; Yu, M. et al. Multifunctional Envelope-Type siRNA Delivery Nanoparticle Platform for Prostate Cancer Therapy. *ACS Nano.* **2017**, *11*, 2618–2627.

Xu, X.; Wu, J.; Liu, Y.; Saw, P. E.; Tao, W.; Yu, M.; Ayyash, D. Multifunctional Envelope-Type siRNA Delivery Nanoparticle Platform for Prostate Cancer Therapy. *ACS Nano* **2017**, *11* (3), 2618–2627.

Xue, X. J.; Wang, F.; Liu, X. G. One-Step, Room Temperature, Colorimetric Detection of Mercury (Hg2þ) Using DNA/ Nanoparticle Conjugates. *J. Am. Chem. Soc.* **2008**, *130*, 3244–3245.

Yaari, Z. et al. Theranostic Barcoded Nanoparticles for Personalized Cancer Medicine. *Nat. Commun.* **2016**, *7*, 13325.

Yin, T. H.; Wang, P.; Li, J. G.; Wang, Y. R.; Zheng, B. W.; Zheng, R. Q. et al. Tumor-Penetrating Codelivery of siRNA and Paclitaxel with Ultrasound-Responsive Nanobubbles Hetero-Assembled from Polymeric Micelles and Liposomes. *Biomaterials* **2014**, *35*, 5932–5943.

Yu, J.; Zhang, Y.; Wang, J.; Wen, D.; Kahkoska, A. R.; Buse, J. B.; Gu, Z. Glucose-Responsive Oral Insulin Delivery for Postprandial Glycemic Regulation. *Nano Res.* **2018**.

Yue, J.; Luo, S. Z.; Lu, M. M.; Shao, D.; Wang, Z.; Dong, W. F. A Comparison of Mesoporous Silica Nanoparticles and Mesoporous Organosilica Nanoparticles as Drug Vehicles for Cancer Therapy. *Chem. Biol. Drug Design* **2018**, *92* (2), 1435–1444.

Yurek, D.; Hasselrot, U.; Sesenoglu-Laird, O.; Padegimas, L.; Cooper, M. Intracerebral Injections of DNA Nanoparticles Encoding for a Therapeutic Gene Provide Partial Neuroprotection in an Animal Model of Neurodegeneration. *Nanomed.: Nanotechnol., Biol. Med.* **2017**, *13* (7), 2209–2217.

Zhang, Y.; Shih, I. M.; Wang, T. L.; Wang, T. H. A Novel Quantum Dot Based Nanoassay for High Resolution Copy Number Variation Detection and DNA Methylation Quantification. *Clin. Cancer Res.* October 1 **2010**, *16* (19 Supplement) A17. doi: 10.1158/DIAG-10-A17

Zhao, Y.; Xing, G.; Chai, Z. Nanotoxicology: Are Carbon Nanotubes Safe? *Nat. Nanotechnol.* 2008, *3* (4), 191.

Zhou, W. et al. Gold Nanoparticles for *in Vitro* Diagnostics. *Chem. Rev.* **2015**, *115*, 10575–10636. doi: 10.1021/acs.chemrev.5b00100.

CHAPTER 6

OVERVIEW AND EMERGENCE OF NANOBIOTECHNOLOGY IN PLANTS

SAHEED ADEKUNLE AKINOLA[1,2], RASHEED OMOTAYO ADEYEMO[1,3], and ISMAIL ABIOLA ADEBAYO[1*]

[1]*Department of Microbiology/Immunology, Faculty of Biomedical Sciences, Kampala International University, Western-Campus, Ishaka-Bushenyi, Uganda*

[2]*Food Security and Safety Niche, Faculty of Natural and Agricultural Sciences, North-West University, Mmabatho, South Africa*

[3]*Department of Paraclinical Sciences, University of Pretoria, Pretoria, South Africa*

Corresponding author. E-mail: ibnahmad507@gmail.com; Ismail.abiola@kiu.ac.ug

ABSTRACT

Manipulation and restructuring of metals into smaller size has effectively transformed the face of nano-size particles using biotechnological methods. This instigates a change in physicochemical and optical properties of valuable materials for better use. Outrightly, phyto-nanotechnological studies on production system of conventional plants allow for easy control of agrochemicals and their side effects. Likewise, the studies of nanoparticle on plants also help in easy manipulation of plant's biomolecules viz. proteins, activators, and nucleotides. Recently, considerable focus has been given to the use of nanotechnology in the discovery of novel particles of specific makeup from plants. Meanwhile, the interplay between plant responses and nanoparticles such as activity and uptake has helped to transform agroecosystem with direct and indirect actions on plants. This reveals the multifarious benefits of nanotechnology over the last decade, and the great promise ahead in the use of biotechnological sciences. Herewith, this mini-review is meant to give a survey on the studies of nanoparticle in plant sciences as a means to intensify its use in environmental- and industrial-based researches.

6.1 INTRODUCTION

The advancement in the search of eco-friendly synthesis routes to improve the benefits of biological systems such as plant tissues, microorganisms, and biomaterials that will help annul the constraints posed by the use of xeno-biotics and chemical compounds has led to the use of nanobiotechnology. Embracing the use of nanoparticles in plant engineering has really helped in the practice of an environment-friendly and long-lasting agroecosystem. Nanobiotechnology focuses mainly on the development of new methods using surface analytical tools to build materials and structures, with better understanding of the modification in their properties and production of smart devices and functional materials (Patra and Baek, 2014). The special properties possessed by nanomaterials enable its use in broad spectrum of newly developed technologies viz. biotechnology, electro-optical devices, chemical industries, catalysts, etc. Nanobiotechnology considered to be a modern concept and now most widely studied in science has formerly been used in the ninth century, where artisans are found using gold and silver to make utensils and pots more attractive and glittering. Meanwhile, the first description of nanoparticle was done by Michael Faraday, showing the effect and relation of metals to light. Besides, the term "nanotechnology" was first coined and used in the year 1959 by Richard Feynman and marked the origin of modern nanobiotechnology (Wang et al., 2016).

The use of plant in nanotechnology has been one of the benign, nontoxic materials for the production of eco-friendly product. Nanomaterials using natural products of green technology have better stability and diverse natures because of the handling and the way they are synthesized. The intensified use of nanoparticles on plants for horticultural and agricultural basis increased gradually over the last decade. Plant biologists have been attracted toward the use of carbon nanotubes (CNTs), due to its intrinsic ability as molecular transporters that penetrate cell membrane of mammalian cells and invari-ably used as gene and drug-delivery vehicles. The mechanism of action of CNTs and their ability to penetrate thick cell walls have been demonstrated by different researchers such as single-walled nanotubes on tobacco cells (Liu et al., 2011) and multiple-walled nanotubes on seed coat of tomato and tobacco plants (Khodakovskaya et al., 2011). Serag et al. (2011a,b) also worked on the penetrating power of CNTs and bring about different mechanisms of cellular penetration into plants. These discoveries have potential to turnaround agricultural practices and be of good help in attaining sustainable plant growth and food safety (Serag et al., 2013).

In this mini-review, we intend to give an overview of multifaceted impact of nanobiotechnology on the study of plant and its help toward enhancing agroecosystem functioning and advancement in plant technology at phenotypic and genomic levels.

6.2 NANOPARTICLE–PLANT DELIVERY OF BIOACTIVE MOLECULES

Several researchers have reported the use of nanoparticles as a means to drive bioactive substances into plant cells. Compounds such as protein, activator, and nucleotides were inserted into plant cells using different methods such as NPs with carbon-coated magnets (González-Melendi et al., 2008; Corredor et al., 2009), magnetic virus-like NPs (Huang and et al., 2011), quantum dots method (Etxeberria et al., 2012), single- or multi-walled CNTs (Liu et al., 2009; Serag et al., 2011a; Khodakovskaya et al., 2011), gold (Au) NPs (Wu et al., 2011; Martin-Ortigosa et al., 2012). According to a research conducted by Torney et al. (2007), mesoporous silica nanoparticle (MSNP) was used to insert DNA and its activator into the cell of isolated plant and healthy leaves of tobacco. This was done by using MSNP to release biomolecules and initiate gene expression in the plant with the aid of uncapped trigger coated with gold attached to the silica NP. In addition, MSNP can also be used to dispatch enzymes and proteins into plant cells, making plant readily available for genome expression and biochemical analysis (Martin-Ortigosa et al., 2014; Guo et al., 2019). In the case of genetically modified plants (GMP), plants have been modified to serve specific functions such as ability to induce resistance against pathogen and stress as a result of plasmid delivery containing different genes based on target function. MSNP can also be used in genome editing as well as to avoid integration of DNA into genome so as to modify traits of plants for future generations (Medintz et al., 2005; Michalet et al., 2005). Nevertheless, NPs possess several benefits ahead of conventional methods of delivery. Those functions include intracellular imaging and labeling; ensure local restructuring of individual tissue, organ, or cells; require a far less DNA for detection of expression; highly efficient and easily operated compared to the conventional methods (Hischemoller et al., 2009; Koo et al., 2015).

6.3 APPLICATION OF NANOTECHNOLOGY TO PLANTS

Manipulated nanoparticles are known to be engineered with distinctive physical and chemical properties such as enhanced reactivity, modified surface structure, and very small. These unique properties are often time used in different commercial and consumer products, drug delivery, microelectronics, catalysts, semiconductors, household products, and the likes. Nanotechnology in plants serves several other functions viz. development of engineered crops structured with the use of nanomaterials of multifunction attributes, specific targeted and time-controlled molecules (Nair et al., 2010). Apart from the specificity, NP also helps in gradual delivery of needed biomolecules of agrochemicals such as herbicides and fertilizer to plant as required. With this, limitations associated with the use of agrochemicals are reduced on environment and plants. Withal, for the control of plant metabolism and genetic modification of plants, target delivery of phytoactive molecules, proteins, and nucleotides helps in easy modification of plant (Nair et al., 2010).

Although the use of NP and nanotechnology in plant science has received considerably low interest so far, in spite of that, they are still found useful in medical and pharmaceutical settings to make delivery of active substances and drugs to a specific site in microbes (especially those with plant growth-promoting traits). This invariable serves as another important means needed to maintain a sustainable agroecosystem and improved yield of farm produce. A considerable number of researchers have tried to introduce this technique (engineered nanoparticle) in human systems to help target a specific drug delivery or eradicate body anomalies such as cancer and abnormal body glucose index that are due to genetic disorder (Gu et al., 2011; Wang et al., 2016).

6.3.1 QUANTIFICATION OF AGROCHEMICAL APPLICATION ON PLANTS

The proposed use of nanofertilizers to help stimulate a better output of plants has been the focus point of botanists over the last decade. Botanists now adopt NP fertilization process to help quantify agrochemical usage. Normally, the release of fertilizer for plant use should be quantified based on plant's need. Gradual introduction of fertilizer prevents loss of soil nutrients due to volatilization (conversion of urea to ammonia) and eutrophication (runoff) of soil nutrient (Hossain et al., 2008; De Rosa et al., 2010). Using aforementioned methods such as quantum dots and MSNP, there is easy control of soil chemicals and they also help to sustain gradual release of chemicals, as needed by plants, and as such agrochemicals are readily available for plants consumption (Liu and Lal, 2014). Apart from fertilizers, pesticides, insecticides, and other agrochemicals are also needed at a considerable amount to ensure effective usage. This helps in plant protection through gradual integration into the cuticular lipids of the insect, and subsequently leading to dehydration. Meanwhile, the choice of NP capping is also very important for easy and proportionate distribution agrochemicals. To understand better the mechanism of absorption and translocation of nanomaterials in plants, Avellan et al. (2019) studied the effect of coating AuNP with either citrate or polyvinylpyrrolidone (PVP). In both combinations, PVP–AuNP introduced to wheat from the leave was well absorbed than citrate-AuNP, but it was not proportionally distributed across the parts of the plant (via leaf–root–rhizosphere transport) compared to citrate-AuNP (Table 6.1).

6.3.2 NANOTECHNOLOGY—A VIABLE MEANS TO IMPROVE PLANT TOLERANCE TO STRESS

Environmental stress caused by either abiotic or biotic factors is strenuous soil anomalies creating setback to the development of plants. Plant resistance to these factors reduces their output and incites the need to develop nanofertilization methods with defense mechanisms that will help in the modification of their gene expression. Modified molecular and biochemical pathways coupled with the effect of antioxidant enzymatic activities turned on by nanobiotechnology will help ameliorate threat posed by soil stress (Saxena et al., 2016).

It has been estimated that one-fifth of the world cultivated land is suffering from salinity stress. Meanwhile, because most plants are halophytes with low salt tolerance, saline soil cripples physiological and biochemical activities of plants, hence, reducing their productivity. Protein biosynthesis, lipid metabolisms, photosynthesis, etc. have been reported as vital processes affected by salinity stress (Rawson et al., 1998).

The use of NPs has been the focus point over recent times with intriguing use in different sectors viz. agriculture, pharmaceutics, etc. In agricultural settings, nanofertilization is used to curb the menace posed by soil stress without accumulating toxic materials, because of its stringent use in regulating the quantity of nutrient applied to the soil (usually based on soil requirement) (Saxena et al., 2016).

6.3.2.1 EFFECT OF SILVER NANOPARTICLES AGNPS ON SOIL STRESS

Silver nanomaterials are essentially needed in a wide range of applications such as commercial products, the use and making of pesticides, coating of household products, food packaging, drug delivery in medicine, electronics, and the host of others (Kim et al., 2012; Shelar and Chavan, 2015). Applied biotechnology using silver nanomaterial is widely used by botanists for crop improvement in agricultural sector. Different researchers positively reported the use of AgNPs on plants, maize, tomato, pear, and cowpea (Karami et al., 2016). The tripartite effect of AgNPs and other selected plant growth-promoting rhizobacteria, namely, *Pseudomonas* sp. and *Bacillus cereus* have been reported by Khan and Bano (2016). Coapplication of these inoculants instigated bioremediation of heavy metals viz. Pb, Cd, Ni, and improved growth of plant (Table 6.1). The test on specific antioxidant enzymes and the ability of AgNPs to mitigate stress induced by heavy metals was reported using *Brassica juncea* in an in-house experiment with inference such as reduced level of proline and better growth of plant noticed across the planting period (Sharma et al., 2012).

6.3.2.2 EFFECT OF SILICON NANOPARTICLES (SINPS) ON SOIL STRESS

Silicon is known to be abundant in the earth crust and its effect in improving the growth and development of agroecosystem is well documented. The uses of SiNPs in mitigating natural or self-induced stress and soil toxicity have been recognized by most researchers but, despite that, the mechanism of action of this useful mineral is yet to be unveiled (Zarafshar et al., 2014). SiNPs easily spread through the cuticle of plants and has the ability to revitalize plant's physiological features viz. water transloca-tion, turgor pressure, xylem humidity, etc. The effect of SiNPs on plant under abiotic stress instigates features such as enhanced nutrient uptake, precipitation of toxic metallic ions, and intensified release of antioxidant enzymes. As noticed in higher plants, features accumulated as a result of SiNPs application will help reinvigorate plant's aptness to different soil stressors (Ashkavand et al., 2015) (Table 6.1). Due to the strength of SiNPs to spread through plant parts, plant-induced resistances are easily triggered with immediate and vast response to soil stress and enhanced phenotypic changes (Zarafshar et al., 2014). The quick response of SiO_4-NPs to soil anomalies have been attributed to easy formation of fine layers at plant cell wall compared to the use of other silicon-related compounds (H_4SiO_4, Na_2SiO_3, etc.). Withal, SiNPs are still without toxic effects as noticed when

Overview and Emergence of Nanobiotechnology in Plants 179

introduced to pear seedling cultivated in a water-stressed soil (Zarafshar et al., 2014). Likewise, in a study reported by Javad et al. (2014) on maize, silicon nanofertilization on different growth stages of *Zea mays* L consistently showed high phenotypic characteristics such as elongated stem and root and enhanced activities photosynthetic pigment as compared to their control (Table 6.1). Even though SiNPs have tremendous contribution in alleviating different soil stressors, little is known about plant interaction with nonmaterial components used in the capping of SiNPs (Saxena et al., 2016).

6.3.2.3 EFFECT OF ALUMINUM NANOPARTICLES (AL_2O_3NPS) ON SOIL STRESS

The physical and chemical properties of aluminum viz. high stiffness and strength, good heat conductivity, and the likes make aluminum a good metallic nanomaterial. But in spite of all its attributes, little work has been done on its use to mitigate soil stress. Rather, researches conducted on the use of aluminum compounds as NPs described AlNPs as a foe rather than friend to the development of agroecosystem. According to Yanık and Vardar (2015), Al_2O_3NPs instigate phytotoxicity in plants leading to alterations in cellular, morphological, and molecular activities of plants.

6.3.2.4 EFFECT OF ZINC NANOPARTICLES (ZNONPS) ON SOIL STRESS

According to Baybordi (2005), nanofertilizers made from micronutrients help lessen the effect of environmental stress such as temperature, salinity, drought, and heavy metals on the growth of plant. Amidst other micronutrients, zinc has been proven to be very important plant growth-promoting mineral and help activate important metabolic activities in proximal plant. As shown on Table 6.1, ZnO-NPs help improve biomass production, tolerance to drought and also promote sprouting by increasing auxin IAA level at the root of plant. A research conducted by Adhikari et al. (2015) showed zinc oxide nanoparticle enhance the growth of *Z. mays* grown under different soil stressors by exhibiting good features such as increased shoot, root, and height of plant. Also, in another study conducted on lettuce and rice exposed to cadmium (Cd) stressed soil, synergistic effect of ZnO-NPs and biochar instigated a decrease in Cd concentration in the soil and a better yield of plant noticed either in the presence or the absence of biochar compared to the control (Ali et al., 2019) (Table 6.1). At a very small amount, ZnO-NPs tend to improve crop growth under the influence of different stressors with little toxicity and hence, considered eco-friendly material for agricultural enhancement.

6.3.2.5 EFFECT OF TITANIUM NANOPARTICLES (TIO$_2$NPS) ON SOIL STRESS

The oxidative–reduction reaction exhibited by TiO$_2$NPs during phytoremediation enables production of hydroxide and superoxidase radicals in the presence of light. But during phytochemical reaction, light sterilization helps improve plant growth and survival. This and many other biochemical activities have been reported by scientists on the use of titanium-coated nanoparticles to improve plant growth and alleviate soil stress (Hu et al., 2020). The research conducted by Jaberzadeh et al. (2013) on wheat grown under drought condition was reported to be influenced by the activities of inoculant comprising TiO$_2$NPs and a host of activities such as water retention rate, improved growth of plant, and the likes were noticed over the growth stages. Likewise, Lei et al. (2008) also showed TiO$_2$NPs induced tolerance of *Spinacia oleracea* to antioxidant stress by regulating several processes such as hydrogen peroxide secretion, reduced superoxidase radicals, and increased enzymes involved in antioxidant activities (catalase, superoxidase dismutase [SOD] and ascorbate peroxidase) of light chemical reaction in plant chloroplast. In addition, SOD activity of onion seedling and a better growth of plant was also noticed with gradual application of TiO$_2$NP in a research conducted by Laware and Raskar (2014) (Table 6.1).

6.4 PLANT'S USE AS GREEN-NP ASSEMBLER

Due to the plethora nature of plants, they are easily used as natural compound to assemble metallic NPs. Plants naturally tolerate toxicity of metallic materials emanating from soil, and also potent enough to withstand hostility of the environment. They maneuver redox states of these minerals to change their toxicity with the aid of enzymes and sequestering proteins. After several reactions and mixing, these compounds are transformed into nontoxic nanoscale fragments (Fig. 6.1). In the case of AuNPs, plants extract is used as an assembler to limit its toxicity and create a niche for its use in greener synthetic application (Oza et al., 2020). With the use of compounds such as alkaloids, ascorbic acid, citric acid, terpenoids, and quinones possessed

TABLE 6.1 Overview of the Use of NPs in Agroecosystem.

Nanoparticles	Plant	Inference	References
TiO$_2$-NPs	Lettuce (*Lactuca sativa* L.)	The use of TiO$_2$-NPs significantly increases the physiological traits of lettuce viz. shoot and root, dry biomass, chlorophyll, and carotenoid content.	Hu et al. (2020)
ZnO-NPs + biochar	Rice (*Oryza sativa* L.)	In a soil contaminated with cadmium (Cd), synergistic effect of ZnO-NPs + biochar decreased Cd concentration and enhanced Zn availability in the root of plant. Besides, physiological properties were also improved in the presence or absence of biochar.	Ali et al. (2019)
PVP-AuNPs and citrate-AuNPs	Wheat (*Triticum aestivum*)	In a research designed to show the effect of different NPs coating. Translocation and coating chemistry of PVP and citrate of AuNPs on wheat were compared. The result showed PVP-coated AuNPs had better absorption but less translocation on wheat compared to citrate coating.	Avellan et al. (2019)
Ag$_2$S-NP	Soybean (*Glycine max* L.)	In an experiment conducted on Ag$_2$S containing NP used on soybean, Ag$_2$S-NP was observed loading excessive Ag in the food web. It was observed that the used NP preserves and accumulates Ag in the food web, for example, in snail via root uptake.	Dang et al. (2019)
Fe-hematite-NP	Chicken pea (*Cicer arietinum*), Black bean (*Phaseolus vulgaris*), etc.	In Fe-hematite-NP using seed presoaked method, visible increase in physiological features, namely, shoot and root length, etc. was noticed across planting period of plants.	Boutchuen et al. (2019)
Mesoporous silica nanoparticle (MSNP)	Tomato (*Lycopersicon esculentum*)	To control *Tuta absoluta* moth (tomato leaf-miner) using innovative methods, pDNA-MSNP was introduced into the plant through spray and injection at the abaxial surface of plant and shoot. Using RT-PCR expression of GUS, the DNA was transcribed and translated into the plant. To test the effectiveness of the complex of pDNA-*GUS*-MSNP and pDNA-Cry1Ab-MSNP, modulated plant species was used to control *T. absoluta*. This research showed easy manipulation of DNA using MSN-mediated gene transient transformation of tomato.	Hajiahmadi et al. (2019)

TABLE 6.1 *(Continued)*

Nanoparticles	Plant	Inference	References
AgNPs + PGPR (*Pseudomonas* sp., *Bacillus cereus*)	Maize (*Zea mays* L.)	The coapplication of AgNP and PGPR organisms induces bioremediation of heavy metals such as Pb, Cd, Ni, and better change in physiological properties of plant.	Khan and Bano (2016)
TiNPs	Wheat (*T. aestivum*)	The effect of foliar application of TiNPs on wheat grown under drought was reported over planting period. The use of titanium oxide as a nanofertilizer increased almost all the agronomic traits, including starch content and glutin.	Jaberzadeh et al. (2013)
Au-MSN	Onion (*Allium cepa*)	Co-delivery of biolistic mediated DNA and protein to plant cell. This provides a better insight into the use of Au-MSN for applied and fundamental research.	Martin-Ortigosa et al. (2012)
Au-MSN	Tobacco (*Nicotiana tabacum*) and maize (*Z. mays* L)	Also, a biolistic delivery of DNA using MSN containing gold nanorods.	Martin-Ortigosa et al. (2012)
CuO-NPs	Soybean (*G. max*) and chicken pea (*C. arietinum*)	The use of CuO-NPs inhibited the growth of test plant which was attributed to accumulation of copper. This can help in search of new directive of research using nano-science.	Adhikari et al. (2012)
SiNPs	Hawthorn (*Crataegus* sp.)	Treatment of *Crataegus* sp. with SiNPs instigate drought resistance with enhanced physiological indexes such as xylem water potential, increased biomass of plant, etc.	Ashkavand et al. (2015)
SiNPs	Maize (*Z. mays* L.)	The application of SiNPs to seed germination induces fascinating physiological features such as relative water content of plant, shoot elongation, increased photosynthetic pigment content, etc. that enables easy tolerance to drought stress.	Javad et al. (2014)
ZnO-NPs	Maize (*Z. mays* L.)	ZnO-NPs introduced into maize enhance the growth, root, shoot and height of plant.	Adhikari et al. (2015)

TABLE 6.1 *(Continued)*

Nanoparticles	Plant	Inference	References
CuO-NPs	Raddish (*Raphanus sativus*), perennial ryegrass, etc.	Nanofertilization using CuO-NP instigates DNA damage by accumulating oxidatively modified mutagenic DNA lesion.	Atha et al. (2012)
ZnO-NPs	Mung (*Vigna radiata*), Gram (*C. arietinum*)	Absorbed NPs help effect change in physiological characteristics of plant such as increased growth of shoot, root, and plant height.	Mahajan et al. (2011)
Multiwall carbon nanotube	Tomato (*L. esculentum*)	Enhanced growth of plant and plant promoting genes such as stress-related genes, water channel genes (LeAqp2 gene) produced by tomato leaves.	Khodakovskaya et al. (2011)
NP template assembly and mutant cowpea	Cowpea (*Vigna unguiculata*)	The synergistic effect of NP templating and mutant plant showed ability to control the size of chlorotic mottle virus used as the mechanism of invasion.	Aniagyei et al. (2009)
TiNPs	Spinach (*Spinacia oleracea*)	The use of titanium oxide nano-anatase to promote spinach growth under antioxidant stress was reported. The research showed that treatment of the plant with TiNPs significantly decrease superoxidase radical accumulation, H_2O_2, and malonyldialdehyde content. Meanwhile, activities of superoxidase dismutase (SOD), catalase, ascorbate peroxide, guaiacol peroxidase, etc. were increased.	Lei et al. (2008)

by plants, the effects of gold toxicity and its ions are easily reduced and recapped form of Au transformed into useful nanomaterial.

Likewise, crapping agents also help to stabilizer gold particles to avoid aggregation (Saxena et al., 2016; Oza et al., 2020). The leaf extract of different plants viz. *Terminalia catappa* (Ankamwar, 2010), *Dracocephalum kotschyi* (Dorosti and Jamshidi, 2016), *Olea europaea* (Khalil et al., 2012), *Mangifera indica* (Philip, 2011), *Abutilon indium* (Mata et al., 2016), *Sesbania grandiflora* (Das and Velusamy, 2014), *Cacumen platycladi* (Wu et al., 2013), *Magnolia kabus* (Song et al., 2009), *Cinnamomum zeylanicum* (Smitha et al., 2009), *Euphorbia hirta* L. (Annamalai et al., 2013), *Azadirachta indica* L (Bindhani and Panigrahi, 2014), *Ficus benghalensis* (Francis et al., 2014), *Ocimum sanctum* (Manoj and Vishwakarma, 2015), and the likes have been useful for nanoparticle synthesis (Table 6.2). The conversion of color, toxicity, and other features during transformation were attributed to the activities of plant's chemical components such as thiamine, terpenoids, and flavonoids. Meanwhile, these compounds (terpenes, etc.) can also be used for capping, stabilizing and as reducing agents.

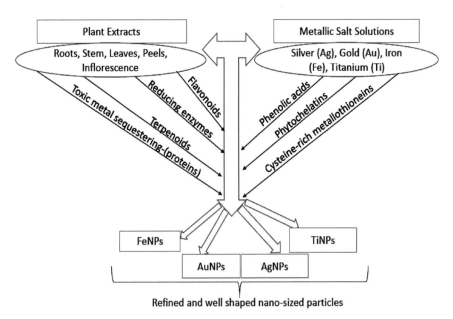

FIGURE 6.1 Sketch illustrating the mechanisms of metal nanoparticle modification using different plant parts.

TABLE 6.2 Overview of Plant Use as Green-NPs Factories and the Potential Use of Refined Form in Medicine, Pharmaceutics, and Crop Enhancement.

Nanoparticles (NPs)	Plant extract	Inference	References
AuNPs	*Pongamia pinnata* leaf extract (PLE)	PLE refined AuNPs was used against human cervical cancer cell line. Synthesized AuNPs induced fascinating results such as altered morphology of the cells, quick recovery from the wounds, etc. Meanwhile, in spite of PLE-AuNPs toxicity against cancer cell, no toxic effect was noticed to embryonic kidney (HEK293) and many other organs.	Khatua et al. (2019)
Fe_3O_4NPs	Corn (*Zea mays*)	Corn extract synthesized FeNPs produces proteasome inhibitory activities (cancer treatment) and antimicrobial effect against multidrug-resistant (MDR) organisms.	Patra et al. (2017)
AgNPs	Purple heart plant (PHP) (*Tradescantia pallida*) extract	Silver NPs synthesized using PHP extract was found effective against enteric and other pathogenic organisms.	Hasnain et al. (2019)
Fe_3O_4NPs	Maize (*Z. mays* L.) + Chinese cabbage (*Brassica rapa* L.) (wastes)	The synergistic effect of wastes derived from the two plants synthesized a nano-sized FeNP with superpragmatic nature. The refined form of FeNPs showed both antioxidative and antimicrobial effects when used against pathogenic fungi and bacteria.	Patra and Baek (2017)
AuNPs	Watermelon (*Citrullus lanatus*) aqueous extract of the rind	The biopotential of AuNPs refined from aqueous extract of watermelon revealed that the synthesized form of AuNPs had antioxidant, antibacterial, and proteasome inhibitory activities against pathogenic organisms (*Listeria monocytogenes*, *Bacillus cereus*, *Salmonella typhimurium*) and it is also viable in the treatment of cancer.	Patra and Baek (2015)
AgNPs	*Citrus limon* + *Citrus sinensis* + *Citrus limetta* (Peel)	Extract from the combination of the peels of citrus plants was used as bioreductant of AgNPs. Tripartite effect showed multifarious activities against different gram +ve and gram −ve pathogenic organisms. Likewise, cytotoxic effect on human lung cell line cancer (A549) was also revealed.	Ahmed et al. (2018); de Barros et al. (2018); Veeraputhiran (2013)

TABLE 6.2 *(Continued)*

Nanoparticles (NPs)	Plant extract	Inference	References
AgNPs	Pomegranate + orange + banana + apple (mixed fruit peel)	The effect of refined form of AgNPs using peels of multiple fruit was studied using UV-visible spectroscopy, XRD, IR and TEM. The antimicrobial and cytotoxicity assay of the synthesized formed showed a considerable reduction in the load of human pathogenic organisms (using agar well diffusion method), and a potential use in the treatment of human breast cancer (MCF-7).	Naganathan and Thirunavukkarasu (2017)
AuNPs	*Abutilon indium* (leaves)	The polyphenolic component of *A. indium* leaf extract was used in the stabilization of AuNPs. The cytotoxicity assay of the refined form showed a viable effect against human colon cancer cells, hence, useful in medicine for the treatment of colon cancer.	Mata et al. (2016)
AuNPs	*Dracocephalum kotschyi* (leaves)	*D. kotschyi* leaf extract mediated AuNPs was refined and tested to determine the cytotoxic and anti-cholinesterase effect. Synthesized AuNPs had effect against cervix cancer (HeLa) and human leukemia (K562). Likewise, it showed satisfactory effect against typed organisms such as *Escherichia coli*, *Pseudomonas aeruginosa*, *Proteus vulgaris*, etc.	Dorosti and Jamshidi (2016)
AgNPs	*Alpinia katsumadai* (seed)	The cytotoxicity, antioxidant and antimicrobial activities of refined AgNPs using *A. katsumadai* seed extract was assessed using EDX, FT-IR, XRD UV-spectrometry, and the likes. The synthesized form of AgNPs showed significant effect on human pathogenic organisms, cytotoxic activities and distinct free radical scavenging.	He et al. (2017)
AgNPs	*Azadirachta indica* (leaves)	Eco-friendly, nontoxic AgNPs was synthesized using *A. indica* leaf extract with refined form of the nanomaterial showing antibacterial effect against pathogens such as *Staphylococcus aureus* and *E. coli*.	Ahmed et al. (2016)

TABLE 6.2 (Continued)

Nanoparticles (NPs)	Plant extract	Inference	References
AgNPs	*Tinospora cordifolia* (leaves)	*T. cordifolia* leaf extract was used as a capping and reducing agent to synthesize eco-friendly AgNPs. The refined form of AgNPs possessed antibacterial and antioxidant activities when tested against pathogens such as *S. aureus* and *Klebsiella* sp. this shows the significance of *T. cordifolia* refined AgNPs in biomedicine.	Selvam et al. (2017)
AgNPs	*Pistacia atlantica* (seed)	Refined AgNPs using extract from the seed of *P. atlantica* yield a considerable nano-size particle and could also be adopted in the treatment of bacterial infections because of its antibacterial properties.	Sadeghi et al. (2015)
AuNPs	*Sesbania grandiflora* (leaves)	Well dispersed and spherical form of AuNPs was formed using *S. grandiflora* leaf extract with refined form reported having catalytic effect on methylene blue and could be useful in degrading chemical dyes.	Das and Velusamy (2014)
AuNPs	*A. indica* (leaves)	Eco-friendly, nontoxic AuNPs were formed using *A. indica* leaf extract as reducing agent. Well refined particle size Au was suggested to be useful in biomedical applications.	Bindhani and Panigrahi (2014)
AuNPs	*Ficus benghalensis* (leaves)	Aqueous gold chloride was reduced using leaf extract of *F. benghalensis* with refined form of AuNPs found effective against plant and human pathogenic organisms.	Francis et al. (2014)
AuNPs	*Hypericum hookerianum* (leaves)	Extracts from leaves of in vitro culture *H. hookerianum* was used as a biogenic approach to refine AuNPs. Synthesized NPs were reported viable against MDR human pathogens such as *P. aeruginosa* and *S. aureus*.	Manoj and Vishwakarma (2015)
AgNPs	*Cocos nucifera* (inflorescence)	Inflorescence of *C. nucifera* was used to synthesize AgNPs with considerable reduction of its particles and antimicrobial efficacy against human pathogens viz. *Klebsiella pneumoniae*, *Bacillus subtilis*, *P. aeruginosa*, *Salmonella paratyphi*.	Mariselvam et al. (2014)

TABLE 6.2 (Continued)

Nanoparticles (NPs)	Plant extract	Inference	References
AuNPs	*Cacumen platycladi* (leaves)	The synthesis of polydispersed fraction of AuNPs using *C. platycladi* leaf extract with size and shape separation achieved using density gradient centrifugation and agarose gel electrophoresis, respectively.	Wu et al. (2013)
AuNPs	*Euphorbia hirta* L. (leaves)	NP size and refined form of AuNPs achieved using *E. hirta* leaf extract ranging between 6 and 71 nm. Refined form of AuNPs reported viable against *K. pneumoniae, E. coli* and *P. aeruginosa.*	Annamalai et al. (2013)
AuNPs	Olive (leaves) + hot water	Hot water and olive leaf extract was used in reshaping AuNPs. Modulated form of the refined NPs was reported to be useful in the therapeutic treatment of human and plant pathogens.	Khalil et al. (2012)
CuNPs	*Calotropis procera* L. (latex)	Biosynthesis of nano-size CuNPs (15 ± 1.7 nm) using *C. procera* latex. Modulated form of CuNPs was effective against cell lines such as A549, BHK21, and HeLa, hence, this shows *C. procera* synthesized CuNPs could be used in cancer treatment.	Harne et al. (2012)
AgNPs	*Trachyspermum ammi* + *Papaver somniferum* (seeds)	Synergistic effect of TA + PS seed extract was used in the formation of biocompatible AgNPs. Constituents of extracts such as codeine, morphine, P-cymene, and thymol derived from the coalescence of the two plants were involved in the restructuring of the compound.	Vijayaraghavan et al. (2012)
AgNPs	*Mangifera indica* (leaves)	The leaf extract of *M. indica* was used to synthesize AgNPs and could be used in agriculture, cosmetics, and medicine.	Philip (2011)
CuNPs	Soybean—*G. max* (seed)	The reduced, uniform, and stable forms of CuNPs were formed using soybeans extract and could be useful in fundamental applications.	Guajardo-Pacheco et al. (2010)
AuNPs	*Terminalia catappa* (leaves)	A reduced, nontoxic, and stabilized form of AuNPs were synthesized using leaf extract of *T. catappa* with the addition of chloroauric acid. This gave the nanomaterial a better outlook with formulated form ranging between 10 and 35 nm.	Ankamwar (2010)

TABLE 6.2 *(Continued)*

Nanoparticles (NPs)	Plant extract	Inference	References
AgNPs	*Carica papaya* (leaves)	In this study, synthesized AgNPs using extracts from the leaf of *C. papaya* were reported highly toxic against MDR human pathogens using standard disc diffusion method.	Jain et al. (2009)
AuNPs	*Magnolia kabus + Diopyros kaki* (leaves)	In this research, Mk + DK leaf extract was used to synthesize gold nanoparticles. The synergistic effect of the two extracts yielded a stable AuNPs that is easily used in agroecosystem, biomedicine, pharmaceutics and in making of household appliances.	Song et al. (2009)
AgNPs	*Jatropha curcas* (seed)	An eco-friendly AgNPs was synthesized using extracts from the seeds of *J. curcas* with refined particles reported to be useful in diverse applications and fundamental studies	Bar et al. (2009)
AuNPs	*Cinnamomum zeylanicum* (leaves)	In this study, *C. zeylanicum* leaf extract was used as a reducing agent for the synthesis well-sized, refined and useful AuNPs for industrial application.	Smitha et al. (2009)

Amongst noble metal NPs, silver is very important with diverse properties such as antimicrobial activity, chemical stability, and catalytic activity (Oza et al., 2020). A whole lot of researchers adopted AgNPs as antimicrobial agent inhibiting pathogenic organisms belonging to the group of fungi, virus, and bacteria. This increased AgNPs usage in biomedical field, agroecosystem, cosmetics, and food industries. Most especially in the treatment of cancerous cells, AgNP has helped in the reduction/spread of cancer cells. The compartmentalization and chelate formation in plants enable detoxification of heavy metals and also play a significant role in bioaccumulation of the ions (Fig. 6.2). Meanwhile, properties such as alkaloids, polysaccharides, protein secondary metabolites, and amino acids help in the reduction of silver particles (Mata et al., 2016). Besides, through tolerance and ability of plant fractions to avoid manipulations, peptides are expressed on the cell surface of metallic ions by binding to the metal to increase their stability (Mata et al., 2016). Nonetheless, different plant extracts have been extensively used to assemble other elements such as copper, zinc, iron, titanium, nickel. Numerous benefits have been deduced from the use of extracts of plant parts such as leaves (Jain et al., 2009; Philip, 2011; Selvam et al., 2017; Ahmed et al., 2018; Khatua et al., 2019), seeds (Bar et al., 2009; Vijayaraghavan et al., 2012; He et al., 2017), fruit and peel (Patra and Baek, 2015; de Barros et al., 2018; Hasnain et al., 2019), and other parts (Mariselvam et al., 2014; Naganathan and Thirunavukkarasu, 2017) adopted as nano-factory to restructure useful metals to an eco-friendly, easily used-up nano-sized particles (Table 6.2).

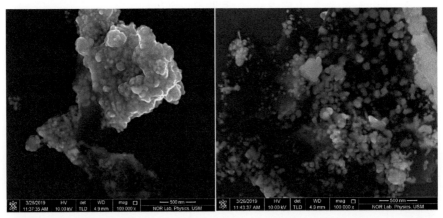

FIGURE 6.2 Refined silver nanoparticles using stem bark extract of *Detarium microcarpum*.

6.5 CONCLUSION AND FUTURE PROSPECT

Apart from agroecosystem that is just gaining ground in the use of nanotechnological methods, the advancement in industrial use of nano-size materials is at its apex over the last two decades. Meanwhile, over the recent times, the alarming increase in the world population with little to feed on has placed immense strain on agroecosystem and there is need to boost farm output. Nevertheless, hazard-free nanotechnology that incorporates either plant or microbes as reducing or capping agents will help ameliorate soil anomalies, reduce the negative effect of synthetic fertilizers, and enhance plant growth. Besides, to maximize the use of nano-sized particles in crop improvement, it is pertinent to understand plant–NPs interactions at both genetic and cellular levels to help reduce phytotoxic effects and make judicious use of the method for food safety.

AUTHORS' CONTRIBUTION

All the authors did sorting and collation of required publications needed for the manuscript. S.A.A. wrote the first draft of the manuscript. R. O. A. and I. A. A. critically assessed it and all authors approved the article for publication.

KEYWORDS

- **phyto-nanotechnological studies**
- **manipulation of plant's biomolecules**
- **discovery of novel particles**
- **plant responses and nanoparticles**
- **agroecosystem**
- **multifarious benefits**

REFERENCES

Adhikari, T.; Kundu, S.; Biswas, A. K.; Tarafdar, J. C.; Rao, A. S. Effect of Copper Oxide Nano Particle on Seed Germination of Selected Crops. *J. Agric. Sci. Technol.* **2012,** *2* (6), 815.

Adhikari, T.; Kundu, S.; Biswas, A.; Tarafdar, J.; Subba Rao, A. Characterization of Zinc Oxide Nano Particles and Their Effect on Growth of Maize (*Zea mays* L.) Plant. *J. Plant Nutr.* **2015**, *38* (10), 1505–1515.

Ahmed, S.; Kaur, G.; Sharma, P.; Singh, S.; Ikram, S. Fruit Waste (Peel) as Bio-Reductant to Synthesize Silver Nanoparticles with Antimicrobial, Antioxidant and Cytotoxic Activities. *J. Appl. Biomed.* **2018**, *16* (3), 221–231.

Ahmed, S.; Saifullah, Ahmad, M.; Swami, B. L.; Ikram, S. Green Synthesis of Silver Nanoparticles Using *Azadirachta indica* Aqueous Leaf Extract. *J. Radiat. Res. Appl. Sci.* **2016**, *9* (1), 1–7.

Ali, S.; Rizwan, M.; Noureen, S.; Anwar, S.; Ali, B.; Naveed, M.; Abd_Allah, E. F.; Alqarawi, A. A.; Ahmad, P. Combined Use of Biochar and Zinc Oxide Nanoparticle Foliar Spray Improved the Plant Growth and Decreased the Cadmium Accumulation in Rice (*Oryza sativa* L.) Plant. *J. Environ. Sci. Pollut. Res.* **2019**, *26* (11), 11288–11299.

Aniagyei, S. E.; Kennedy, C. J.; Stein, B.; Willits, D. A.; Douglas, T.; Young, M. J.; De, M.; Rotello, V. M.; Srisathiyanarayanan, D.; Kao, C. C. Synergistic Effects of Mutations and Nanoparticle Templating in the Self-Assembly of Cowpea Chlorotic Mottle Virus Capsids. *J. Nano Lett.* **2009**, *9* (1), 393–398.

Ankamwar, B. Biosynthesis of Gold Nanoparticles (Green-Gold) Using Leaf Extract of *Terminalia catappa*. *J. Chem.* **2010**, *7* (4), 1334–1339.

Annamalai, A.; Christina, V.; Sudha, D. e. a. Green Synthesis, Characterization and Antimicrobial Activity of AuNPs Using *Euphorbia hirta* L. Leaf Extract. *Colloids Surf. B Biointerf.* **2013**, *108*, 60–65.

Ashkavand, P.; Tabari, M.; Zarafshar, M.; Tomaskova, I.; Struve, D. Effect of SiO2 Nanoparticles on Drought Resistance in Hawthorn Seedlings. *Forest Res. Papers* **2015**, *76* (4), 350–359.

Atha, D. H.; Wang, H.; Petersen, E. J.; Cleveland, D.; Holbrook, R. D.; Jaruga, P.; Dizdaroglu, M.; Xing, B.; Nelson, B. C. Copper Oxide Nanoparticle Mediated DNA Damage in Terrestrial Plant Models. *J. Environ. Sci. Technol.* **2012**, *46* (3), 1819–1827.

Avellan, A.; Yun, J.; Zhang, Y.; Spielman-Sun, E.; Unrine, J. M.; Thieme, J.; Li, J.; Lombi, E.; Bland, G.; Lowry, G. V. Nanoparticle Size and Coating Chemistry Control Foliar Uptake Pathways, Translocation, and Leaf-to-Rhizosphere Transport in Wheat. *J. ACS Nano.* **2019**, *13* (5), 5291–5305.

Bar, H.; Bhui, D.; Sahoo, G. et al. Green Synthesis of Silver Nanoparticles Using Seed Extract of *Jatropha curcas*. *Colloids Surf. A Physicochem. Eng. Asp.* **2009**, *348*, 212–216.

Baybordi, A. Effect of Zinc, Iron, Manganese and Copper on Wheat Quality Under Salt Stress Conditions. *J. Water Soil* **2005**, *140*, 150–170.

Bindhani, B. K.; Panigrahi, A. K. Green Synthesis of Gold Nanoparticles Using Neem (*Azadirachta indica* L.) Leaf Extract and Its Biomedical Applications. *J. Int. J. Adv. Biotechnol. Res.* **2014**, *5*, 457–464.

Boutchuen, A.; Zimmerman, D.; Aich, N.; Masud, A. M.; Arabshahi, A.; Palchoudhury, S. Increased Plant Growth with Hematite Nanoparticle Fertilizer Drop and Determining Nanoparticle Uptake in Plants Using Multimodal Approach. *J. Nanomater.* **2019**.

Corredor, E.; Testillano, P. S.; Coronado, M.-J.; González-Melendi, P.; Fernández-Pacheco, R.; Marquina, C.; Ibarra, M. R.; de la Fuente, J. M.; Rubiales, D.; Pérez-de-Luque, A. Nanoparticle Penetration and Transport in Living Pumpkin Plants: In Situsubcellular Identification. *J. BMC Plant Biol.* **2009**, *9* (1), 45.

Overview and Emergence of Nanobiotechnology in Plants

Dang, F.; Chen, Y.-Z.; Huang, Y.-N.; Hintelmann, H.; Si, Y.-B.; Zhou, D.-M. Discerning the Sources of Silver Nanoparticle in a Terrestrial Food Chain by Stable Isotope Tracer Technique. *J. Environ. Sci. Technol.* **2019**, *53* (7), 3802–3810.

Das, J.; Velusamy, P. Catalytic Reduction of Methylene Blue Using Biogenic Gold Nanoparticles from *Sesbania grandiflora* L. *J. Taiwan Inst. Chem. Eng.* **2014**, *45* (5), 2280–2285.

de Barros, C. H. N.; Cruz, G. C. F.; Mayrink, W.; Tasic, L. Bio-Based Synthesis of Silver Nanoparticles from Orange Waste: Effects of Distinct Biomolecule Coatings on Size, Morphology, and Antimicrobial Activity. *J. Nanotechnol., Sci. App.* **2018**, *11*, 1.

DeRosa, M. et al. Nanotechnology in Fertilizers. *Nat. Nanotechnol.* **2010**, *5*, 91.

Dorosti, N.; Jamshidi, F. Plant-Mediated Gold Nanoparticles by *Dracocephalum kotschyi* as Anticholinesterase Agent: Synthesis, Characterization, and Evaluation of Anticancer and Antibacterial Activity. *J. Appl. Biomed.* **2016**, *14* (3), 235–245.

Etxeberria, E.; Pozueta-Romero, J.; Fernández, E. B. Fluid-Phase Endocytosis in Plant Cells. In *Endocytosis in Plants*; Springer, 2012; pp 107–122.

Francis, G.; Thombre, R.; Parekh, F.; Lekshminarayan, P. Bioinspired Synthesis of Gold Nanoparticles Using *Ficus benghalensis* (Indian Banyan) Leaf Extract. *J. Chem. Sci. Trans.* **2014**, *3*, 470–474.

González-Melendi, P. et al. Nanoparticles as Smart Treatment-Delivery Systems in Plants: Assessment of Different Techniques of Microscopy for Their Visualization in Plant Tissues. *Ann. Bot.* **2008**, *101*, 187–195.

Gu, Z.; Biswas, A.; Zhao, M.; Tang, Y. Tailoring Nanocarriers for Intracellular Protein Delivery. *J. Chem. Soc. Rev.* **2011**, *40* (7), 3638–3655.

Guajardo-Pacheco, M. J.; Morales-Sánchez, J.; González-Hernández, J.; Ruiz, F. Synthesis of Copper Nanoparticles Using Soybeans as a Chelant Agent. *J. Mater. Lett.* **2010**, *64* (12), 1361–1364.

Guo, H.; Ma, C.; Thistle, L.; Huynh, M.; Yu, C.; Clasby, D.; Chefetz, B.; Polubesova, T.; White, J. C.; He, L. Transformation of Ag Ions into Ag nanoparticle-loaded AgCl Microcubes in the Plant Root Zone. *J Environ. Sci. Nano* **2019**, *6* (4), 1099–1110.

Hajiahmadi, Z.; Shirzadian-Khorramabad, R.; Kazemzad, M.; Sohani, M. M. Enhancement of Tomato Resistance to *Tuta absoluta* Using a New Efficient Mesoporous Silica Nanoparticle-Mediated Plant Transient Gene Expression Approach. *J. Sci. Horticult.* **2019**, *243*, 367–375.

Harne, S.; Sharma, A.; Dhaygude, M.; Joglekar, S.; Kodam, K.; Hudlikar, M. Novel Route for Rapid Biosynthesis of Copper Nanoparticles Using Aqueous Extract of *Calotropis procera* L. Latex and Their Cytotoxicity on Tumor Cells. *J. Colloids Surf. B: Biointerf.* **2012**, *95*, 284–288.

Hasnain, M. S.; Javed, M. N.; Alam, M. S.; Rishishwar, P.; Rishishwar, S.; Ali, S.; Nayak, A. K.; Beg, S. Purple Heart Plant Leaves Extract-Mediated Silver Nanoparticle Synthesis: Optimization by Box-Behnken Design. *J. Mater. Sci. Eng.: C* **2019**, *99*, 1105–1114.

He, Y.; Wei, F.; Ma, Z.; Zhang, H.; Yang, Q.; Yao, B.; Huang, Z.; Li, J.; Zeng, C.; Zhang, Q. Green Synthesis of Silver Nanoparticles Using Seed Extract of *Alpinia katsumadai*, and Their Antioxidant, Cytotoxicity, and Antibacterial Activities. *J. RSC Adv.* **2017**, *7* (63), 39842–39851.

Hischemoller, A. et al. In-vivo Imaging of the Uptake of Upconversion Nanoparticles by Plant Roots. *J. Biomed. Nanotechnol.* **2009**, *5*, 278–284.

Hossain, K. et al. Adsorption of Urease on PE-MCM-41 and Its Catalytic Effect on Hydrolysis of Urea. *Colloids Surf. B Biointerf.* **2008**, *62*, 42–50.

Hu, J.; Wu, X.; Wu, F.; Chen, W.; Zhang, X.; White, J.; Li, J.; Yi, W.; Liu, J.; Wang, X. TiO_2 nanoparticle exposure on lettuce (*Lactuca sativa* L.): Dose-Dependent Deterioration of Nutritional Quality. *J. Environ. Sci.: Nano* **2020**.

Huang, X. et al. Magnetic Virus-Like Nanoparticles in *N. benthamiana* Plants: A New Paradigm for Environmental and Agronomic Biotechnological Research. *ACS Nano* **2011**, *5*, 4037–4045.

Jaberzadeh, A.; Moaveni, P.; Moghadam, H. R. T.; Zahedi, H. Influence of Bulk and Nanoparticles Titanium Foliar Application on Some Agronomic Traits, Seed Gluten and Starch Contents of Wheat Subjected to Water Deficit Stress. *J. Notulae Botanicae Horti Agrobotanici Cluj-Napoca* **2013**, *41* (1), 201–207.

Jain, D.; Daima, H.; Kachhwaha, S.; Kothari, S. Synthesis of Plant-Mediated Silver Nanoparticles Using Papaya Fruit Extract and Evaluation of Their Antimicrobial Activities. **2009**, *4*, 557–563.

Javad, S.; Javad, K.; Sasan, M.; Majid, S. R. J. M. Evaluating SiO_2 Nanoparticles Effects on Developmental Characteristic and Photosynthetic Pigment Contents of *Zea mays* L. *Bull. Env. Pharmacol. Life Sci.* **2014**, *3* (6), 194–201.

Karami, M.; Heidari, R.; Rahmani, F.; Solnamaz, N. Effect of Chemical Synthesis Silver Nanoparticles on Germination Indices and Seedlings Growth in Seven Varieties of *Lycopersicon esculentum* Mill (tomato) Plants. *J. Clust. Sci.* **2016**, *27*, 327.

Khalil, M. M.; Ismail, E. H.; El-Magdoub, F. Biosynthesis of Au Nanoparticles Using Olive Leaf Extract: 1st Nano Updates. *Arab. J. Chem.* **2012**, *5* (4), 431–437.

Khan, N.; Bano, A. Role of Plant Growth Promoting Rhizobacteria and Ag-nano Particle in the Bioremediation of Heavy Metals and Maize Growth under Municipal Wastewater Irrigation. *Int. J. Phytoremediat.* **2016**, *18* (3), 211–221.

Khatua, A.; Prasad, A.; Priyadarshini, E.; Patel, A. K.; Naik, A.; Saravanan, M.; Barabadi, H.; Paul, B.; Paulraj, R.; Meena, R. Emerging Antineoplastic Plant-Based Gold Nanoparticle Synthesis: A Mechanistic Exploration of their Anticancer Activity toward Cervical Cancer Cells. *J. Clust. Sci.* **2019**, 1–12.

Khodakovskaya, M. V.; de Silva, K.; Nedosekin, D. A.; Dervishi, E.; Biris, A. S.; Shashkov, E. V.; Galanzha, E. I.; Zharov, V. P. Complex Genetic, Photothermal, and Photoacoustic Analysis of Nanoparticle-Plant Interactions. *J. Proc. Natl. Acad. Sci.* **2011**, *108* (3), 1028–1033.

Kim, S. W.; Jung, J. H.; Lamsal, K.; Kim, Y. S.; Min, J. S.; Lee, Y. S. Antifungal Effects of Silver Nanoparticles (AgNPs) against Various Plant Pathogenic Fungi. *J. Mycobiol.* **2012**, *40* (1), 53–58.

Koo, Y. et al. Fluorescence Reports Intact Quantum Dot Uptake Into Roots and Translocation to Leaves of *Arabidopsis thaliana* and Subsequent Ingestion by Insect Herbivores. *Environ. Sci. Technol.* **2015**, *49*, 626–632.

Laware, S.; Raskar, S. Effect of Titanium Dioxide Nanoparticles on Hydrolytic and Antioxidant Enzymes during Seed Germination in Onion. *Int. J. Curr. Microbiol. App. Sci.* **2014**, *3* (7), 749–760.

Lei, Z.; Mingyu, S.; Xiao, W.; Chao, L.; Chunxiang, Q.; Liang, C.; Hao, H.; Xiaoqing, L.; Fashu, H. Antioxidant Stress Is Promoted by Nano-Anatase in Spinach Chloroplasts under UV-Beta Radiation. *Biol. Trace Elem. Res.* **2008**, *121*, 69–79.

Liu, R.; Lal, R. Synthetic Apatite Nanoparticles as a Phosphorus Fertilizer for Soybean (*Glycine max*). *Sci. Rep.* **2014**, *4*, 5686.

Liu, Z.; Tabakman, S.; Welsher, K.; Dai, H. Carbon Nanotubes in Biology and Medicine: In Vitro and In Vivo Detection, Imaging and Drug Delivery. *J. Nano Res.* **2009,** *2* (2), 85–120.

Liu, Z.; Yang, K.; Lee, S.-T. Single-Walled Carbon Nanotubes in Biomedical Imaging. *J. Mater. Chem.* **2011,** *21* (3), 586–598.

Mahajan, P.; Dhoke, S.; Khanna, A. Effect of Nano-ZnO Particle Suspension on Growth of Mung (*Vigna radiata*) and Gram (*Cicer arietinum*) Seedlings Using Plant Agar Method. *J. Nanotechnol.* **2011.**

Manoj, L.; Vishwakarma, V. Green Synthesis and Spectroscopic Characterisations of Gold Nanoparticles Using In vitro Grown Hypericin Rich Shoot Cultures of *Hypericum hookerianum. Int. J. Chem. Tech. Res.* **2015,** *8,* 194–199.

Mariselvam, R.; Ranjitsingh, A.; Nanthini, A. U. R.; Kalirajan, K.; Padmalatha, C.; Selvakumar, P. M. Green Synthesis of Silver Nanoparticles from the Extract of the Inflorescence of *Cocos nucifera* (Family: Arecaceae) for Enhanced Antibacterial Activity. *J. Spectrochim. Acta, A: Mol. Biomol. Spectrosc.* **2014,** *129,* 537–541.

Martin-Ortigosa, S. et al. Mesoporous Silica Nanoparticle Mediated Intracellular Cre Protein Delivery for Maize Genome Editing via loxP Site Excision. *Plant Physiol.* **2014,** *164,* 537–547.

Martin-Ortigosa, S.; Valenstein, J. S.; Lin, V. S. Y.; Trewyn, B. G.; Wang, K. Gold Functionalized Mesoporous Silica Nanoparticle Mediated Protein and DNA Codelivery to Plant Cells via the Biolistic Method. *J. Adv. Funct. Mater.* **2012,** *22* (17), 3576–3582.

Martin-Ortigosa, S.; Valenstein, J. S.; Sun, W.; Moeller, L.; Fang, N.; Trewyn, B. G.; Lin, V. S. Y.; Wang, K. Parameters Affecting the Efficient Delivery of Mesoporous Silica Nanoparticle Materials and Gold Nanorods into Plant Tissues by the Biolistic Method. *J. Small* **2012,** *8* (3), 413–422.

Mata, R.; Nakkala, J. R.; Sadras, S. R. Polyphenol Stabilized Colloidal Gold Nanoparticles from *Abutilon indicum* Leaf Extract Induce Apoptosis in HT-29 Colon Cancer Cells. *J. Colloids Surf. B: Biointerf.* **2016,** *143,* 499–510.

Medintz, I. et al. Quantum Dot Bioconjugates for Imaging, Labelling and Sensing. *Nat. Mater.* **2005,** *4,* 435–446.

Michalet, X. et al. Quantum Dots for Live Cells, in Vivo Imaging, and Diagnostics. *Science* **2005,** *307.*

Naganathan, K.; Thirunavukkarasu, S. Green Way Genesis of Silver Nanoparticles Using Multiple Fruit Peels Waste and Its Antimicrobial, Anti-Oxidant and Anti-Tumor Cell Line Studies. Paper Read at IOP Conference Series: Materials Science and Engineering, 2017.

Nair, R.; Varghese, S. H.; Nair, B. G.; Maekawa, T.; Yoshida, Y.; Kumar, D. S. Nanoparticulate Material Delivery to Plants. *J/ Plant Sci.* **2010,** *179* (3), 154–163.

Oza, G.; Reyes-Calderón, A.; Mewada, A.; Arriaga, L. G.; Cabrera, G. B.; Luna, D. E.; Iqbal, H. M.; Sharon, M.; Sharma, A. Plant-Based Metal and Metal Alloy Nanoparticle Synthesis: A Comprehensive Mechanistic Approach. *J. Mater. Sci.* **2020,** 1–22.

Patra, J. K.; Ali, M. S.; Oh, I.-G.; Baek, K.-H. Proteasome Inhibitory, Antioxidant, and Synergistic Antibacterial and Anticandidal Activity of Green Biosynthesized Magnetic Fe_3O_4 Nanoparticles Using the Aqueous Extract of Corn (*Zea mays* L.) Ear Leaves. *J. Artif. Cells, Nanomed., Biotechnology* **2017,** *45* (2), 349–356.

Patra, J. K.; Baek, K.-H. Green Nanobiotechnology: Factors Affecting Synthesis and Characterization Techniques. *J. Nanomater.* **2014.**

Patra, J. K.; Baek, K.-H. Novel Green Synthesis of Gold Nanoparticles Using *Citrullus lanatus* Rind and Investigation of Proteasome Inhibitory Activity, Antibacterial, and Antioxidant Potential. *Int. J. Nanomed.* **2015,** *10*, 7253.

Patra, J. K.; Baek, K.-H. Green Biosynthesis of Magnetic Iron Oxide (Fe_3O_4) Nanoparticles Using Aqueous Extract of Food Processing Wastes Under Photo-Catalyzed Condition and Investigation of Their Antimicrobial and Antioxidant Activity. *J. Photochem. Photobiol. B: Biol.* **2017,** *173*, 291–300.

Philip, D. *Mangifera indica* Leaf-Assisted Biosynthesis of Well-Dispersed Silver Nanoparticles. *J. Spectrochimica Acta Part A: Mol. Biomol. Spectrosc.* **2011,** *78* (1), 327–331.

Rawson, H.; Long, M.; Munns, R. Growth and Development in NaCl Treated Plants. *J Plant Physiol.* **1998,** *15*, 519–527.

Sadeghi, B.; Rostami, A.; Momeni, S. Facile Green Synthesis of Silver Nanoparticles Using Seed Aqueous Extract of *Pistacia atlantica* and Its Antibacterial Activity. *J. Spectrochim. Acta Part A: Mol. Biomol. Spectrosc.* **2015,** *134*, 326–332.

Saxena, R.; Tomar, R. S.; Kumar, M. Exploring Nanobiotechnology to Mitigate Abiotic Stress in Crop Plants. *J. Pharm. Sci. Res.* **2016,** *8* (9), 974.

Selvam, K.; Sudhakar, C.; Govarthanan, M.; Thiyagarajan, P.; Sengottaiyan, A.; Senthilkumar, B.; Selvankumar, T. Eco-friendly Biosynthesis and Characterization of Silver Nanoparticles Using *Tinospora cordifolia* (Thunb.) Miers and Evaluate Its Antibacterial, Antioxidant Potential. *J. Radiat. Res. Appl. Sci.* **2017,** *10* (1), 6–12.

Serag, M. F.; Kaji, N.; Gaillard, C.; Okamoto, Y.; Terasaka, K.; Jabasini, M.; Tokeshi, M.; Mizukami, H.; Bianco, A.; Baba, Y. Trafficking and Subcellular Localization of Multiwalled Carbon Nanotubes in Plant Cells. *J. ACS Nano* **2011,** *5* (1), 493–499.

Serag, M. F.; Kaji, N.; Venturelli, E.; Okamoto, Y.; Terasaka, K.; Tokeshi, M.; Mizukami, H.; Braeckmans, K.; Bianco, A.; Baba, Y. J. A. n. Functional Platform for Controlled Subcellular Distribution of Carbon Nanotubes. *ACS Nano* **2011,** *5* (11), 9264–9270.

Serag, M. F.; Kaji, N.; Habuchi, S.; Bianco, A.; Baba, Y. Nanobiotechnology Meets Plant Cell Biology: Carbon Nanotubes as Organelle Targeting Nanocarriers. *J. RSC Adv.* **2013,** *3* (15), 4856–4862.

Sharma, P.; Bhatt, D.; Zaidi, M.; Saradhi, P. P.; Khanna, P.; Arora, S. Silver Nanoparticle-Mediated Enhancement in Growth and Antioxidant Status of *Brassica juncea*. *J. Appl. Biochem. Biotechnol.* **2012,** *167* (8), 2225–2233.

Shelar, G. B.; Chavan, A. M. Myco-synthesis of Silver Nanoparticles from *Trichoderma harzianum* and Its Impact on Germination Status of Oil Seed. *J. Biolife* **2015,** *3* (1), 109–113.

Smitha, S.; Philip, D.; Gopchandran, K. Green Synthesis of Gold Nanoparticles Using *Cinnamomum zeylanicum* Leaf Broth. *Spectrochim. Acta Part A Mol. Biomol. Spectrosc.* **2009,** *74*.

Song, J.; Jang, H.-K.; Kim, B. Biological Synthesis of Gold Nanoparticles Using *Magnolia kobus* and *Diopyros kaki* Leaf Extracts. *Process Biochem.* **2009,** *44*, 1133–1138.

Torney, F. et al. Mesoporous Silica Nanoparticles Deliver DNA and Chemicals Into Plants. *Nat. Nanotechnol.* **2007,** *2*, 295–300.

Veeraputhiran, V. Bio-Catalytic Synthesis of Silver Nanoparticles. *J. Int. J. Chem. Tech. Res.* **2013,** *5* (5), 255–2562.

Vijayaraghavan, K.; Nalini, S. K.; Prakash, N. U.; Madhankumar, D. One Step Green Synthesis of Silver Nano/Microparticles Using Extracts of *Trachyspermum ammi* and *Papaver somniferum*. *J. Colloids Surf. B: Biointerf.* **2012,** *94*, 114–117.

Wang, P.; Lombi, E.; Zhao, F.-J.; Kopittke, P. M. Nanotechnology: A New Opportunity in Plant Sciences. *J. Trends Plant Sci.* **2016,** *21* (8), 699–712.

Wu, P.; Gao, L.; Zou, D.; Li, S. An Improved Particle Swarm Optimization Algorithm for Reliability Problems. *J. ISA Trans.* **2011,** *50* (1), 71–81.

Wu, W.; Huang, J.; Wu, L.; Sun, D.; Lin, L.; Zhou, Y.; Wang, H.; Li, Q. Two-Step Size-and Shape-Separation of Biosynthesized Gold Nanoparticles. *J. Separat. Purif. Technol.* **2013,** *106*, 117–122.

Yanık, F.; Vardar, F. Toxic Effects of Aluminum Oxide (Al_2O_3) Nanoparticles on Root Growth and Development in *Triticum aestivum. Water, Air Soil Pollut.* **2015,** *226*, 296.

Zarafshar, M.; Akbarinia, M.; Askari, H.; Hosseini, S. M.; Rahaie, M.; Struve, D.; Striker, G. Morphological, Physiological and Biochemical Responses to Soil Water Deficit in Seedlings of Three Populations of Wild Pear Tree (*Pyrus boisseriana*) Biotechnology. *Agron. Soc. Environ.* **2014,** *18*, 353–366.

CHAPTER 7

NANOFERTILIZERS AND NANOPESTICIDES: APPLICATION AND IMPACT ON AGRICULTURE

ZEENAT JAVEED[1], UMAIR RIAZ[1*], GHULAM MURTAZA[2],
SHAHZADA MUNAWAR MEHDI[3], MUHAMMAD IDREES[4],
QAMAR UZ ZAMAN[5], and WAQAS KHALID[6]

[1]*Soil and Water Testing Laboratory for Research, Bahawalpur, Pakistan*

[2]*Assistant Soil Fertility Office, Bahawalpur, Pakistan*

[3]*Rapid Soil Fertility Survey and Soil Testing Institute, Lahore, Punjab, Pakistan*

[4]*Department of Computer Science and Engineering, University of Engineering and Technology, Lahore, Narowal Campus, Pakistan*

[5]*Department of Environmental Sciences, University of Lahore, Lahore Campus, Pakistan*

[6]*Department of Biomedical Engineering Technology, NFC Institute of Engineering and Technology Multan, Pakistan*

*Corresponding author.
E-mail: umairbwp3@gmail.com, umair.riaz@uaf.edu.pk*

ABSTRACT

Nanotechnology is a growing field of multidisciplinary development with numerous applications in the medical, electrical, pharmaceutical, and agricultural fields. Nanomaterials range in size from 1 to 100 nm, only those whose size in this range is called nanomaterials. Different properties of nanomaterials, for example, biological, physical, and optical are due to their small particle size and large geographical area. A major factor limiting the production of plants is the pests of plants. Nanotechnology is emerging as nanoencapsulation, and the use of nanostructure catalyst has emerged as one of the most common pest control methods. Nanofertilizer are carriers of synthetic ingredients formed using nano-sized substrates of 1–100 nm. Due to their broad geographical location, nanoparticles can absorb nutrients excessively and release slowly in such a way that it resembles the absorption of nutrients as a plant requirement without the adverse effects of conventional fertilizers. Nanofertilizers and nanocomposites also have their utility in the control of releasing nutrients from the fertilizer, which improves the efficiency of the nutrients while preventing the organic ions from being consumed or lost in the environment. Nanoparticles, when used as nanopesticides, can help fight crop diseases by targeting pathogens. It is an agrochemical mixture used to reduce the problems produced due to the use of conventional pesticides. Nanosilver is the nanoparticle widely used in biological systems. The fried nanosilver removes unwanted germs from growing the soil and prevents a few other plant diseases. The use of various nanofertilizers plays a significant role in increasing crop yields, which will decrease the expenditure for fertilizer for crop production and reduce the risk of contamination. In this chapter, the available data relating to the scope and application of nanotechnology has been extensively analyzed for their present and future applications in the fight against the challenges of soil nutrition and crop protection.

7.1 INTRODUCTION

World agricultural areas face many challenges, such as crop production, nutrient uptake, biodiversity loss, biodiversity scarcity, arable land scarcity, water scarcity, and labor shortages of people from agriculture (FAO, 2009; Godfray et al., 2010). Information provided by the FAO shows that declining and obliterating areas and water presents critical difficulties in delivering adequate nourishment and other agricultural items to help jobs and address the issues of the developing global population (UN, 2015). In nanotechnology, the structure of ultrasmall particles with various properties, for example, surface-area-to-volume proportion and improved optoelectronic and physicochemical properties, contrasted with vast numbers of their counterparts (UN, 2015) is presently developing as a promising system to advance plant development and profitability (Gogos et al., 2012). This idea is part of an ever-changing agricultural science of precision, where

Nanofertilizers and Nanopesticides

technology is used by the farmers to use water, fertilizers, and other inputs efficiently. Nanotechnology and its derivatives are explored for their various applications in the agricultural sector, such as pesticides and pollutants, dealing out with agricultural goods with better storage life, pesticides for the delivery of biomolecules, fortification, water purification, nutrient recycling (Saharan et al., 2014). In this study, we focus on quick fertilizers delivery, called "nanofertilizer." The design of nanotechnology has led us to the point that we will create future foods such as consumer choice with a strong flavor, long shelf life, enhanced nutrient content, and texture. Keen safe bundling will be utilized to wrap nourishment; this bundling will have the option to identify impurity and plundering. Nanotechnology is an interdisciplinary research field that opens new bearings in the fields of medication, innovation, pharmaceuticals, and horticulture. The "term nanomaterials" are utilized to portray materials that are in the scope of 1 to 100 nm; this tiny particle size with broad region making them have distinctive nanomaterials properties, for example, physical, physical, and natural properties. Nanotechnology is a rising multidisciplinary look into the field, which fuses the structure, advancement, application, and plan of materials and gadgets at the subatomic level at the nanometer scale, for example, the base breadth runs in size from 1 to 100 nm, a million meters (Fakruddin et al., 2012).

Developing the world relies on the agricultural sector, with over 60% of its population directly or indirectly dependent on their income (Brock et al., 2011; Rai and Ingle, 2012; Qamar et al., 2014). Though, in this fast era of technological development, farmers must face challenges like the sustainable utilization of natural resources, the elimination of nutrients through the year, and environmental issues such as running and collecting fertilizers and pesticides. It is therefore needed for the hour to use such technology that can transform and rebuild agricultural strategies to produce its product and make it cost valid with the delivery of the right amount of timely input (Prasad et al., 2012). A significant reason for diminishing harvest yields is a plant pest. The most commonly used pest control is the utilization of pesticides in higher quantities, which finally adds extra value to crop production. At the point when these pesticides are utilized, a lot of them can cause ecological contamination and pollution. Therefore, to utilize pesticides at the most reduced conceivable expense to decrease production and environmental costs (Sharon et al., 2010). It very well may be accomplished by expanding the span of pesticide stockpiling through the necessary activity or the required operation. The persistence of pesticides in the early stages of plant growth helps to reduce the number of pests below the limit, and consequently, better control over the long term. The nanotechnology, "nanoencapsulation" strategy, can be utilized to improve the insecticidal worth. In the nano-exemplification technique, the dynamic nanosized pest spray is fixed by a sac or flimsy-walled shell. A useful methodology in this regard is the "controlled release of active ingredient" that will significantly improve the efficiency and decrease of pesticide exposure and the related environmental risks. For example, the "halloysite" (clay nanotubes) was developed as an active, cost-effective pesticide. This will lessen a significant number of pesticides, for example, expanding the discharge time and better contact with plants, decreasing the expense of pesticides to a more prominent degree, and less impact on the environment (Allen, 1994). One development in this regard is the possible use of structured nanocatalysts that will increase the effectiveness of pesticides and will also reduce the number of pesticides needed (Joseph and Morrison, 2006). Liu et al. (2006) revealed that porous hollow silica nanoparticles (PHSNs) combined with validamycin (a pesticide) could be used successfully to control the pesticides. Nano-silica has already been tested for agricultural pest control. Physisorption is a nano-silica action mode. It is absorbed by the lipids of decorative insects which is why it leads to the death of the insect (Ulrichs et al., 2005).

7.2 IMPLICATIONS OF NANOMATERIAL AS FERTILIZER AND PESTICIDE IN AGRICULTURE

Extensive use of nanotechnology has been identified in heavy industry (Lo et al., 2010), the medical industry (Zhou et al., 2004), energy (Das et al., 2007), consumer goods (Schneider, 2010), and environment (Shi et al., 2007). Nanotechnology has a significant contribution to the medical field, but as well it has more productive use in the agriculture sector nowadays, and it is estimated to grow considerably. The agricultural producing sector has been recognized as the most prospective industry to create substantial investments around the world, so for future investments, this productive sector has been considerably used. Nanotechnology is capable of transforming the agriculture sector over the use of cell-based disease therapies, short-term disease detection, increased plant performance, absorption of nutrients, etc. Smart sensors, fast and secure delivery systems, will support the agricultural industry struggle off against insects and other plant pathogens. To increase the efficiency of pesticides and herbicides, future indices will be presented that will allow their lower use and effective results. Nanotechnology in the future will also indirectly protect the environment by using other nonrenewable

Nanofertilizers and Nanopesticides   203

(renewable) materials, and filters or catalysts to reduce the pollution and detoxify of remaining pollutants. Nanotechnology is another feasible and straightforward way to achieve this goal. There are some reasons in the use of nanotechnology, which is gaining the high intentions of scientists toward the range of fertilizers used as nanoparticles for use in agricultural and agricultural sciences (Buentello et al., 2005).

7.2.1 NANOFERTILIZER

Nanofertilizer technology is a relatively new way, and now it is pioneering days, and there are few books available. Optimal fertilizer efficiency will not exceed 30%–35%, 18%–20%, and 35%–40% N, P, and K, respectively. It is notable that the description of "nanofertilizer" is being answered. In many studies, related to nanoapplication of fertilizer in agriculture as a nanofertilizer is being used in both materials to form a thickness of in the range of 1 to 100 nm in at least one place (e.g., Zinc Oxide nanoparticles) and those present at a larger size greater than 100 nm but that have been customized by materials of nanoscale (e.g., massive fertilizer made of nanomaterial). Thus, in this study, the word "nanofertilizer" means to that bulk and nonmaterial used as fertilizers. In this regard their distinctive characters, nanoparticles can affect the metabolic activities of a plant to vary degree in comparison to conventional materials and thus have the ability to synthesize basic elements, such as phosphorus, in the rhizosphere. Three different categories are (1) nonmaterials made of macronutrients; (2) nonmaterials made of micronutrients, and (3) nonmaterials that hold carriers, macronutrients. The first two phases form nanomaterials as the materials themselves while they are added in the third phase. Some researchers refer to the first two phases as nanofertilizers, and the third phase is called a "neutralizer" or "nanomaterial-fermented fertilizer." Additional categories include "vanadium," which promotes plant growth and nanomaterials that act as carriers of micronutrients.

Emerging research, awareness, and the availability of useful natural zeolites have stimulated enthusiasm for the development of zeolite, which generally based on nanofertilizers. In one experiment, a patented nanocomposite containing N, P, K, B, Zn, Fe, Mn, and Cu with mannose and amino acids was publicized to enhance the absorption and consumption of plant nutrients (Jinghua, 2004). In another report, it is exhibited that the balanced use of nitrogen, phosphorus, potassium and sulfur, zinc, Boron would significantly enhance the formation of plumes in red and lateral soils (Bhattacharya et al., 2004). The sufficient amount of NPK fertilizer increased green and black yields by 13% and 38% more. Liu et al. (2006) has revealed in his study that organic substances (polystyrene) that bind to endogenous clay components form nano cement and subnanocomposites that can regulate the release of nutrients from the organic surface. The nanoparticles can also be used in the management of the physical properties of nutrients. Subramanian et al. (2008) depicted in his work that nanofertilizers and nanocomposites can be used to control the release of fertilizer to improve the efficiency of nutrients while preventing the ions of nutrients from degrading or degrading. Recently, Sharmila Rahale (2011) considered the pattern of nutrient deletions produced by a nitrogen-laden nanofertilizer. Data have shown that nano-number fertilizers (zeolite and montmorillonite in size 30–40 nm) can release nitrogen longer (>1000 h) than conventional fertilizers (<500 h). Nanofertilizer technology enhances crop fertilizer use efficiency. There is a growing interest in the use of nanomaterial in agriculture over the years due to current public concerns about the adverse effects of chemical fertilizers on environment (Ramesh et al., 2010).

7.2.2 NANOPESTICIDES

It is essential to differentiate the various ways through which the word nanopesticides work for the entire release of the active ingredient, which will control the plant diseases that have been discussed by Kah et al. (2013). The only size-based description (e.g., most often 100 nm as the maximum size, for example,, according to the EU, 2011) may be too simple as, on the other hand, the 100 nm size limit does not provide many novel ideas.

The high proportion of nanoformulations for insecticidal purposes can be partially explained by the active ingredients (AI) of most common insects with aqueous solubility and consequently require a system for their effective delivery in the field (Adak et al., 2012). Another emerging feature of pest-related research is the possibility of using other AI pests that do not harm nonconcentrated organisms and may limit the development of resistance. Many of these unique AI systems are unstable and require protection from premature deterioration, which can be detected by nanoformulations (Loha et al., 2012).

Nanoparticles can help fight plant diseases by targeting viral infections (Philip, 2011). Other nanoparticles are nanotypes of carbon, silver, silica, and alumina cultures used to control plant diseases. Singh et al. (2014) reported that nanosilver is the most exploitative nanoparticles in the biological system. The nanocatalyst removes unwanted bacteria from growing the soil and prevents a few other plant diseases (Bhattacharyya et al., 2010; Singh et al.,

2014). In agriculture, pesticides or weeds are used to control pests or weeds to increase crop yields. However, they also harm soil health. Nanopesticide is an agrochemical compact that is used to overcome problems caused by conventional pesticides (Sasson et al., 2007). Several types of materials viz. survivors, organic polymers, and mineral nanoparticles fall in the size range of nanometer are used in the manufacture of nanopesticides (Alfadul et al., 2017). The new generation of nanopesticides will be more effective against pests and harm other essential soil pests (Kah et al., 2013).

7.3 RESTRICTIONS IN CONVENTIONAL CULTURE FOR CULTIVATION OVER NANOTECHNOLOGY

Agriculture needs new help for innovation to cope with the growing demand for food while declining its impact on the environment. Many research trials must be conducted in developing competing vital products and have the potential to make modern cultivation sustainable. Proper use of nanotechnologies can participate as a major function in achieving the target.

Nanofertilizers have an advantage over traditional fertilizers as they increase soil fertility, yields, and crop quality limits are unhealthy and are harmless to the environment and people, reducing costs and increasing profitability. Nanoformulations enhance nutrients using effectiveness and reduce the cost of environmental protection (Naderi and Abedi, 2012). They have improved the nutritional value of crops and flavor quality. Iron utilization improves and increases the protein content of wheat cereals (Abecassis et al., 2008). Improve plant growth through disease resistance and improve crop stability by fighting crop rotation. Tarafdar et al. (2012) also suggested that a balanced fertilizer in a plant can be achieved through nanotechnology.

Modern agriculture is increasingly chemically sophisticated with the use of multiple doses of pesticides, diseases, weeds, and natural nutrient management to obtain the perfect product for each area without regard to natural resources and the environment (Singh, 2017). Many studies have reported that nanofertilizer had a significant effect on seed germination and seedling growth, which enhances seed vigor (Prasad et al., 2014). Nanofertilizer can quickly absorb the seed and enhance the availability of nutrients in growing plants leading to the length and length of shoot and root length. At present, agricultural fertilizer contributes to 50% of agricultural production, but increased consumption of high-quality fertilizer does not guarantee improved productivity but leads to many problems such as soil degradation

206 Diverse Applications of Nanotechnology in the Biological Sciences

and water pollution. Increase fertilizer consumption by solutions to optimize and decrease doses, it is revealed that fertilizer alone contributes 50% to crop production (Qamar et al., 2014).

- High travel costs of fertilizer because they require high quantities.
- Reduce the amount of fertilizer.
- Much of the contamination of organic material by using excess volumes of crop production.
- Addition of value to traditional fertilizer and reduce capacity in each area.
- Lack of nutrients in the soil.
- Combine the use of macro and micronutrient sources.

This analysis considered only nanoformulations of active ingredients at present. The general purpose of air regeneration is to gradually release the portable or unstable AI after its use in the wild. In such cases, the extraction of the active ingredient from the nanocarrier system is a necessary process that controls the natural end of the nanopesticides (Kah and Hofmann, 2014).

7.4 NANOFERTILIZERS AND NUTRIENT USE EFFICIENCY

The nanofertilizer has a high surface area, mostly due to the small size of the particles that provide much space to facilitate the various metabolic processes resulting in the assembly of many photosynthates. Nanofertilizer has a large surface area and less particle size less than the size of the roots and leaves of the plant that can increase plant penetration from the implanted area and improve the absorption and utilization of nanofertilizer fertilizers. The reduction in particle size leads to more surface area and particle size in each fertilizer that provides an additional opportunity for nanofertilization resulting in more absorption and absorption of nutrients (Liscano et al., 2000).

Fertilizers with nanoparticles will enhance the accessibility and absorption of nutrients in plant tissues (Tarafdar et al., 2012). Zeolite-based nanofertilizer can release nutrients slowly from a plant to increase plant nutrients while growth time that prevents nutrient loss from denigration, leaching, volatilization, and fixation in soils, especially nitrate and ammonical forms of nitrogen. Size of particle less than 100 nm nanoparticles can be used as fertilizer to treat environment-friendly nutrients and significantly decrease environmental hazardous (Joseph and Morrisson, 2006). The primary purpose of high-quality fertilizer is primarily their absorption capacity, more surface area, and small size, which usually give the same effect found in the bulk form. In part, it is because nanoparticles exhibit a large surface area over average size. Therefore, the active site is in a more representative area in terms of nanoparticles compared to conventional sizes. The surface area of the units increases with decreasing size, and the nonlinear strength of the particles is their size function. A related study was found by Liscano et al. (2000). Many research works exhibited that nanofertilizer had a significant impact on germination percentage of seeds and growth of seedlings, which exposed the influence of nanofertilizer. Nanofertilizer can easily penetrate plants and increase the availability of nutrients in growing plants leading to height and height of roots and root length. Nanofertilizers have played a significant part in crop production, and many research studies have shown that nanofertilizer improves the growth, yield, and quality of fruit yields as high-quality food used by humans and animals. This translates into development in three major production areas. Many studies show that the use of nanofertilizers significantly increases crop yields over control. The unique use of nanoparticles as fertilizer significantly increases the yield of crops (Ulrichs et al., 2005). Nanofertilizer provides more surface area and greater availability of nutrients to plants that help improve these plant quality categories (such as protein, oil content, and sugar content) by improving the reaction rate or process initiation in the plant system. Nano-Fe_2O_3 improves photosynthesis and the healthy growth of unusual plants (Wang et al., 2016). Some nutrients are responsible for disease prevention in the plant due to the high availability of nanonutrients in plants that protect against disease. Nano-ZnO also helps to grow under pressure conditions (Singh, 2017).

7.5 CONCLUSION AND FUTURE PROSPECTIVE

The use of various nanofertilizers plays a significant role in increasing crop yields, which will diminish the total capita of fertilizer for crop production and reduce the risk of contamination. The use of nanofertilizers in agriculture should have a significant impact on society. The efficiency of organic fertilizers in crop production can be increased by the efficient application of nanopesticides and nanofertilizers. Nanofertilizer improves crop production and provides high insertion capacity and concentration but also has a detrimental effect on crop quality if the collection exceeds the maximum limit leading to increased yield and yield. Many studies are in progress to develop nanocomposites to provide all the necessary nutrients correctly with an excellent delivery system that can assist in the delivery of plant nutrients; the need to study nutritional requirements in plant systems, assessing the impact of nanofertilizers on soil and soil is beneficial, healthy, and nutritious solutions. Learn the need to increase absorption and the amount of nanofertilizer for various crops and local control of nanofertilizers in agriculture precisely these are just a few of the problems you need to have in order to attain a better result from nanoparticles in crop production.

KEYWORDS

- **nanofertilizer**
- **agricultural technologies**
- **nanopesticides**
- **nanocomposites**
- **pollution**

REFERENCES

Abecassis, J.; David, C.; Fontaine, L.; Taupier-Létage, B.; Viaux, P. A Multidisciplinary Approach to Improve the Quality of Organic Wheat-Bread Chain. **2008,** 201–206.

Adak, T.; Kumar, J.; Shakil, N. A.; Walia, S. Development of Controlled Release Formulations of Imidacloprid Employing Novel Nano-Ranged Amphiphilic Polymers. *J. Environ. Sci. Health B* **2012,** *47* (3), 217–225.

Alfadul, S.M.; Altahir, O. S.; Khan, M. Application of Nanotechnology in the Field of Food Production. *Acad. J. Sci. Res.* **2017,** *5* (7), 143–154.

Allen, R. Agriculture during the Industrial Revolution. *Econ. History Britain Since* **1994,** *1700* (3), 96–123.

Bhattacharya, S. S.; Debkanta, G. N.; Mandal, C.; Majumdar, K. Effect of Balanced Fertilization on Pulse Crop Production in Red and Lateritic Soils. *Better Crops.* **2004,** *88* (4), 52–57.

Bhattacharyya, A.; Bhaumik, A.; Rani, P. U.; Mandal, S.; Epidi, T. T. Nanoparticles – a Recent Approach to Insect Pest Control. *Afr. J. Biotechnol.* **2010,** *9* (24), 3489–3493.

Brock, D. A.; Douglas, T. E.; Queller, D. C.; Strassmann, J. E. Primitive Agriculture in a Social Amoeba. *Nature* **2011,** *469* (7330), 393–396.

Buentello, S.; Persad, D. L.; Court, E. B.; Martin, D. K.; Daar, A. S.; Peter, A. Nanotech and the Developing World. *Pl. Med.* **2005,** *2*, 97.

Calder, A. J.; Dimkpa, C. O.; McLean, J. E.; Britt, D. W.; Johnsonc, W.; Anderson, A. J. Soil Components Mitigate the Antimicrobial Effects of Silver Nanoparticles towards a

Nanofertilizers and Nanopesticides

Beneficial Soil Bacterium, Pseudomonas Chlororaphis O$_6$. *Sci. Total Environ.* **2012**, *429*, 215–222.

Chuprova, V. V.; Ulyanova, O. A.; Kulebakin, V. G. The Effect of Bark–Zeolite Fertilizers on Mobile Humus Substances of Chernozem and on Biological Productivity of Corn. *Poster Presented Euro Soil, Freiburg, Germany* 2004; pp 4–12.

Colman, B. P.; Arnaout, C. L.; Anciaux, S.; Gunsch, C. K.; Hochella, M. F. Jr.; Kim, B. Low Concentrations of Silver Nanoparticles in Biosolids Cause Adverse Ecosystem Responses under Realistic Field Scenario. *PLoS One* **2013**, *8*, 571–589.

Das, S.; Gates, A. J.; Abd, H. A.; Rose, G. S.; Picconatto, C. A.; Ellenbogen, J. C. Designs for Ultra-Tiny, Special-Purpose Nano Electronic Circuits. *IEEE Trans. Circ. Syst.* **2007**, *154*, 2528–2540.

DeRosa, M. C.; Monreal, C.; Schnitzer, M.; Walsh, R.; Sultan, Y. Nanotechnology in Fertilizers. *Nat. Nanotechnol.* **2010a**, *5*, 91.

DeRosa, M. C.; Monreal, C.; Schnitzer, M.; Walshand, R.; Sultan, Y. Nanotechnology in Fertilizers. *Nat. Nanotechnol.* **2010b**, *5*, 540–547.

Dimkpa, C. O.; Latta, D. E.; McLean, J. E.; Britt, D. W.; Boyanov, M. I.; Anderson, A. J.. Fate of CuO and ZnO Nano-and Microparticles in the Plant Environment. *Environ. Sci. Technol.* **2013**, *47*, 4734–4742.

Du, W.; Sun, Y.; Ji, R.; Zhu, J.; Wu, J.; Guo, H. TiO$_2$ and ZnO Nanoparticles Negatively Affect Wheat Growth and Soil Enzyme Activities in Agricultural Soil. *J. Environ. Monit.* **2011**, *13*, 822–828.

Fakruddin, M.; Hossain, Z.; Afroz, H. Prospects and Applications of Nanobiotechnology: A Medical Perspective. *J. Nanobiotechnol.* **2012**, *10* (1), 1–8.

Food and Agriculture Organization of the United Nation (FAO). *Global Agriculture Towards 2050*; FAO: Rome, Italy, 2009.

Frenk, S.; Ben-Moshe, T.; Dror, I.; Berkowitz, B.; Minz, D. Effect of Metal Oxide Nanoparticles on Microbial Community Structure and Function in Two Different Soil Types. *PLoS One* **2013**, *8*, 84441.

Ge, Y.; Priester, J. H.; Van De Werfhorst, L. C.; Schimel, J. P.; Holden, P. A. Potential Mechanisms and Environmental Controls of TiO$_2$ Nanoparticle Effects on Soil Bacterial Communities. *Environ. Sci. Technol.* **2013**, *47*, 14411–14417.

Ge, Y.; Schimel, J. P.; Holden, P. A. Evidence for Negative Effects of TiO$_2$ and ZnO Nanoparticles on Soil Bacterial Communities. *Environ. Sci. Technol.* **2011**, *45*, 1659–1664.

Ge, Y.; Schimel, J. P.; Holden, P. A. Identification of Soil Bacteria Susceptible to TiO$_2$ and ZnO Nanoparticles. *Appl. Environ. Microbiol.* **2012**, *78*, 6749–6758.

Godfray, H. C. J.; Beddington, J. R.; Crute, I. R.; Haddad, L.; Lawrence, D.; Muir, J. F.; Pretty, J.; Robinson, S.; Thomas, S. M.; Toulmin, C. Food Security: The Challenge of Feeding 9 Billion People. *Science* **2010**, *327*, 812–818.

Gogos, A.; Knauer, K.; Bucheli, T. D. Nanomaterials in Plant Protection and Fertilization: Current State, Foreseen Applications, and Research Priorities. *J. Agric. Food Chem.* **2012**, *60*, 9781–9792.

Hänsch, M.; Emmerling, C. Effects of Silver Nanoparticles on the Microbiota and Enzyme Activity in Soil. *J. Plant Nutr. Soil Sci.* **2010**, *173*, 554–558.

Hillie, T. *Nano Computers and Swarm Intelligence*; London: ISTE, 2007; p 26.

Jinghua, G. Synchrotron Radiation, Soft X-ray Spectroscopy and Nano-materials. *J. Nanotech.* **2004**, *1* (1–2), 193–225.

Jo´sko, I.; Oleszczuk, P. Influence of Soil Type and Environmental Conditions on ZnO, TiO$_2$ and Ni Nanoparticles Phytotoxicity. *Chemosphere* **2013**, *92*, 91–99.

Joseph, T.; Morrison, M. Nanotechnology in Agriculture and Food: A Nanoforum Report.; 2006a of publication; *Nanoforum.org.*

Joseph, T.; Morrisson, M. *Eur. Nanotechnol. Gateway*, 2006b.

Kah, M. Nanopesticides and Nanofertilizers: Emerging Contaminantsor Opportunities for Risk Mitigation. *Front. Chem.* **2015**, *3*, 64–64.

Kah, M.; Beulke, S.; Tiede, K.; Hofmann, T. Nano-Pesticides: State of Knowledge, Environmental Fate, and Exposure Modeling. *Crit. Rev. Environ. Sci. Technol.* **2013**, *43* (16), 1823–1867.

Kah, M.; Hofmann, T. Nanopesticide Research: Current Trends and Future Priorities. *Environ. Int.* **2014**, *63*, 224–235.

Kaushik, P.; Shakil, N. A.; Kumar, J.; Singh, M. K.; Yadav, S. K. Development of Controlled Release Formulations of Thiram Employing Amphiphilic Polymers and Their Bioefficacy Evaluation in Seed Quality Enhancement Studies. *J. Environ. Sci. Health B* **2013**, *48* (8), 677–685.

Kim, S.; Kim, J.; Lee, I. Effects of Zn and ZnO Nanoparticles and Zn2+on Soil Enzyme Activity and Bioaccumulation of Zn in *Cucumis sativus*. *Chem. Ecol.* **2011**, *27*, 49–55.

Liscano, J. F.; Wilson, C. E.; Norman, R. J.; Slaton, N. A. *AAES Res. Bull.* **2000**, *963*, 1–31.

Liu, F.; Wen, L. X.; Li, Z. Z.; Yu, W.; Sun, H. Y. Porous Hollow Silica Nanoparticles as Controlled Delivery System for Water-Soluble Pesticide. *Mater. Res. Bull.*, **2006**, *41* (12), 2268–2275.

Lo, P. K.; Karam, P.; Sleiman, H. F. Loading and Selective Release of Cargo in DNA Nano Tubes with Longitudinal Variation. *Nat. Chem.*, **2010**.

Loha, K. M.; Shakil, N. A.; Kumar, J.; Singh, M. K.; Adak, T.; Jain, S. Release Kinetics of Beta-Cyfluthrin from Its Encapsulated Formulations in Water. *J. Environ. Sci. Health B* **2011**, *46* (3).

Loha, K. M.; Shakil, N. A.; Kumar, J.; Singh, M.; Srivastava, C. Bio-Efficacy Evaluation of Nanoformulations of Beta-Cyfluthrin against *Callosobruchus maculatus* (Coleoptera: Bruchidae). *J. Environ. Sci. Health B* **2012**, *47* (7), 687–691.

Naderi, M. R.; Abedi, A. *J. Nanotech.* **2012**, *11* (1), 18–26.

National Institute of Food and Agriculture (NIFA), United States Department of Agriculture (USDA). *Nanotechnology Program*, 2017.

Pankaj Shakil, N. A.; Kumar, J.; Singh, M. K.; Singh, K. Bioefficacy Evaluation of Controlled Release Formulations Based on Amphiphilic Nano-Polymer of Carbofuran against Meloidogyne Incognita Infecting Tomato. *J. Environ. Sci. Health B* **2012**, *47* (6), 520–528.

Pérez-de-Luque, A.; Rubiales, D. Nanotechnology for Parasitic Plant Control. *Pest Manage. Sci.* **2009**, *65* (5), 540-545.

Philip, D. Mangifera Indica Leaf- Assisted Biosynthesis of Well Dispersed Silver Nanoparticles. *Acad. J. Sci. Res.; Spectrochim. Acta Part A.* **2011**,*78* (1), 327–331.

Prasad, R.; Bagde, U.; Varma, A. An Overview of Intellectual Property Rights in Relation to Agricultural Biotechnology. *Afr. J. Biotechnol.* **2012**, *11* (73), 13746–13752.

Prasad, R.; Kumar, V.; Prasad, K. S. Nanotechnology in Sustainable Agriculture: Present Concerns and Future Aspects. *Afr. J. Biotechnol.* **2014**, *13* (6), 705–713.16.

Qamar, Z.; Nasir, I. A.; Husnain, T. In-Vitro Development of Cauliflower Synthetic Seeds and Conversion to Plantlets. *Adv. Life Sci.* **2014**, *1* (2), 34–41.

Nanofertilizers and Nanopesticides 211

Rai, M.; Ingle, A. Role of Nanotechnology in Agriculture with Special Reference to Management of Insect Pests. *Appl. Microbiol. Biotechnol.* **2012,** *94* (2), 287–293.

Ramesh, K.; Biswas, A. K.; Soma Sundaram, J.; Subba Rao, A. Nanoporous Zeolites in Farming: Current Status and Issues Ahead. *Curr. Sci.* **2010,** *99* (6), 25.

Saharan, V.; Khatik, R.; Kumari, M.; Raliya, R.; Nallamuthu, I.; Pal, A. Proceedings of the 4th Annual International Conference on Advances in Biotechnology (BioTech 2014); Dubai, United Arab Emirates, March 10–11, 2014; p 23.

Sarkar, D. J.; Kumar, J.; Shakil, N. A.; Walia, S. Release Kinetics of Controlled Release Formulations of Thiamethoxam Employing Nano-Ranged Amphiphilic PEG and Diacid Basedblock Polymers in Soil. *J. Environ. Sci. Health A Tox Hazard Subst. Environ. Eng.* **2012,** *47* (11), 1701–1712.

Sasson, Y.; Levy-Ruso, G.; Toledano, O.; Ishaaya, I. *Nano-Suspensions: Emerging Novel Agrochemical Formulations Insecticides Design Using Advanced Technologies*; Springer-Verlag: Berlin Heidelberg, 2007; pp 1–39.

Schlich, K.; Hund-Rinke, K. Influence of Soil Properties on the Effect of Silver Nanomaterials on Microbial Activity in Five Soils. *Environ. Pollut.* **2015,** *196*, 321–330.

Schneider, A. Amid Nanotech's Dazzling Promise, Health Risks Grow. http://www.aolnews.com/nanotech/article/amidnantechs-dazzling-promise-health risks-grow/19401235, 2010; p 24.

Sharmila, R.; Nutrient Release Pattern of Nanofertilizer Formulation. Ph.D (Agri.) Thesis. Tamilnadu Agricultural University, Coimbatore, 2011.

Sharon, M.; Choudhary, A. K.; Kuma, R. Nanotechnology in Agricultural Diseases and Food Safety. *J. Phytol.,* **2010,** *2* (4), 83–92.

Sheng, Z.; Liu, Y. Effects of Silver Nano Particles on Waste Water Biofilms. *Water Res.* **2011,** *45*, 6039–6050.

Shi, X.; Wan, S.; Meshinchi, S.; Van Antwerp, M. E.; Bi, X.; Lee, I.; Baker, J. R. Dendrimer – Entrapped Gold Nanoparticles as a Platform for Cancer Cell Targeting and Imaging. *Small,* **2007,** *3*, 1245–1252.

Shoults, W.; Reinsch, W. A.; Tsyusko, B. C.; Bertsch, O. V.; Lowry, P. M.; Unrine, J. M. Role of Particle Size and Soil Type in Toxicity of Silver Nanoparticles to Earthworms. *Soils Sci. Soc. Am. J.* **2011,** *75*, 365–377.

Simonin, M.; Guyonnet, J. P.; Martins, J. M.; Ginot, M.; Richaume, A. Influence of Soil Properties on the Toxicity of TiO_2 Nanoparticles on Carbon Mineralization and Bacterial Abundance. *J. Hazard. Mater.* **2015,** *283*, 529–535.

Singh, M. D. Nano-Fertilizers Is a New Way to Increase Nutrients Use Efficiency in Crop Production. *Int. J. Agric. Sci.* **2017.**

Singh, S.; Singh, B. K.; Yadav, S. M.; Gupta, A. K. Applications of Nanotechnology and Their Role in Disease Management. *Res. J. Nanosci. Nanotech.* **2014.**

Subramanian, K. S.; Paulraj, C.; Natarajan, S. Nanotechnological Approaches in Nutrient Management. *Nanotechnol. App. Agric.* **2008,** 37–42.

Tarafdar, J. C.; Raliya, R.; Tathore, I. *J. Bionanosci.* **2012a,** *6*, 84–89.

Tarafdar, J. C.; Xiang, Y.; Wang, W. N.; Dong, Q.; Biswas, P. *Appl. Biol. Res.* **2012b,** *14*, 138–144.

Ulrichs, C.; Mewis, I.; Goswami, A. Crop Diversification Aiming Nutritional Security in West Bengal—Biotechnology of Stinging Capsules in Nature's Waterblooms. *Ann. Tech. Issue State Agric. Technol. Serv. Assoc.* **2005,** 1–18.

United Nations (UN). *World Population Projected to Reach 9.7 Billion by 2050*; UN: New York, 2015.

Vittori Antisari, L.; Carbone, S.; Gatti, A.; Vianello, G.; Nannipieri, P. Toxicity of Metal Oxide (CeO2, Fe3O4, SnO2) Engineered Nanoparticles on Soil Microbial Biomass and Their Distribution in Soil. *Soil Biol. Biochem.* **2013**, *60*, 87–94.

Wang, Y.; Hu, J.; Dai, Z.; Li, J.; Huang, J. In Vitro Assessment of Physiological Changes of Watermelon (*Citrullus lanatus*) Upon Iron Oxide Nanoparticles Exposure. *Plant Physio. Biochem.* **2016,** *108*, 353–360.

Zhao, L.; Hernandez-Viezcas, J. A.; Peralta-Videa, J. R.; Bandyopadhyay, S.; Peng, B.; Munoz, B. ZnO Nanoparticle Fate in Soil and Zinc Bioaccumulation in Corn Plants (Zea mays) Influenced by Alginate. *Environ. Sci. Process. Impacts* **2013,** *15*, 260–266.

Zhou, B.; Hermans, S.; Gabor, A. Nanotechnology in Catalysis, 2004. ISBN: 978-0-306-555.

CHAPTER 8

GREENER METHODS OF NANOPARTICLE SYNTHESIS

MOHD YOUSUF RATHER[1,2], SOMAIAH SUNDARAPANDIAN[1*], and MOHAMMED LATIF KHAN[2]

[1]Department of Ecology and Environmental Sciences, Pondicherry University, Puducherry, India

[2]Department of Botany, Dr. Harisingh Gour Vishwavidyalaya (A Central University), Sagar, Madhya Pradesh, India

[*]Corresponding author. E-mail: smspandian65@gmail.com

ABSTRACT

Nanotechnology is regarded as the future of all technologies and the subject is evolving as a multidisciplinary research concept. Nanomaterials have currently found applications in almost all fields of science and technology. The wide use of metal nanoparticles (NPs) in the daily life of humans and their surrounding environment has urged the researchers to develop NP synthesis protocols that are economical, energy-efficient, and which can overcome the toxicity caused by chemically synthesized NPs. Therefore, nanotechnology and green chemistry combined and proposed the green methods for synthesizing such NP. Hence, this work explores the different approaches and methods that are used for NP production and it provides a brief explanation of how green methods are advantageous over other methods. Additionally, the influence of various factors such as temperature, reaction time, pH of the reaction medium, the volume of extract used, and the metal salt concentration used on the NP morphology and properties was investigated. The applications of green synthesized NPs have been highlighted as well and mostly these NPs are used in the applications meant for sustainable development. The work concludes that green methods of NP synthesis are the simplest,

214 Diverse Applications of Nanotechnology in the Biological Sciences

cheapest, and most effective methods for manufacturing NPs at a larger scale with no adverse environmental factors.

8.1 INTRODUCTION

Nanotechnology is an art of synthesizing and manipulating particles or structures of nanoscale, that is, at least one dimension below 100 nm size (Masciangioli and Zhang, 2003) and it has fascinated a lot of researchers in course of time because of its diverse range of applications (Lohrasbi et al., 2019). The National Nanotechnology Initiative (NNI) of the United States has defined the nanoscience and nanotechnology as, "*Nanoscience involves research to discover new behaviors and properties of materials with dimensions at the nanoscale which ranges roughly from 1 to 100 nanometers (nm). Nanotechnology is the way discoveries made at the nanoscale are put to work. Nanotechnology is more than throwing together a batch of nanoscale materials – it requires the ability to manipulate and control those materials in a useful way*" (NNI Strategic Plan, 2007). Hence, nanotechnology is the science, which deals with the synthesis of nanomaterials and their potential applications for human use.

The nanotechnology concept is a very old concept dating back to the ninth century but the awareness about the subject was not proper. Mesopotamia artisans used gold nanoparticles (NPs) (GNPs) and silver NPs (SNPs) for producing the glittering effect on the surface of pots and it is evident on the Lycurgus Cup back (Horikoshi and Serpone, 2013). The industrial use of colloidal particles dates back to 1676 and the gold colloidal particles were used for the application in 1718 (Helcher, 1718). In 1857, Michael Faraday published a paper "Experimental relations of gold (and other metals) to light" that is the first scientific explanation of NP properties (Faraday, 1857). The NPs were hypothesized as "magic bullets" as early as 1906 by the Nobel Prize winner P. Ehrlich (Himmelweit, 2017). Richard Feynman's talk of 1959 entitled "There's plenty of space at the bottom" that describes "the molecular machines built with atomic precision" was reflected as the first insight on nanotechnology. The talk spawned the thought and theory of nanoscience and the control and manipulation of atoms and molecules gave rise to nanotechnology (Das and Chatterjee, 2019).

World focus was turned toward NP use in drug delivery in the 1950s and 1960s, and Professor Peter Paul Speiser was one of the pioneers. His research group focused on the investigation of the oral administration of polyacrylic beads, and then on microcapsules. Late in the 1960s, his group produced the first NPs for the application of drug delivery. This was the beginning of NP research for the application of drug delivery and it was followed by advancement in transport of drugs through the blood–brain barrier. Theoretically, the revolution of nanotechnology is believed to have begun in the 1980s as K. Eric Drexler published the first report on nanotechnology (Drexler, 1981). With regular developments in the techniques used in nanotechnology over a while, it has achieved the status of being named as the future of all technologies. Recently, nanotechnology has been evolving as a multidisciplinary research concept (Nasrollahzadeh et al., 2019). It is developed rapidly and the last decade has witnessed considerable development in this field (Das and Chatterjee, 2019).

8.2 NANOPARTICLES

NPs are generally defined as the materials with 1–100 nm dimensions (Dey et al., 2018). Some important definitions of NPs are as follows (Horikoshi and Serpone, 2013; Das and Chatterjee, 2019):

- International Organization for Standardization (ISO), "*A particle spanning 1–100 nm (diameter)*"
- National Institute of Occupational Safety and Health (NIOSH), "*A particle with a diameter between 1 and 100 nm and a fiber spanning the range 1–100 nm*"
- American Society of Testing and Materials (ASTM), "*An ultrafine particle whose length in two or three places is 1–100 nm*"
- British Standards Institution (BSI), "*All the three dimensions of Nano object are in the nanoscale range*"
- Scientific Committee on Consumer Products (SCCP), "*At least one side is in the nanoscale range*"
- Bundesanstalt für Arbeitsschutz und Arbeitsmedizin (BAuA), "*All the fields or diameters are in the nanoscale range*"

The NPs reveal entirely novel physiochemical properties in contrast to their bulk material such as they possess an extremely small size and high surface-to-volume ratio (Siddiqui et al., 2015; Santhoshkumar et al., 2017). They display different properties as well, for example, titanium dioxide is white-colored in bulk and it has no color at nanosize. In general, the properties like color, weight, strength, electrical conductivity, and chemical

216   Diverse Applications of Nanotechnology in the Biological Sciences

properties of NPs completely differ from that of bulk matter. Because of nanoscale dimensions and different variety of properties, they are used in numerous applications in every field of science and technology and this has also resulted in their increased need and rate of production. They also attract scientific interest because they fill the gap between the atomic and bulk structures (Rauwel et al., 2015). These properties depend on the shape, size, and chemical surroundings (Chang, 2005). The chemical reactivity of the substance depends heavily on the surface-area-to-volume ratio and hence the tunable particle size and morphology make their synthesis interesting. Keeping in view the demand for NPs and the promise nanotechnology holds for its potential applications in various fields, the synthesis of nanomaterials is gaining thrust for a few bygone decades.

8.3 APPROACHES OF NANOPARTICLE SYNTHESIS

In general, two approaches for NP synthesis have been defined: the top-down and the bottom-up approaches (Abdelghany et al., 2018). The two approaches of NP synthesis are demonstrated in Figure 8.1 and all the methods of NP synthesis follow either of two approaches.

8.3.1 TOP-DOWN APPROACH

The NPs are prepared by size reduction in the top-down approach. It includes the breaking down of bulk materials into fine particles (Husein and Nassar, 2008) with the help of suitable lithographic technique (Meyers et al., 2006). By the application of multiple external huge forces, for example, temperature, cutting, stress, crushing, grinding, strain, cryo-grinding, physical pressure, disruption, homogenization, and degradation, the pulverization of bulk material to fine particulate material is done (Lade and Shanware, 2020). The top-down approach creates some imperfections in the NP surface (Nasrollahzadeh et al., 2019).

8.3.2 BOTTOM-UP APPROACH

In the bottom-up approach, the NPs are synthesized from smaller entities like atoms and molecules (Mukherjee et al., 2001). The approach utilizes the reducing and capping agents for synthesizing NPs, for example, the

Greener Methods of Nanoparticle Synthesis

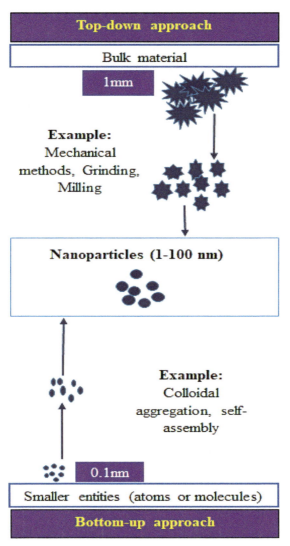

FIGURE 8.1 Approaches of nanoparticle synthesis.

actual entities could be the electrons of the metal ions and the electrons of the reducing/capping agents (Lade and Shanware, 2020). It includes the nucleation and growth of atoms or molecules (Husein and Nassar, 2008). As of now, due to the simplicity of bottom-up approaches, it has dominated the top-down approaches of NP synthesis (Handoko et al., 2019).

8.4 METHODS OF NANOPARTICLE SYNTHESIS

Physical, chemical, and biological methods are exploited for the synthesis of various metal NPs (MNPs) (Iravani et al., 2014; Gade et al., 2014). Some of the methods of NP synthesis are presented in Figure 8.2.

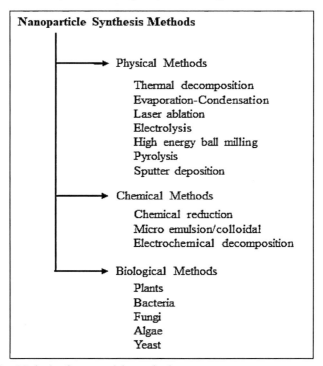

FIGURE 8.2 Methods of nanoparticle synthesis.

8.4.1 PHYSICAL METHODS

The physical methods fall in the category of top-down approach and NPs are synthesized from pulverization of bulk materials with the help of external forces. These methods are used to synthesize NPs at a large scale (Jin et al., 2003; Oliveira et al., 2005; Merga et al., 2007). Some of the methods are laser ablation (Scaramuzza et al., 2015), arc discharge method (Tsuji et al., 2002), high-energy ball milling (Kuang et al., 2015), metal sputtering (Tsuji et al., 2003), electro-spraying (Foroutan et al., 2015), atomization (Okuyama and Lenggoro, 2003), evaporation–condensation (Abbasi et al., 2016), annealing (Nakade et al., 2002), and laser pyrolysis (Bouhadoun et

Greener Methods of Nanoparticle Synthesis

al., 2015). These methods are time and energy-consuming because they take too much time to attain thermal stability and hence consume a significant amount of energy (Kawasaki and Nishimura, 2006). The instruments needed for physical synthesis are very expensive and their maintenance is costly as well. Additionally, the NPs synthesized by physical processes possess some surface defects and contamination.

8.4.2 CHEMICAL METHODS

Chemical methods of NPs synthesis fall under bottom-up techniques. The NP synthesis by chemical methods is a multistep process that includes reduction, nucleation, growth, coarsening, and agglomeration (Cushing et al., 2004). Some of the chemical methods are microemulsion techniques, sol–gel process, chemical vapor deposition process, pyrolysis, hydrothermal synthesis, etc. and the reducing agents used in these processes are elemental hydrogen, ascorbate, gallic acid, sodium citrate, hydroquinone, sodium borohydride, etc. (Tarasenko et al., 2006). Chemicals like sodium borohydride, hydroxylamine, poly-N-vinylpyrrolidone, and polyvinyl alcohol are reported of damaging human health as well as the environment (Chhipa, 2019). Some chemicals used for stabilization of synthesized NPs are alcohols, acids, thiols, and amines. The use of capping agents and surfactants is mandatory in chemical synthesis methods and hence the use of chemicals cannot be avoided in these methods of NP synthesis (Lade and Shanware, 2020). These reducing agents and surfactants used are also toxic to the human and the environment (Rauwel et al., 2015; Abegunde et al., 2019). Sometimes, the key product and by-products generated in these methods tend to cause contamination of terrestrial and aquatic ecosystems, contributing inevitably to their environmental accumulation (Lade and Shanware, 2020).

8.4.3 BIOLOGICAL METHODS

Multiple disadvantages are related to physical and chemical methods; however, the problems have been answered by the biological methods. Biological methods are of considerable importance in the NP synthesis area. These methods use the matter or living things as reducing agents; therefore, they culminate the expensive nature and imperfections in the morphology of NPs caused by the physical methods and also avoid the use of toxic chemicals (Bharimalla, 2018). The green reducing agents include extracts of plants, bacteria, etc. that are environmentally safe, nonhazardous to humans, and biodegradable. The adaptation of biological methods of NP synthesis has become common, as it is clear from the reports that the SNPs synthesis has been attempted by using organisms from four kingdoms of living organisms out of five which include Monera, Protista, Fungi, and Plantae (Mohanpuria et al., 2008). In most cases, both the reducing and stabilizing agents are present within the same biological extracts and hence there is no need for adding extra chemicals (Pal et al., 2019).

The wide use of MNPs in the daily life of humans and their surrounding environment has urged the researchers to develop NP synthesis protocols that are economical and don't require the use of toxic chemicals. The biological methods of NP synthesis are one of them. The biological methods of NP synthesis are human and environmentally safe and therefore are also termed as *"green methods."* Green methods follow the principles of nature and involve natural phenomena for reducing metal salts to NPs, hence they are used as an alternative method for physical and chemicals approaches of NP synthesis (Hemmati et al., 2019).

8.5 GREEN SYNTHESIS

To overcome the toxicity, which can be caused to the environment and the humans by chemically synthesized NPs (Patil and Kim, 2017), the considerable requirement of safer fabrication of NP has encouraged researchers for the development of new and eco-friendly synthesis methods (Khalil et al., 2017). Therefore, nanotechnology combined with the "Green Chemistry" principles that state as *"Utilization of a set of principles that reduces or eliminates the use or generation of hazardous substances in the design, manufacture, and application of chemical products"* (Anastas and Warner, 1998) has rapidly developed as a multidisciplinary research area. The green synthesis emphasizes on nonhazardous, clean, and environment-friendly procedures and techniques for nanomaterial synthesis. GNPs (Dey et al., 2018), iron (Rather and Sundarapandian, 2020), palladium (Yang et al., 2010), SNPs (Rather et al., 2020), zinc (Elumalai et al., 2015), and several other metals have been synthesized by biological methods and have been successfully used for various applications (Pal et al., 2019). The green methods are simple, efficient, eco-friendly, and economical in competition to other methods and the synthesis is easily saleable up to industrial level (Iravani et al., 2014). The biological methods are called green methods because the initial, as well as the by-products of the process, are biodegradable and eco-friendly. Various routes for green NP synthesis are explained in more detail.

8.5.1 ACTINOMYCETES

Actinomycetes belong to phylum *Actinobacteria* and they share various characteristics with fungi. They are used for producing two-thirds of the overall antimicrobial compounds in usage today (Bentley et al., 2002). NPs can be synthesized using actinomycetes both intracellularly and extracellularly and studies on ultrasmall size, stable, and polydisperse MNPs synthesis by Actinomycetes have been reported (Manimaran and Kannabiran, 2017). Synthesis of monodisperse spherical GNPs stable for about six months was reported from *Thermomonospora* sp. (Ahmad et al., 2003). The *Streptomyces* sp. (Alani et al., 2012), *Streptomyces* HBUM 171191 (Waghmare et al., 2011), *Rhodococcus* NCIM 2891 (Otari et al., 2012), and *Thermoactinomyces* sp. (Deepa et al., 2013) were also reported for SNPs synthesis. The *Rhodococcus* sp. (Ahmad et al., 2003), *Streptomyces hygroscopicus* (Waghmare et al., 2014) were used for GNPs synthesis. The electrostatic attraction between the negatively charged carboxylate groups presents in the enzymes of the cell wall of mycelia and the positively charged metal ions is responsible for the intracellular NP synthesis (Sunitha et al., 2013). For the extracellular synthesis of NPs, the enzymes involved in the nitrogen cycle are proposed to be responsible for the electron shuttle enzymatic reduction of metal ions (Karthik et al., 2014).

8.5.2 ALGAE

Algae, a diverse category of aquatic organisms varying from microscopic (picoplankton) to macroscopic (Rhodophyta), are extensively used for NP synthesis (Ahmad et al., 2019). GNPs were synthesized from a blue-green algae *Spirulina platensis* (Suganya et al., 2015), a brown-seaweed *Sargassum wightii* (Singaravelu et al. 2007), and diatoms *Diadesmis gallica* and *Navicula atomus* (Schröfel et al., 2011). *S. wightii* was reported for the production of SNPs (Govindaraju et al., 2009), *Cystophora moniliformis* (Prasad et al., 2013), *Caulerpa racemosa* (Kathiraven et al., 2015), *Chaetomorpha linum* (Kannan et al., 2013), *Gelidium amansii* (Pugazhendhi et al., 2018), and *Chaetoceros salina, Chaetoceros calcitrans, Tetraselmis gracilis,* and *Isochrysis galbana* (Merin et al., 2010). In addition to these, SNPs were also synthesized from polysaccharide extracts of *Jania rubins, Pterocladia capillacae, Ulva fasciata,* and *Colpomenia sinuosa* species (El-Rafie et al., 2013). Using algae for NP synthesis is advantageous because the surface of the algal cells is negatively charged which accelerates the nucleation and crystal growth and they can also be used for cost-effective large-scale synthesis (Sharma et al., 2019). It was also proposed that the enzymes and functional groups of algal cell wall of algae are accountable for the metal ion reduction by forming complexes with them and thereby forming MNPs (Gade et al., 2014).

8.5.3 BACTERIA

Bacterial species produce various inorganic materials intracellularly or extracellularly (Pal et al., 2019). The bacterial species can detoxify heavy metals as their system reduces the metal ions to insoluble nontoxic metal products (Kishen et al., 2020). SNPs synthesis from *Pseudomonas stutzeri* AG259 bacterial strain segregated from a silver mine was the first evidence (Prabhu and Poulose, 2012). In recent times, however, bacteria-mediated NP synthesis has developed as a rapidly growing field of green nanotechnology globally. It was reported that many microorganisms can produce inorganic nanostructures and MNPs with similar properties to chemically synthesized materials (Arya, 2010). The causative factor behind microbe-mediated NP production is reported to be nitrate reductase. Many researchers supported nitrate reductase as the principal factor for the extracellular synthesis of NPs. To quote: the *Rhodopseudomonas capsulate* secretes the cofactors NADH and NADH-dependent enzymes that are responsible for the extracellular synthesis of GNPs (He et al., 2007). The bioreduction of gold ions has been shown to be triggered by the electron transfer from the NADH. The electrons are accepted by the gold ions and are subsequently reduced to their seed particles, followed by aggregation to form GNPs. The biosynthesis of GNPs by bacterium *Stenotrophomonas maltophilia* and their stabilization also involves NADPH-dependent reductase enzyme that is responsible for conversion of gold ions to GNPs (Nangia et al., 2009). Rai et al. (2009) also reinforced the fact that large quantities of enzymes are secreted by biological agents that hydrolyze the metals and reduce them.

8.5.4 FUNGI

Fungi have been used for synthesizing metallic NPs because they show resistance and ability to bioaccumulate metals, strong binding potential, and intracellular uptake of metals comparable to bacteria (Ahmad et al., 2003).

Greener Methods of Nanoparticle Synthesis

The NP fabrication by fungal extracts is further beneficial than utilizing other microbes because they grow quicker and are modest to treat and produce than bacteria under laboratory conditions (Chhipa, 2019). Fungal species may also tolerate different conditions like agitation and flow pressure in reaction instruments (Pal et al., 2019). Fungal-mediated NP synthesis mechanism is different in that fungal species secrete enormous quantities of enzymes that facilitate the reduction of metal ions to NPs (Mandal et al., 2006). The fungal species secrete the enzymes to protect themselves from metal salt to which they are exposed and subsequently the enzymes reduce the metal ions to MNPs (Kishen et al., 2020). Some fungi such as *Alternaria* sp., *Bipolaris* sp., *Colletotrichum* sp., *Humicola* sp., *Neurospora* sp., *Penicillium* sp., *Phanerochaete* sp., *Schizophyllum* sp., and *Trichoderma* sp. were used for the NPs synthesis but *Aspergillus* and *Fusarium* families were explored more (Chhipa, 2019). The first fungus used for NP synthesis was *Verticillium* (Mukherjee et al., 2001). Extract of *Volvariella volvacea* was used for the synthesis of SNPs, GNPs, and SNP–GNPs extracellularly (Philip, 2009). The fungi produce napthoquinones and anthraquinones (Medentsev and Akimenko, 1998; Bell et al., 2003) that act as reducing agents.

8.5.5 PLANTS

NP synthesis by this greener method is a novel technique. The process involves the use of plants or parts of the plant to reduce the metal salt into MNPs (Rauwel and Rauwel, 2017). In recent years, these processes have appealed to the attention of researchers as the synthesis processes are one-step, nonpathogenic, economic, and eco-friendly procedures (Rather and Sundarapandian, 2020). It is the simplest, cheapest, and fastest method of NP synthesis. The plant extract–mediated NP synthesis is even conducted at ambient room conditions. The use of plant extracts in NP synthesis has been proven advantageous over other biological methods by many authors (Shankar et al., 2016; Hemmati et al., 2019). Additionally, the biomolecules responsible for reducing the metal salt also functionalize the NP surface (Nasrollahzadeh et al., 2019). Various plant parts that have been explored for NP synthesis are bark, callus, flower, fruit, leaves, root, seeds, stem, and tendril. The reducing agents are phytochemicals, for example, amino acids, alkaloids, proteins, polyols, polysaccharides, terpenoids, saponins, polyphenols, and vitamins present within the plant extract (Kharissova et al., 2013; Roy and Das, 2015). Plant extracts have been used for the synthesis of various MNPs and metal oxide NPs like SNPs (Rather et al., 2020), gold (Dey et al., 2018), iron oxide (Rather and Sundarapandian, 2020), palladium (Edayadulla et al., 2015), lithium (Álvarez et al., 2015), zinc oxide (Mehr et al., 2018), titanium dioxide (Chowdhury et al., 2016), copper oxide (Mehr et al., 2018), cerium dioxide (Darroudi et al., 2014), etc. and even nanocomposites (Venkateswarlu and Yoon, 2015) and quantum dots (Sachdev and Gopinath, 2015) have been successfully synthesized using plant extracts.

8.5.6 YEASTS

Yeast, a single-celled eukaryotic organism is primarily used for the synthesis of semiconductor materials (Pal et al., 2019). Yeasts are also preferred for the large-scale synthesis of NPs (Kishen et al., 2020). As many as 1500 species of yeast have been identified (Yurkov et al., 2011). Several species have been used for NP synthesis like MKY3 yeast strain for SNPs (Kowshik et al., 2002), *Saccharomyces cerevisiae* broth for SNPs (Jha and Prasad, 2008) and manganese dioxide (Salunke et al., 2015), *S. cerevisiae* strain AP22 and CCFY-100 (Sen et al., 2011), *Saccharomyces boulardii* (Kaler et al., 2013), and marine yeast *Yarrowia lipolytica* (Agnihotri et al., 2009; Pimprikar et al., 2009; Nair et al., 2013) for gold, etc. The growth of yeasts can be controlled under laboratory conditions, and they grow rapidly even in simple nutrients which gives them an advantage over the use of bacteria in NP synthesis (Kumar et al., 2011).

8.5.7 BIOMOLECULAR TEMPLATES

Several biomolecular templates like cell membranes, nucleic acids, etc. are reportedly used for NP generation. One of the excellent templates is DNA because it is attracted to transition metal ions (Pal et al., 2019). DNA as a biotemplate was used for the synthesis of GNPs (Kundu and Jayachandran, 2008; Zinchenko et al., 2014). Synthesis of highly stable SNPs of the wire-like structure was reported from the DNA template as well (Majumdar et al., 2013). The DNA has been used as reducing as well as capping agent for the production of gold, palladium, and cadmium sulfide nanowires (Kundu et al., 2013). Similarly, biological membranes like rubber membrane made from *Hevea brasiliensis* have been used for GNPs synthesis (Cabrera et al., 2013). Viruses as templates have also been described for NPs synthesis (Pokorski and Steinmetz, 2010), for instance, tobacco mosaic virus (TMV) for iron oxides, SiO_2, PbS, and CdS (Shenton et al., 1999). Zinc sulfide and

Greener Methods of Nanoparticle Synthesis 225

cadmium sulfide quantum dots were synthesized from M13 bacteriophage (Lee et al., 2002; Mao et al., 2003).

8.6 PROCESS OF NANOPARTICLE SYNTHESIS BY GREEN METHODS

NP synthesis by biological methods commonly involves reactant 1 (metal salt solution) and reactant 2 (extract from a biological source), and the process is completed in two steps, that is, nucleation and stabilization (Edayadulla et al., 2015). The reactants interact together forming metal complexes that are reduced to seed particles subsequently. These seed particles undergo nucleation and aggregation followed by stabilization by biomolecules present within the extract (Keihan et al., 2017). A range of biological extracts have been studied for the synthesis of homogenous, monodisperse, and stable NPs and it was reported that biomolecules vitamins, polysaccharides, amino acids, polyphenols, peptides, and proteins are majorly responsible for NPs synthesis (Chhipa, 2019). The whole process operates at certain parameters and the synthesis is significantly enhanced by stabilizing agents (Lade and Shanware, 2020). The capping agents prevent the agglomeration and mass synthesis of NPs. The optimal results can be achieved at specific process factors like temperature, reaction time, the pH of reaction medium, the volume of extract used, the metal salt concentration used, etc. In general, green nanotechnology with the aid of biotechnological techniques uses reducing agents from biological sources, so the NPs synthesized by these methods are human and environment-friendly (Hussain et al., 2016).

8.7 FACTORS AFFECTING NANOPARTICLE SYNTHESIS

NP synthesis is a very critical process and numerous factors are responsible for the varying quality and yield of NPs. The factors like the concentration of leaf extract, the concentration of metal salt, temperature, pH, reaction time, etc. are imperative in the synthesis process. These parameters are adjusted for achieving the NPs of the required properties. Some of the factors that affect the NP shape, size, morphology, and properties are described briefly.

8.7.1 TEMPERATURE

The NP synthesis by green synthesis is normally conducted at room temperature but by increasing the temperature, the rate of reaction can be enhanced. For example, the rapid reduction of silver ions to SNPs was achieved on increasing the reaction temperature (Lade and Patil, 2017). Philip et al. (2009) and Dwivedi and Gopal (2010) also found that the rate of reaction increases as soon as the temperature as increases. Reaction temperature also influences the shape of the NP, for example, at lower temperature, triangle-shaped NPs were obtained while as more spherical ones were obtained at higher temperatures from *Cymbopogon flexuosus* leaf extract (Raju et al., 2011). The size can be controlled by the temperature of the process as well, as and the size of NPs at higher temperatures was found small (Lade and Patil, 2017). It has been reported that the bimetallic NPs (silver and gold) synthesized from leaf extract of *Anacardium occidentale* required less amount of extract at higher temperatures and more amount at a lower temperature (Sheny et al., 2011). The synthesis of SNPs by *Melia azedarach* at different temperatures was reported and increased production of NPs was observed with increasing temperatures (Sukirtha et al., 2012).

8.7.2 REACTION TIME

The morphology of NP depends upon the reaction time or contact time (Darroudi et al., 2011). It has been found that longer reaction time or contact time between reducing agents and metal salt results in the aggregation of the NPs and this might reduce the potential of NPs as well (Baer, 2011). SNPs synthesized using an extract of *Capsicum annuum* with a reaction time of 5, 9, and 13 h had a size of 10, 25, and 40 nm, respectively, which depicts that increase in incubation time can lead to an increase in the NP size (Li et al., 2007). With an increase in incubation time, an increase in the absorption peak was reported in synthesized gold and SNPs (Dubey et al., 2010). The optimum incubation time for synthesis and stability of SNPs using leaf extract of mangosteen was observed to be 1 h (Veerasamy et al., 2011).

8.7.3 THE pH OF THE REACTION MEDIUM

The pH is an important parameter that is responsible for different shapes and sizes of NP synthesized (Armendariz et al., 2004) as this alters the electrical loads of biomolecules that may influence the capping and stabilization process and subsequently affecting the growth of the NPs (Vadlapudi and Kaladhar, 2014). The conformation of proteins or peptides specifically influenced by shifts in the protonation condition of the amino acids generates NPs of different shapes and dimensions (Gericke and Pinches, 2006). Different pH of the reaction mixture produces NPs of different morphologies, for example, acidic pH of reaction mixture produced NPs of larger size compared to the higher pH (Dubey et al., 2010). Similarly, large and relatively small-sized GNPs were synthesized at pH 2 and pH 3–4 from *Avena sativa* (Armendariz et al., 2004). It was found that at pH 5 and above, spherical NPs are achieved from bark extract of *Cinnamon zeylanicum* and it was explained that the smaller size at high pH could be due to the availability of more functional groups to bind with metal ions (Sathishkumar et al., 2009). The different biological extracts would have different pH, so it is important to optimize the pH of the reaction mixture for the optimized synthesis of NPs.

8.7.4 VOLUME/CONCENTRATION OF EXTRACT

Volume/concentration of biological extract is also crucial for NP synthesis and it affects the rate and time of reaction. The volume of extract used greatly influences the average particle size of synthesized NPs and the volume up to a certain quantity is only helpful in NP formation (Rather et al., 2020). Different concentrations of fruit extract of *Tanacetum vulgare* were used for silver and GNPs synthesis and the NPs of larger size were achieved at lower concentrations of extracts (Dubey et al., 2010). A similar decrease in particle size of silver and GNPs with an increase in leaf extract concentration was found on using *Chenopodium album* leaf extract (Dwivedi and Gopal, 2010). The volume and concentration indicate the content of reducing biomolecules (e.g., anthocyanin's, polysaccharides, phenols, tannins, and polyphenols in leaf extract) present in the extract which significantly contribute to the reduction of metal ions to NPs and hence its optimization becomes necessary for manufacturing NPs (Lade and Shanware, 2020). It has been also reported that synthesized SNPs by low concentrated *Carica papaya* peel extract were found uniformly distributed but aggregation and destabilization were observed when the extract concentration was further increased (Balavijayalakshmi and Ramalakshmi, 2017).

8.7.5 METAL SALT CONCENTRATION

It has been reported that the metal concentration also affects the NP formation. Several different metal concentrations for NP synthesis were used and 1-mM metal salt concentration yielded better results (Gondwal and Pant, 2013). On increasing the metal salt concentration, the metal ion reduction will be less effective that could be visible through salt accumulation (Lade and Shanware, 2020). The SNPs were synthesized at different metal salt concentrations and the size of NPs was observed to be increasing with an increased metal salt concentration (Dubey et al., 2010). They also found that the size in the case of GNPs reduces when the metal salt concentration was increased from 1 to 2 mM but it increased when the metal concentration was further increased to 3 mM. Similar results were reported for silver and GNPs synthesized from *Chenopodium album* leaf extract (Dwivedi and Gopal, 2010). They also observed the faster rate of NP synthesis at higher metal salt concentrations.

8.8 ADVANTAGES OF GREEN METHODS

The green methods of NP synthesis are mostly preferred because of their eco-friendly nature. These processes require less human work compared to physical and chemical methods. The chemical synthesis methods usually require chemicals for generating NPs that are sometimes toxic to human health and the environment, and the green synthesis methods replace these chemicals with biomolecules. Some of the important advantages of green methods over physical and chemical methods are:

- The green methods are easy to handle.
- The procedures are one-pot, economic, and environment-friendly.
- For the synthesis process, the methods and instrumentation used are very simple.
- The reducing agents, that is, biological extracts can be prepared anytime and in any quantities.
- NPs with different morphologies can be synthesized.
- Mostly NPs are synthesized at room temperature
- Both the reducing and capping/stabilizing agents responsible for NP synthesis are present within the same extract.

8.9 LIMITATIONS OF GREEN METHODS

Multiple advantages of green methods of NP synthesis have been provided till now in this chapter but there are some limitations attached to these methods as well. Some of them are (Das and Chatterjee, 2019):

- Resource availability such as cultures of microbial growth
- Uniformity in biomolecules in different extracts
- Separation of NPs when they are synthesized intracellularly

Greener Methods of Nanoparticle Synthesis

- Heterogeneous NP production
- The conversion rate of reactants to products is less
- Reproducibility of the results
- Large-scale applicability

8.10 APPLICATIONS OF GREEN NPS

The property of NPs to change their color in the solution was described by Mie in 1908 (Hussain et al., 2016). NPs have displayed new features that are different from the atom-based properties and some of these essential characteristics are large surface-area-to-volume ratio, large surface energy, specific electronic structure, plasmon excitation, and quantum confinement (Ahmed et al., 2016). Because of these unique physiochemical properties, they possess countless uses and applications. They have been found more useful in many applications in the field of science like biology, chemistry, medical sciences, pharmaceutics, energy science, chemical sensing and imaging, information storage, catalysis, photonics, environmental remediation, biological labeling, and drug delivery (Vanaja and Annadurai, 2013; Shankar et al., 2016; Foo et al., 2018; Anastopoulos et al., 2018). Other fields include electronics (Maekawa et al., 2012), cosmetics (Azzazy et al., 2012), chemical industry (Virkutyte and Varma, 2011), food industry (Sanguansri and Augustin, 2006), etc. NPs are generally studied for medicine (Burgess, 2009; Blasiak et al., 2013), agriculture (Chen and Yada, 2011; Khot et al., 2012), and food industry (Sanchez-Garcia et al., 2010; Martirosyan and Schneider, 2014). They also offer a novel potential for the treatment of wastewater contaminated with various pollutants like heavy metals, toxic metal ions, microorganisms, and organic and inorganic solutes. Hence, nanotechnology is regarded as one of the most promising technologies that could resolve multiple problems, including water quality and purification (Sánchez et al., 2011; Rather and Sundarapandian, 2020). Agriculture confronted significant challenges in the 21st century owing to climate change, overuse of chemicals, and enhanced tolerance of phytopathogens and pests, and nanotechnology offered solutions to such problems (Solanki et al., 2015; Zielonka and Klimek-Ochab, 2017). In agriculture, NPs have been used as nanofertilizers, nanoherbicides, and nanosensors for pathogen detection, nanofungicides, nanopesticides, etc. They are readily accepted in food-packaging industries also as they are reported to increase the shelf life of several packaged food products (Chhipa, 2019).

8.11 CONCLUSION

NPs are used in every field in the daily life of humans and their surrounding environment. Hence, to overcome the toxicity which can be caused to the environment and the humans by chemically synthesized NPs, researchers have been urged to develop alternate synthesis protocols. Green synthesis methods have been developed as a significant means for the synthesis of biologically safe and cost-effective NPs. The processes are simple, easy, economical, and time-efficient. The biggest advantage of greener methods is that it is a single-pot method and both reducing and capping agents are present within the same biological extract. Different factors like temperature, reaction time, pH, metal concentration, etc. influence the NP synthesis process and hence these factors must be considered for the optimization of the results. The green synthesized NPs have been explored for a diversity of applications and mostly the results have been reported as significant. The field of green nanotechnology is still in its initial stages and a lot needs to be explored in the coming years. For example, the green methods have been widely used for monometallic or bimetallic NP synthesis and the processes should be explored for the synthesis of nanocomposites as well. Elaborated studies both by independent scientists and governments on the effect of green NPs on larger animals, including humans, are advised.

ACKNOWLEDGMENTS

The authors would like to acknowledge the financial support provided by the Department of Biotechnology, Government of India (grant number No. BT/PR12899/NDB/39/506/2015 dt. 20/06/2017)

KEYWORDS

- **biosynthesis**
- **green chemistry**
- **nanoparticle**
- **plant extract**
- **synthesis methods**

REFERENCES

Abbasi, E.; Milani, M.; Fekri, Aval S.; Kouhi, M.; Akbarzadeh, A.; Tayefi Nasrabadi, H.; Nikasa, P.; Joo, S. W.; Hanifehpour, Y.; Nejati-Koshki, K.; Samiei M. Silver Nanoparticles: Synthesis Methods, Bio-applications and Properties. *Crit. Rev. Microbiol.* **2016,** *42,* 173–180.

Abdelghany, T. M.; Al-Rajhi, A. M.; Al Abboud, M. A.; Alawlaqi, M. M.; Magdah, A. G.; Helmy, E. A.; Mabrouk, A. S. Recent Advances in Green Synthesis of Silver Nanoparticles and Their Applications: About Future Directions. A Review. *BioNanoScience* **2018,** *8* (1), 5–16.

Abegunde, O. O.; Akinlabi, E. T.; Oladijo, O. P.; Akinlabi, S.; Ude, A. U. Overview of thin film deposition techniques. *AIMS Mater. Sci.* **2019,** *6* (2), 174–199.

Agnihotri, M.; Joshi, S.; Kumar, A. R.; Zinjarde, S.; Kulkarni, S. Biosynthesis of Gold Nanoparticles by the Tropical Marine Yeast *Yarrowia lipolytica* NCIM 3589. *Mater. Lett.* **2009,** *63,* 1231–1234.

Ahmad, A.; Mukherjee, P.; Senapati, S.; Mandal, D.; Khan, M. I.; Kumar, R.; Sastry, M. Extracellular Biosynthesis of Silver Nanoparticles Using the Fungus *Fusarium oxysporum*. *Colloids Surf. B.* **2003,** *28* (4), 313–318.

Ahmad, A.; Senapati, S.; Khan, M. I.; Kumar, R.; Ramani, R.; Srinivas, V.; Sastry, M. Intracellular Synthesis of Gold Nanoparticles by a Novel Alkalotolerant Actinomycete, *Rhodococcus* Species. *Nanotechnology* **2003a,** *14,* 824–828.

Ahmad, A.; Senapati, S.; Khan, M. I.; Kumar, R.; Sastry, M. Extracellular Biosynthesis of Monodisperse Gold Nanoparticles by a Novel Extremophilic Actinomycete, *Thermomonospora* sp. *Langmuir* **2003b,** *19* (8), 3550–3553.

Ahmad, S.; Munir, S.; Zeb, N.; Ullah, A.; Khan, B.; Ali, J.; Bilal, M.; Omer, M.; Alamzeb, M.; Salman, S. M.; Ali, S. Green Nanotechnology: A Review on Green Synthesis of Silver Nanoparticles—An Ecofriendly Approach. *Int. J. Nanomed.* **2019,** *14,* 5087.

Ahmed, S.; Ahmad, M.; Swami, B. L.; Ikram, S. A Review on Plants Extract Mediated Synthesis of Silver Nanoparticles for Antimicrobial Applications: A Green Expertise. *J. Adv. Res.* **2016,** *7* (1), 17–28.

Alani, F.; Moo-Young, M.; Anderson, W. Biosynthesis of Silver Nanoparticles by a New Strain of *Streptomyces sp.* Compared with *Aspergillus fumigatus*. *World. J. Microbiol. Biotechnol.* **2012,** *28* (3), 1081–1086.

Álvarez, R. A.; Cortez-Valadez, M.; Britto-Hurtado, R.; Bueno, L. O. N.; Flores-Lopez, N. S.; Hernández-Martínez, A. R.; Gámez-Corrales, R.; Vargas-Ortiz, R.; Bocarando-Chacon, J. G.; Arizpe-Chavez, H.; Flores-Acosta, M. Raman Scattering and Optical Properties of Lithium Nanoparticles Obtained by Green Synthesis. *Vib. Spectrosc.* **2015,** *77,* 5–9.

Anastas, P. T.; Warner, J. C. Principles of Green Chemistry. *Green Chem.: Front.* **1998,** *640.*

Anastopoulos, I.; Hosseini-Bandegharaei, A.; Fu, J.; Mitropoulos, A. C.; Kyzas, G. Z. Use of Nanoparticles for Dye Adsorption. *J. Dispersion Sci. Technol.* **2018,** *39* (6), 836–847.

Armendariz, V.; Herrera, I.; Jose-Yacaman, M.; Troiani, H.; Santiago, P.; Gardea-Torresdey, J. L. Size Controlled Gold Nanoparticle Formation by *Avena sativa* Biomass: Use of Plants in Nanobiotechnology. *J. Nanopart. Res.* **2004,** *6* (4), 377–382.

Arya, V. Living Systems: Ecofriendly Nanofactories. *Dig. J. Nanomater. Bios.* **2010,** *5* (1), 9–11.

Azzazy, H. M.; Mansour, M. M.; Samir, T. M.; Franco, R. Gold Nanoparticles in the Clinical Laboratory: Principles of Preparation and Applications. *Clin. Chem. Lab. Med.* **2012,** *50* (2), 193–209.

Baer, D. R. Surface Characterization of Nanoparticles. *J. Surf. Anal.* **2011,** *17* (3), 163–169.

Balavijayalakshmi, J.; Ramalakshmi, V. *Carica papaya* Peel Mediated Synthesis of Silver Nanoparticles and Its Antibacterial Activity against Human Pathogens. *J. Appl. Res. Technol.* **2017,** *15* (5), 413–422.

Bell, A. A.; Wheeler, M. H.; Liu, J.; Stipanovic, R. D.: Puckhaber, L. S.; Orta, H. Polyketide Toxins of *Fusarium oxysporum, F. vasinfectum. Pest Manage. Sci.* 2003, *59* (6–7), 736–747.

Bentley, S. D.; Chater, K. F.; Cerdeno-Tarraga, A. M.; Challis, G. L.; Thomson, N. R.; James, K. D.; Harris, D. E.; Quail, M. A.; Kieser, H.; Harper, D.; Bateman, A. Complete Genome Sequence of the Model Actinomycete *Streptomyces coelicolor* A3 (2). *Nature* **2002,** *417* (6885), 141.

Bharimalla, A. K. Chapter 4. In: Training-Manual Advances in Applications of Nanotechnology, ICAR, Central Institute for Research on Cotton Technology (ICAR-CIRCOT) D.A.R.E, Ministry of Agriculture & Farmers Welfare, Govt of India, **2018,** 28–38.

Blasiak, B.; Van Veggel, F. C.; Tomanek, B. Applications of Nanoparticles for MRI Cancer Diagnosis and Therapy. *J. Nanomater.* **2013,** *2013*.

Bouhadoun, S.; Guillard, C.; Dapozze, F.; Singh, S.; Amans, D.; Bouclé, J.; Herlin-Boime, N. One Step Synthesis of N-doped and Au-loaded TiO_2 Nanoparticles by Laser Pyrolysis: Application in Photocatalysis. *Appl. Catal. Environ.* **2015,** *174*, 367–375.

Burgess, R. Medical Applications of Nanoparticles and Nanomaterials. *Stud. Health Technol. Inform.* **2009,** *149*, 257–283.

Cabrera, F. C.; Mohan, H.; Dos Santos, R. J.; Agostini, D. L.; Aroca, R. F.; Rodríguez-Pérez, M. A.; Job, A. E. Green Synthesis of Gold Nanoparticles with Self-Sustained Natural Rubber Membranes. *J. Nanomater.* **2013,** *2013*.

Chang, K. Tiny Is Beautiful, Translating "Nano" into Practical. *NY Times*, **2005,** *22*.

Chen, H.; Yada, R. Nanotechnologies in Agriculture: New Tools for Sustainable Development. *Trends Food Sci. Technol.* **2011,** *22* (11), 585–594.

Chhipa, H. Mycosynthesis of Nanoparticles for Smart Agricultural Practice: A Green and Eco-friendly Approach. In *Green Synthesis, Characterization and Applications of Nanoparticles*; Elsevier 2019; pp 87–109.

Chowdhury, I. H.; Ghosh, S.; Naskar, M. K. Aqueous-Based Synthesis of Mesoporous TiO_2 and Ag–TiO_2 Nanopowders for Efficient Photodegradation of Methylene Blue. *Ceram. Int.* **2016,** *42* (2), 2488–2496.

Cushing, B. L.; Kolesnichenko, V. L.; O'Connor, C. J. Recent Advances in the Liquid-Phase Syntheses of Inorganic Nanoparticles. *Chem. Rev.* **2004,** *104* (9), 3893–3946.

Darroudi, M.; Ahmad, M. B.; Zamiri, R.; Zak, A. K.; Abdullah, A. H.; Ibrahim, N. A. Time Dependent Effect in Green Synthesis of Silver Nanoparticles. *Int. J. Nanomed.* **2011,** *6*, 677.

Darroudi, M.; Sarani, M.; Oskuee, R. K.; Zak, A. K.; Amiri, M. S. Nanoceria: Gum Mediated Synthesis and in Vitro Viability Assay. *Ceram. Int.* **2014,** *40* (2), 2863–2868.

Das, M.; Chatterjee, S. Green Synthesis of Metal/Metal Oxide Nanoparticles toward Biomedical Applications: Boon or Bane. In *Green Synthesis, Characterization and Applications of Nanoparticles*; Elsevier, 2019; pp 265–301.

Deepa, S.; Kanimozhi, K.; Panneerselvam, A. Antimicrobial Activity of Extracellularly Synthesized Silver Nanoparticles from Marine Derived Actinomycetes. *Int. J. Curr. Microbiol. App. Sci.* **2013,** *2*, 223–230.

Dey, A.; Yogamoorthi, A.; Sundarapandian, S. Green Synthesis of Gold Nanoparticles and Evaluation of its Cytotoxic Property against Colon Cancer Cell Line. *RJLBPCS* **2018,** *4* (1), 17.

Drexler, K. E. Molecular Engineering: An Approach to the Development of General Capabilities for Molecular Manipulation. *PNAS* **1981,** *78* (9), 5275–5278.

Dubey, S. P.; Lahtinen, M.; Sillanpää, M. Tansy Fruit Mediated Greener Synthesis of Silver and Gold Nanoparticles. *Process. Biochem.* **2010,** *45* (7), 1065–1071.

Dwivedi, A. D.; Gopal, K. Biosynthesis of Silver and Gold Nanoparticles Using *Chenopodium album* Leaf Extract. *Colloids Surf. A.* **2010,** *369* (1–3), 27–33.

Edayadulla, N.; Basavegowda, N.; Lee, Y. R. Green Synthesis and Characterization of Palladium Nanoparticles and Their Catalytic Performance for the Efficient Synthesis of Biologically Interesting Di (indolyl) Indolin-2-ones. *J. Ind. Eng. Chem.* **2015,** *21,* 1365–1372.

El-Rafie, H. M.; El-Rafie, M.; Zahran, M. K. Green Synthesis of Silver Nanoparticles Using Polysaccharides Extracted from Marine Macro Algae. *Carbohydr. Polym.* **2013,** *96* (2), 403–410.

Elumalai, K.; Velmurugan, S.; Ravi, S.; Kathiravan, V.; Ashokkumar, S. Retracted: Green Synthesis of Zinc Oxide Nanoparticles Using *Moringa oleifera* Leaf Extract and Evaluation of Its Antimicrobial Activity. *Spectrochim. Acta, A.* **2015,** *143,* 158–164.

Faraday, M. X. The Bakerian Lecture—Experimental Relations of Gold (and Other Metals) to Light. *Philos. Trans. R. Soc. London Ser. A.* **1857,** *147,* 145–181.

Foo, M. E.; Anbu, P.; Gopinath, S. C.; Lakshmipriya, T.; Lee, C. G.; Yun, H. S.; Uda, M. N. A.; Yaakub, A. R. W. Antimicrobial Activity of Functionalized Single-Walled Carbon Nanotube with Herbal Extract of *Hempedu bumi. Surf. Interf. Anal.* **2018,** *50* (3), 354–361.

Foroutan, F.; Jokerst, J. V.; Gambhir, S. S.; Vermesh, O.; Kim, H. W.; Knowles, J. C. Sol–Gel Synthesis and Electrospraying of Biodegradable (P_2O_5) 55–(CaO) 30–(Na_2O) 15 Glass Nanospheres as a Transient Contrast Agent for Ultrasound Stem Cell Imaging. *ACS Nano.* **2015,** *9* (2), 1868–1877.

Gade, A.; Gaikwad, S.; Duran, N.; Rai, M. Green Synthesis of Silver Nanoparticles by *Phoma glomerata. Micron* **2014,** *59,* 52–59.

Gericke, M.; Pinches, A. Microbial Production of Gold Nanoparticles. *Gold Bull.* **2006,** *39* (1), 22–28.

Gondwal, M.; Pant, G. J. N. Biological Evaluation and Green Synthesis of Silver Nanoparticles Using Aqueous Extract of *Calotropis procera. IJPBS.* **2013,** *4* (4), 635–643.

Govindaraju, K.; Kiruthiga, V.; Kumar, V. G.; Singaravelu, G. Extracellular Synthesis of Silver Nanoparticles by a Marine Alga, *Sargassum wightii* Grevilli and Their Antibacterial Effects. *J. Nanosci. Nanotechnol.* **2009,** *9* (9), 5497–5501.

Handoko, C. T.; Huda, A.; Gulo, F. Synthesis Pathway and Powerful Antimicrobial Properties of Silver Nanoparticle: A Critical Review. *Asian J. Sci. Res.* **2019,** *12,* 1–17.

He, S.; Guo, Z.; Zhang, Y.; Zhang, S.; Wang, J.; Gu, N. Biosynthesis of Gold Nanoparticles Using the Bacteria *Rhodopseudomonas capsulata. Mater. Lett.* **2007,** *61* (18), 3984–3987.

Helcher, H. H. *Aurum Potabile Oder Gold Tinstur*; Herbord Klossen, J., Eds. Breslau and Leipzig, 1718.

Hemmati, S.; Rashtiani, A.; Zangeneh, M. M.; Mohammadi, P.; Zangeneh, A.; Veisi, H. Green Synthesis and Characterization of Silver Nanoparticles Using Fritillaria Flower Extract and Their Antibacterial Activity against Some Human Pathogens. *Polyhedron* **2019,** *158,* 8–14.

Himmelweit, F. *The Collected Papers of Paul Ehrlich: In Four Volumes Including a Complete Bibliography*; Elsevier, 2017.

Horikoshi, S.; Serpone, N. Introduction to Nanoparticles. *Microwaves Nanopart. Synth. Fundamentals App.* **2013,** *24,* 1–24.

Husein, M. M.; Nassar, N. N. Nanoparticle Preparation Using the Single Microemulsions Scheme. *Curr. Nanosci.* **2008,** *4* (4), 370–380.

Hussain, I.; Singh, N. B.; Singh, A.; Singh, H.; Singh, S. C. Green Synthesis of Nanoparticles and Its Potential Application. *Biotechnol. Lett.* **2016,** *38* (4), 545–560.

Iravani, S.; Korbekandi, H.; Mirmohammadi, S. V.; Zolfaghari, B. Synthesis of Silver Nanoparticles: Chemical, Physical and Biological Methods. *Res. Pharm. Sci.* **2014,** *9* (6), 385.

Jha, A. K.; Prasad, K. Yeast Mediated Synthesis of Silver Nanoparticles. *Int. J. Nanosci. Nanotechnol.* **2008,** *4,* 17–22.

Jin, R.; Cao, Y. C.; Hao, E.; Métraux, G. S.; Schatz, G. C.; Mirkin, C. A. Controlling Anisotropic Nanoparticle Growth through Plasmon Excitation. *Nature* **2003,** *425* (6957), 487.

Kaler, A.; Jain, S.; Banerjee, U. C. Green and Rapid Synthesis of Anticancerous Silver Nanoparticles by *Saccharomyces boulardii* and Insight into Mechanism of Nanoparticle Synthesis. *BioMed. Res. Int.* **2013,** *2013,* 1–8.

Kannan, R. R. R.; Arumugam, R.; Ramya, D.; Manivannan, K.; Anantharaman, P. Green Synthesis of Silver Nanoparticles Using Marine Macroalga *Chaetomorpha linum. Appl. Nanosci.* **2013,** *3* (3), 229–233.

Karthik, L.; Kumar, G.; Vishnu Kirthi, A.; Rahuman, A. A.; Rao, K. V. B. *Streptomyces* sp. LK3 Mediated Synthesis of Silver Nanoparticles and Its Biomedical Application. *Bioprocess. Biosyst. Eng.* **2014,** *37,* 261–267.

Kathiraven, T.; Sundaramanickam, A.; Shanmugam, N.; Balasubramanian, T. Green Synthesis of Silver Nanoparticles Using Marine Algae *Caulerpa racemosa* and Their Antibacterial Activity against Some Human Pathogens. *Appl. Nanosci.* **2015,** *5* (4), 499–504.

Kawasaki, M.; Nishimura, N. 1064-nm Laser Fragmentation of Thin Au and Ag Flakes in Acetone for Highly Productive Pathway to Stable Metal Nanoparticles. *Appl. Surf. Sci.* **2006,** *253* (4), 2208–2216.

Keihan, A. H.; Veisi, H.; Veasi, H. Green Synthesis and Characterization of Spherical Copper Nanoparticles as Organometallic Antibacterial Agent. *Appl. Organomet. Chem.* **2017,** *31* (7), 3642.

Khalil, A. T.; Ovais, M.; Ullah, I.; Ali, M.; Shinwari, Z. K.; Maaza, M. Biosynthesis of Iron Oxide (Fe_2O_3) Nanoparticles via Aqueous Extracts of *Sageretia thea* (Osbeck.) and Their Pharmacognostic Properties. *Green Chem. Lett. Rev.* **2017,** *10* (4), 186–201.

Kharissova, O. V.; Dias, H. R.; Kharisov, B. I.; Pérez, B. O.; Pérez, V. M. The Greener Synthesis of Nanoparticles. *Trends. Biotechnol.* **2013,** *31* (4), 240–248.

Khot, L. R.; Sankaran, S.; Maja, J. M.; Ehsani, R.; Schuster, E. W. Applications of Nanomaterials in Agricultural Production and Crop Protection: A Review. *Crop Prot.* **2012,** *35,* 64–70.

Kishen, S.; Mehta, A.; Gupta, R. Biosynthesis and Applications of Metal Nanomaterials. In *Green Nanomaterials*; Springer: Singapore, 2020; pp 139–157.

Kowshik, M.; Ashtaputre, S.; Kharrazi, S.; Vogel, W.; Urban, J.; Kulkarni, S. K.; Paknikar, K. Extracellular Synthesis of Silver Nanoparticles by a Silver-Tolerant Yeast Strain MKY3. *Nanotechnology* **2002,** *14,* 95–100.

Kuang, L.; Mitchell, B. S.; Fink, M. J. Silicon Nanoparticles Synthesised through Reactive High- Energy Ball Milling: Enhancement of Optical Properties from the Removal of Iron Impurities. *J. Exp. Nanosci.* **2015,** *10* (16), 1214–1222.

Greener Methods of Nanoparticle Synthesis

Kumar, D.; Karthik, L.; Kumar, G.; Roa, K. B. Biosynthesis of Silver Nanoparticles from Marine Yeast and Their Antimicrobial Activity against Multidrug Resistant Pathogens. *Pharmacologyonline* **2011**, *3*, 1100–1111.

Kundu, S.; Jayachandran, M. The Self-Assembling of DNA-Templated Au Nanoparticles into Nanowires and Their Enhanced SERS and Catalytic Applications. *RSC Adv.* **2013**, *3* (37), 16486–16498.

Kundu, S.; Maheshwar, V.; Saraf, R. F. Photolytic Metallization of Au Nanoclusters and Electrically Conducting Micrometer Long Nanostructures on a DNA Scaffold. *Langmuir* **2008**, *24* (2), 551–555.

Lade, B. D.; Gogle, D. P. Chapter 10: Nanobiopesticide: Synthesis and Applications in Plant Safety. In *Nanobiotechnology Applications in Plant Protection, the Nanotechnology in the Life Sciences*; Abd-Elsalam, K., Prasad, R., Eds., Vol. 2; Springer Nature, USA, 2019; pp 169–189.

Lade, B. D.; Patil, A. S. Silver Nano Fabrication Using Leaf Disc of *Passiflora foetida* linn. *Appl. Nanosci.* **2017**, *7* (5), 181–119.

Lee, S. W.; Mao, C.; Flynn, C. E.; Belcher, A. M. Ordering of Quantum Dots Using Genetically Engineered Viruses. *Science* **2002**, *296* (5569), 892–895.

Li, S.; Shen, Y.; Xie, A.; Yu, X.; Qiu, L.; Zhang, L.; Zhang, Q. Green Synthesis of Silver Nanoparticles Using *Capsicum annuum* L. Extract. *Green. Chem.* **2007**, *9* (8), 852–858.

Lohrasbi, S.; Kouhbanani, M. A. J.; Beheshtkhoo, N.; Ghasemi, Y.; Amani, A. M.; Taghizadeh, S. Green Synthesis of Iron Nanoparticles Using *Plantago major* Leaf Extract and Their Application as a Catalyst for the Decolorization of Azo Dye. *BioNanoScience*, **2019**, 1–6.

Maekawa, K.; Yamasaki, K.; Niizeki, T.; Mita, M.; Matsuba, Y.; Terada, N.; Saito, H. Drop-on Demand Laser Sintering with Silver Nanoparticles for Electronics Packaging. *IEEE Trans. Compon. Packag. Manuf. Technol.* **2012**, *2* (5), 868–877.

Majumdar, D.; Singha, A.; Mondal, P. K.; Kundu, S. DNA-mediated Wirelike Clusters of Silver Nanoparticles: An Ultrasensitive SERS Substrate. *ACS Appl. Mater. Interfaces* **2013**, *5* (16), 7798–7807.

Mandal, D.; Bolander, M. E.; Mukhopadhyay, D.; Sarkar, G.; Mukherjee, P. The Use of Microorganisms for the Formation of Metal Nanoparticles and Their Application. *Appl. Microbiol. Biotechnol.* **2006**, *69* (5), 485–492.

Manimaran, M.; Kannabiran, K. Actinomycetes-Mediated Biogenic Synthesis of Metal and Metal Oxide Nanoparticles: Progress and Challenges. *Lett. Appl. Microbiol.* **2017**, *64* (6), 401–408.

Mao, C.; Flynn, C. E.; Hayhurst, A.; Sweeney, R.; Qi, J.; Georgiou, G.; Iverson, B.; Belcher, A. M. Viral Assembly of Oriented Quantum Dot Nanowires. *Proc. Natl. Acad. Sci. U. S. A.* **2003**, *100* (12), 6946–6951.

Martirosyan, A.; Schneider, Y. J. Engineered Nanomaterials in Food: Implications for Food Safety and Consumer Health. *Int. J. Environ. Res. Public Health.* **2014**, *11* (6), 5720–5750.

Masciangioli, T.; Zhang, W. X. Peer Reviewed: Environmental Technologies at the Nanoscale. **2003**, 102A–108A.

Medentsev, A. G.; Akimenko, V. K. Naphthoquinone Metabolites of the Fungi. *J. Phytochem.* **1998**, *47* (6), 935–959.

Mehr, E. S.; Sorbiun, M.; Ramazani, A.; Fardood, S. T. Plant-Mediated Synthesis of Zinc Oxide and Copper Oxide Nanoparticles by Using *Ferulago angulata* (schlecht) Boiss Extract and Comparison of Their Photocatalytic Degradation of Rhodamine B (RhB) under Visible Light Irradiation. *J. Mater. Sci. – Mater. Electron.* **2018**, *29* (2), 1333–1340.

Merga, G.; Wilson, R.; Lynn, G.; Milosavljevic, B. H.; Meisel, D. Redox Catalysis on "Naked" Silver Nanoparticles. *J. Phys. Chem. C.* **2007,** *111* (33), 12220–12226.

Merin, D. D.; Prakash, S.; Bhimba, B. V. Antibacterial Screening of Silver Nanoparticles Synthesized by Marine Micro Algae. *Asian Pac. J. Trop. Med.* **2010,** *3* (10), 797–799.

Meyers, M. A.; Mishra, A.; Benson, D. J. Mechanical Properties of Nanocrystalline Materials. *Prog. Mater. Sci.* **2006,** *51* (4), 427–556.

Mohanpuria, P.; Rana, N. K.; Yadav, S. K. Biosynthesis of Nanoparticles: Technological Concepts and Future Applications. *J. Nanopart. Res.* **2008,** *10* (3), 507–517.

Mukherjee, P.; Ahmad, A.; Mandal, D.; Senapati, S.; Sainkar, S. R.; Khan, M. I.; Ramani, R.; Parischa, R.; Ajayakumar, P. V.; Alam, M.; Sastry, M. Bioreduction of $AuCl_4^-$ Ions by the Fungus, *Verticillium sp.* and Surface Trapping of the Gold Nanoparticles Formed. *Angew. Chem. Int. Ed.* **2001,** *40* (19), 3585–3588.

Nair, V.; Sambre, D.; Joshi, S.; Bankar, A.; Ravi Kumar, A.; Zinjarde, S. Yeast-Derived Melanin Mediated Synthesis of Gold Nanoparticles. *J. Bionanosci.* **2013,** *7,* 159–168.

Nakade, S.; Matsuda, M.; Kambe, S.; Saito, Y.; Kitamura, T.; Sakata, T.; Wada, Y.; Mori, H.; Yanagida, S. Dependence of TiO_2 Nanoparticle Preparation Methods and Annealing Temperature on the Efficiency of Dye-Sensitized Solar Cells. *J. Phys. Chem. B.* **2002,** *106* (39), 10004–10010.

Nangia, Y.; Wangoo, N.; Goyal, N.; Shekhawat, G.; Suri, C. R. A Novel Bacterial Isolate *Stenotrophomonas maltophilia* as Living Factory for Synthesis of Gold Nanoparticles. *Microb. Cell Fact.* **2009,** *8* (1), 1.

Nasrollahzadeh, M.; Atarod, M.; Sajjadi, M.; Sajadi, S. M.; Issaabadi, Z. Plant-Mediated Green Synthesis of Nanostructures: Mechanisms, Characterization, and Applications. *Interf. Sci. Technol.,* **2019,** *28,* 199–322.

Okuyama, K.; Lenggoro, I. W. Preparation of Nanoparticles via Spray Route. *Chem. Eng. Sci.* **2003,** *58* (3–6), 537–547.

Oliveira, M. M.; Ugarte, D.; Zanchet, D.; Zarbin, A. J. Influence of Synthetic Parameters on the Size, Structure, and Stability of Dodecanethiol-Stabilized Silver Nanoparticles. *J. Colloid Interf. Sci.* **2005,** *292* (2), 429–435.

Otari, S. V.; Patil, R. M.; Nadaf, N. H.; Ghosh, S. J.; Pawar, S. H. Green Biosynthesis of Silver Nanoparticles from an Actinobacteria *Rhodococcus* sp. *Mater. Lett.* **2012,** *72,* 92–94.

Pal, G.; Rai, P.; Pandey, A. Green Synthesis of Nanoparticles: A Greener Approach for a Cleaner Future. In *Green Synthesis, Characterization and Applications of Nanoparticles*; Elsevier, 2019; pp 1–26.

Patil, M. P.; Kim, G. D. Eco-friendly Approach for Nanoparticles Synthesis and Mechanism behind Antibacterial Activity of Silver and Anticancer Activity of Gold Nanoparticles. *Appl. Microbiol. Biotechnol.* **2017,** *101* (1), 79–92.

Philip, D. Biosynthesis of Au, Ag and Au–Ag Nanoparticles Using Edible Mushroom Extract. *Spectrochim. Acta. A: Mol. Biomol. Spectrosc.* **2009,** *73* (2), 374–381.

Pimprikar, P.; Joshi, S.; Kumar, A.; Zinjarde, S.; Kulkarni, S. Influence of Biomass and Gold Salt Concentration on Nanoparticle Synthesis by the Tropical Marine Yeast *Yarrowia lipolytica* NCIM 3589. *Colloids. Surf. B.* **2009,** *74,* 309–316.

Pokorski, J. K.; Steinmetz, N. F. The Art of Engineering Viral Nanoparticles. *Mol. Pharm.* **2010,** *8* (1), 29–43.

Prabhu, S.; Poulose, E. K. Silver Nanoparticles: Mechanism of Antimicrobial Action, Synthesis, Medical Applications, and Toxicity Effects. *Int. Nano Lett.* **2012,** *2* (1), 32.

Prasad, T. N.; Kambala, V. S. R.; Naidu, R. Phyconanotechnology: Synthesis of Silver Nanoparticles Using Brown Marine Algae *Cystophora moniliformis* and Their Characterisation. *J. Appl. Phycol.* **2013,** *25* (1), 177–182.

Pugazhendhi, A.; Prabakar, D.; Jacob, J. M.; Karuppusamy, I.; Saratale, R. G. Synthesis and Characterization of Silver Nanoparticles Using *Gelidium amansii* and Its Antimicrobial Property against Various Pathogenic Bacteria. *Microb. Pathog.* **2018,** *114*, 41–45.

Rai, M.; Yadav, A.; Bridge, P.; Gade, A.; Rai, M.; Bridge, P. D. Myconanotechnology: A New and Emerging Science. *Appl. Mycol.* **2009,** 258–267.

Raju, D.; Mehta, U. J.; Hazra, S. Synthesis of Gold Nanoparticles by Various Leaf Fractions of *Semecarpus anacardium* L. Tree. *Trees.* **2011,** *25* (2), 145–151.

Rather, M. Y.; Shincy, M.; Sundarapandian, S. Silver Nanoparticles Synthesis Using *Wedelia urticifolia* (Blume) DC. Flower Extract: Characterization and Antibacterial Activity Evaluation. *Microsc. Res. Tech.* **2020.** https://doi.org/10.1002/jemt.23499

Rather, M. Y.; Sundarapandian, S. Magnetic Iron Oxide Nanorod Synthesis by *Wedelia urticifolia* (Blume) DC. Leaf Extract for Methylene Blue Dye Degradation. *Appl. Nanosci.* **2020.** https://doi.org/10.1007/s13204-020-01366-2

Rauwel, P.; Küünal, S.; Ferdov, S.; Rauwel, E. A Review on the Green Synthesis of Silver Nanoparticles and Their Morphologies Studied via TEM. *Adv. Mater. Sci. Eng.* **2015,** *2015*.

Rauwel, P.; Rauwel, E. Emerging Trends in Nanoparticle Synthesis Using Plant Extracts for Biomedical Applications. *GJN* **2017,** *1*, 555–562.

Roy, S.; Das, T. K. Plant Mediated Green Synthesis of Silver Nanoparticles—A Review. *Int. J. Plant Biol. Res.* **2015,** *3* (3), 1044–1055.

Sachdev, A.; Gopinath, P. Green Synthesis of Multifunctional Carbon Dots from Coriander Leaves and Their Potential Application as Antioxidants, Sensors and Bioimaging Agents. *Analyst* **2015,** *140* (12), 4260–4269.

Salunke, B. K.; Sawant, S. S.; Lee, S. I.; Kim, B. S. Comparative study of MnO_2 Nanoparticle Synthesis by Marine Bacterium *Saccharophagus degradans* and Yeast *Saccharomyces cerevisiae*. *Appl. Microbiol. Biotechnol.* **2015,** *99* (13), 5419–5427.

Sánchez, A.; Recillas, S.; Font, X.; Casals, E.; González, E.; Puntes, V. Ecotoxicity of, and Remediation with, Engineered Inorganic Nanoparticles in the Environment. *TrAC Trends Anal. Chem.* **2011,** *30* (3), 507–516.

Sanchez-Garcia, M. D.; Lopez-Rubio, A.; Lagaron, J. M. Natural Micro and Nanobiocomposites with Enhanced Barrier Properties and Novel Functionalities for Food Biopackaging Applications. *Trends Food Sci. Technol.* **2010,** *21* (11), 528–536.

Sanguansri, P.; Augustin, M. A. Nanoscale Materials Development–a Food Industry Perspective. *Trends Food Sci. Technol.* **2006,** *17* (10), 547–556.

Santhoshkumar, J.; Rajeshkumar, S.; Kumar, S. V. Phyto-Assisted Synthesis, Characterization and Applications of Gold Nanoparticles–A Review. *Biochem. Biophys. Rep.* **2017,** *11*, 46–57.

Sathishkumar, M.; Sneha, K.; Won, S. W.; Cho, C. W.; Kim, S.; Yun, Y. S. *Cinnamon zeylanicum* Bark Extract and Powder Mediated Green Synthesis of Nano-Crystalline Silver Particles and Its Bactericidal Activity. *Colloids Surf. B.* **2009,** *73* (2), 332–338.

Scaramuzza, S.; Agnoli, S.; Amendola, V. Metastable alloy Nanoparticles, Metal-Oxide Nanocrescents and Nanoshells Generated by Laser Ablation in Liquid Solution: Influence of the Chemical Environment on Structure and Composition. *Phys. Chem. Chem. Phys.* **2015,** *17* (42), 28076–28087.

Schröfel, A.; Kratošová, G.; Bohunická, M.; Dobročka, E.; Vávra, I. Biosynthesis of Gold Nanoparticles Using Diatoms-Silica-Gold and EPS-Gold Bionanocomposite Formation. *J. Nanopart. Res.* **2011,** *13* (8), 3207–3216.

Sen, K.; Sinha, P.; Lahiri, S. Time Dependent Formation of Gold Nanoparticles in Yeast Cells: A Comparative Study. *Biochem. Eng. J.* **2011,** *55,* 1–6.

Shankar, P. D.; Shobana, S.; Karuppusamy, I.; Pugazhendhi, A.; Ramkumar, V. S.; Arvindnarayan, S.; Kumar, G. A Review on the Biosynthesis of Metallic Nanoparticles (Gold and Silver) Using Bio-Components of Microalgae: Formation Mechanism and Applications. *Enzyme Microb. Technol.* **2016,** *95,* 28–44.

Sharma, D.; Kanchi, S.; Bisetty, K. Biogenic Synthesis of Nanoparticles: A Review. *Arabian J. Chem.* **2019,** *12* (8), 3576–3600.

Shenton, W.; Douglas. T.; Young, M.; Stubbs, G.; Mann, S. Inorganic–Organic Nanotube Composites from Template Mineralization of Tobacco Mosaic Virus. *Adv. Mater.* **1999,** *11* (3), 253–256.

Sheny, D.; Mathew, J.; Philip, D. Phytosynthesis of Au, Ag and Au–Ag Bimetallic Nanoparticles Using Aqueous Extract and Dried Leaf of *Anacardium occidentale. Spectrochim. Acta Part A.* **2011,** *79* (1), 254–262.

Siddiqui, M. H.; Al-Whaibi, M. H.; Firoz, M.; Al-Khaishany, M. Y. Role of Nanoparticles in Plants. In *Nanotechnology and Plant Sciences*; Springer, 2015; pp 19–35.

Singaravelu, G.; Arockiamary, J. S.; Kumar, V. G.; Govindaraju, K. A Novel Extracellular Synthesis of Monodisperse Gold Nanoparticles Using Marine Alga, *Sargassum wightii* Greville. *Colloids Surf. B.* **2007,** *57* (1), 97–101.

Solanki, P.; Bhargava, A.; Chhipa, H.; Jain, N.; Panwar, J. Nano-Fertilizers and Their Smart Delivery System. In *Nanotechnologies in Food and Agriculture*; Springer, 2015, pp 81–101.

Suganya, K. U.; Govindaraju, K.; Kumar, V. G.; Dhas, T. S.; Karthick, V.; Singaravelu, G.; Elanchezhiyan, M. Blue Green Alga Mediated Synthesis of Gold Nanoparticles and Its Antibacterial Efficacy against Gram Positive Organisms. *Mater. Sci. Eng. C.* **2015,** *47,* 351–356.

Sukirtha, R.; Priyanka, K. M.; Antony, J. J.; Kamalakkannan, S.; Thangam, R.; Gunasekaran, P.; Krishanan, M.; Achiraman, S. Cytotoxic Effect of Green Synthesized Silver Nanoparticles Using *Melia azedarach* against in Vitro HeLa Cell Lines and Lymphoma Mice Model. *Process. Biochem.* **2012,** *47* (2), 273–279.

Sunitha, A.; Isaac, R. S. R.; Geo, S.; Sornalekshmi, S.; Rose, A.; Praseetha, P. K. Evaluation of Antimicrobial Activity of Biosynthesized Iron and Silver Nanoparticles Using the Fungi *Fusarium oxysporum* and Actinomycetes sp. on Human Pathogens. *Nano. Biomed. Eng.* **2013,** *5,* 39–45.

Tarasenko, N. V.; Butsen, A. V.; Nevar, E. A.; Savastenko, N. A. Synthesis of Nanosized Particles during Laser Ablation of Gold in Water. *Appl. Surf. Sci.* **2006,** *252* (13), 4439–4444.

The National Nanotechnology Initiative (NNI). Strategic Plan. 2007. http://www.nano.gov/NNI_Strategic_Plan_2007.pdf.

Tsuji, T.; Iryo, K.; Watanabe, N.; Tsuji, M. Preparation of Silver Nanoparticles by Laser Ablation in Solution: Influence of Laser Wavelength on Particle Size. *Appl. Surf. Sci.* **2002,** *202* (1–2), 80–85.

Tsuji, T.; Kakita, T.; Tsuji, M. Preparation of Nano-Size Particles of Silver with Femtosecond Laser Ablation in Water. *Appl. Surf. Sci.* **2003,** *206* (1–4), 314–320.

Vadlapudi, V.; Kaladhar, D. S. V. G. K. Review: Green Synthesis of Silver and Gold Nanoparticles. *Middle East J. Sci. Res.* **2014,** *19* (6), 834–842.

Greener Methods of Nanoparticle Synthesis 239

Vanaja, M.; Annadurai, G. *Coleus aromaticus* Leaf Extract Mediated Synthesis of Silver Nanoparticles and Its Bactericidal Activity. *Appl. Nanosci.* **2013,** *3* (3), 217–223.

Veerasamy, R.; Xin, T. Z.; Gunasagaran, S.; Xiang, T. F. W. Yang, E. F. C.; Jeyakumar, N.; Dhanaraj, S. A. Biosynthesis Of Silver Nanoparticles Using Mangosteen Leaf Extract and Evaluation of Their Antimicrobial Activities. *J. Saudi. Chem. Soc.* **2011,** *15* (2), 113–120.

Venkateswarlu, S.; Yoon, M. Rapid Removal of Cadmium Ions Using Green-Synthesized Fe_3O_4 Nanoparticles Capped with Diethyl-4-(4 amino-5-mercapto-4 H-1,2,4-triazol-3-yl) Phenyl Phosphonate. *RSC Adv.* **2015,** *5* (80), 65444–65453.

Virkutyte, J.; Varma, R. S. Green Synthesis of Metal Nanoparticles: Biodegradable Polymers and Enzymes in Stabilization and Surface Functionalization. *Chem. Sci.* **2011,** *2* (5), 837–846.

Waghmare, S. S.; Deshmukh, A. M.; Kulkarni, W.; Oswaldo, L. A. Biosynthesis and Characterization of Manganese and Zinc Nanoparticles. *Univ. J. Environ. Res. Technol.* **2011,** *1,* 64–69.

Waghmare, S. S.; Deshmukh, A. M.; Sadowski, Z. Biosynthesis, Optimization, Purification and Characterization of Gold Nanoparticles. *Afr. J. Microbiol. Res.* **2014,** *8,* 138–146.

Yang, X.; Li, Q.; Wang, H.; Huang, J.; Lin, L.; Wang, W.; Sun, D.; Su, Y.; Opiyo, J. B.; Hong, L.; Wang, Y. Green Synthesis of Palladium Nanoparticles Using Broth of *Cinnamomum camphora* Leaf. *J. Nanopart. Res.* **2010,** *12* (5), 1589–1598.

Yurkov, A. M.; Kemler, M.; Begerow, D. Species Accumulation Curves and Incidence Based Species Richness Estimators to Appraise the Diversity of Cultivable Yeasts from Beech Forest Soils. *PLoS One* **2011,** *8* (8).

Zielonka, A.; Klimek-Ochab, M. Fungal Synthesis of Size-Defined Nanoparticles. *Adv. Nat. Sci. Nanosci. Nanotechnol.* **2017,** *8* (4), 043001.

Zinchenko, A.; Miwa, Y.; Lopatina, L. I.; Sergeyev, V. G.; Murata, S. DNA Hydrogel as a Template for Synthesis of Ultrasmall Gold Nanoparticles for Catalytic Applications. *ACS Appl. Mater. Interf.* **2014,** *6* (5), 3226–3232.

CHAPTER 9

NANOSCIENCE IN BIOTECHNOLOGY

CHARU GUPTA[1*], MIR SAJAD RABANI[2], MAHENDRA K GUPTA[2], SHIVANI TRIPATHI[2], and ANJALI PATHAK[2]

[1]*School of Studies in Microbiology, Jiwaji University, Gwalior, Madhya Pradesh, India*

[2]*Microbiology Research Lab, School of Studies in Botany, Jiwaji University, Gwalior, Madhya Pradesh, India*

Corresponding author. E-mail: gcharu720@gmail.com

ABSTRACT

Nanoscience is a new burgeoning technology in diverse regimes of life sciences. Nanomedicines or nanobiopharmaceuticals (antimicrobial and antibiofilm agents), environmental system (e.g., water purification systems), food industries (for better food productions), fuel cells, solar cells, fuels, sporting goods, privacy and security (nanoelectronics), textile (fabric), etc. are myriad applications that emphasize potentials of nanobiotechnology in distinct sector. Nanoparticles (NPs) are synthesized with highly studied metals and metal salts such as copper, cadmium, palladium, silver, platinum, gold, zinc oxide, cadmium sulfide, and titanium oxide by biological (bacteria, fungi, algae, plant) as well as by chemical method. From the last few decades, nanobiotechnology has earned considerable attention; therefore, different nanomaterials have been accomplished and found useful in remediation processes for eliminating environmental pollution. The elimination of these noxious pollutants from contaminated sites is a need of an hour to forbid the negative impact on human health and the natural habitat. Due to the exclusive properties of nanomaterials like magnetic, antimicrobial, anticancer catalytic activity, and optical properties, these metal NPs have been explored extensively. Their large surface-area-to-volume ratio is the most important property, which increases their interaction with other molecules. In recent years, nanobiotechnology has become an emergent technology due to their remarkable and fascinating properties.

9.1 INTRODUCTION

Nanobiotechnology or bionanotechnology has emerged as a novel and flourishing science that attracted the scientific mind in the recent past. The term nanobiotechnology refers to the involvement of nanosciences and biotechnology (Scott et al., 2002; Ehud, 2007). Nanoscience is an innovative field and due to its countless application in various disciplines of science like physics, biotechnology, medical science, microbiology, chemical science, material science, and nanotechnology, it needs to be explored for the benefits of mankind and environment. Nano is a Greek word that means dwarf (Rai et al., 2008). Norio Taniguchi, a researcher at the University of Tokyo, Japan, introduced the word "Nanotechnology" (Taniguchi, 1974). Physicist and Nobelist Professor Richard Feynman during a talk entitled "there's plenty of room at the bottom" gave the concept of nanobiotechnology (Feynman, 1959).

Nanoparticles (NPs) are the ultrafine metallic materials. They exhibit various shapes like rod, spherical, triangular, etc. and have size of less than 100 nm (Sau and Rogach, 2010). The exploitation of NPs in the production area is due to their large surface-area-to-volume ratio together with some exclusive properties, including:

i.   Chemical properties such as photocatalytic capacity as a semiconductor (Tong et al., 2012); rising optical performance (Kelly et al., 2003) and high resistance to corrosion (Hamdy and Butt, 2007);

ii. Physical properties such as tribological increase to the thermal resistance (Miyake et al., 2013), superconductivity (Shi et al., 2012; Iijima, 2002), superparamagnetism (Vatta et al., 2009), and ultra-hardness (Lamni et al., 2005); and

iii. Biological properties such as environmental remediation (Bodzek and Konieczny, 2011), antimicrobial coatings (Singh and Nalwa, 2011), and antimicrobial properties (Chen and Chiang, 2008).

Different physical and chemical methods are extensively applied for the synthesis of metal NP. However, these methods provide better sized synthesized NPs with high productivity, but because of being less compatible, high energy requirements, higher capital cost, use of toxic

Nanoscience in Biotechnology   243

chemicals, and production of noxious wastes, they are found unfavorable (Hosseini and Sarvi, 2015). Therefore, the biological method for NP synthesis could be a better replacement for chemical and physical methods and due to its eco-friendly, biocompatible, and cost-efficient approach (Ahmed and Ikram, 2016). Microbes are considered as green eco-friendly nanofactories. Thus, microbial NP synthesis constructs a new burgeoning field, that is, nanobiotechnology, which is the interconnection of microbiology, nanotechnology, and biotechnology. Interaction of microbes and metals has been extensively used for biomineralization, bioleaching, and bioremediation, but bionanotechnology is still flourishing (Narayanan and Sakthivel, 2010). Microorganisms and plants have been widely exploited for the biosynthesis of metal NPs, such as titanium (Ti), gold (Au), copper (Cu), silver (Ag), and zinc (Zn) (Thakkar et al., 2010).

Nanobiotechnology has vast applications in the textile, biosensor, cosmetics, nanomedicine, food industry, nanobioremediation, and so on. NPs possess unique properties such as high reactivity, large surface area, and small size; therefore, it has become the interesting area of research in the present era. These nanomaterials play a pivotal role in bioremediation of the contaminated environment. This suggested that nanobiotechnology can be used for the human and environment welfare. However, a lot of research is to be done for its wider applications.

9.2 CLASSIFICATION OF NANOPARTICLES

NPs are referred to as a conglomerate of atoms having dimension of 1–100 nm. The morphology and shape of NPs play a significant role in their functioning to a certain extent (Buzea et al., 2007). On the basis of their dimension, composition, morphology, agglomeration, and uniformity, NPs can be classified as follows.

Based on dimension:

One-dimensional nanoparticles—These include thin layers or surface plating and are used in respective fields like biological and chemical sensors, electronics, optical and magneto-optic devices, information storage systems, fiber-optic systems (Seshan, 2002)

Two-dimensional nanoparticles—These include long nanoclusters and thick membranes with nanopores, for example, nanowires, fibrils, fibers, and carbon nanotubes.

Three-dimensional nanoparticles—These include small nanoclusters and membranes with nanopores, for example, nanocrystals or quantum dots, liposomes, dendrimers (Pal et al., 2011).

Based on morphology: They are flat, crystalline, and spherical.

Based on composition: They can be classified as single form and composite.

Based on uniformity and agglomeration: On the basis of uniformity and agglomeration, they are classified as agglomerates and dispersed (Anila Fariq et al., 2017).

9.3 SYNTHESIS OF NANOPARTICLES

Considering their diverse applications in the field of nanoscience, various techniques have been developed for the synthesis of metal NPs (Reddy, 2006). Two different approaches have been adopted for the synthesis of these NPs:

- **Top-down method**—Slicing of massive pieces into nanosized particles (Husen and Siddiqi, 2014) by different ablations, for example, mechanical grinding, thermal decomposition, sputtering, etching, cutting, and diffusion. But the surface structural defects are the main demerits of this approach. Further, physicochemical properties of metallic NPs are notably influenced by these defects (Siddiqi et al., 2018).

- **Bottom-up method**—This approach is extensively used for the fabrication of NPs in which atoms or molecules (liquid or gaseous state) synthesize nanostructures (Husen and Siddiqi, 2014) by involving a reducing agent and enzymes. The bottom-up approach can be achieved by methods such as chemical reduction, electrochemical synthesis, seeded growth method, biological methods, and polyol synthesis method (Siddiqi et al., 2018) (Fig. 9.1).

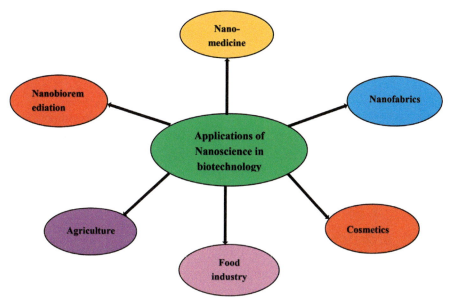

FIGURE 9.1 Two methods for synthesizing nanoparticles: (A) top-down approach and (B) bottom-up approach.

9.3.1 CHEMICAL SYNTHESIS

Chemical synthesis of metallic NPs can be done by any of the processes such as chemical reduction, photochemical, electrochemical, pyrolysis- and irradiation-assisted hybrid chemical approaches (Zhang et al., 2007; Maretti et al., 2009). Chemical synthesis process comprises mainly three components: (1) reducing agents, (2) stabilizing/capping agents, and (3) metal precursors.

Chemical reduction method is the most frequently used approach for synthesizing NPs. The reducing agents such as alcohol, poly(*N*-vinylpyrrolidone) (PVP) (Kim, 2007), ascorbic acid, sodium citrate, or sodium borohydride (Cao and Hu, 2009), tetra-*n*-tetra-fltetra-009e (TFATFB), CTAB (Hanauer et al., 2007), and *N,N*-dimethylformamide (DMF) (Pastoriza-Santos and Liz-Marzan, 2000) facilitate the reduction of metal particles into NPs. Electrochemical method of synthesizing NPs stimulates chemical reactions with the help of an applied voltage in an electrolyte solution (Sau and Rogach, 2010). In the photochemical synthesis method, silver (Ag) salt can be photoreduced with citrate or a polymer, irradiated by various sources of light, for instance, ultraviolet radiation (Sato-Berru et al., 2009). Further, traces of these toxic chemicals such as sodium borohydride or hydrazine

246 Diverse Applications of Nanotechnology in the Biological Sciences

are highly sensitive, nonbiodegradable and cause a hazardous effect on the environment.

9.3.2 PHYSICAL SYNTHESIS

Physical methods applied for the synthesis include thermal decomposition, cocondensation, evaporation/condensation, laser ablation, inert gas condensation, ultraviolet irradiation, arc discharge, electrolysis, diffusion, etc. (Jung et al., 2006). However, high energy requirement is a demerit of physical synthesis of NPs (Asanithi, 2012).

9.3.3 BIOLOGICAL SYNTHESIS

Mostly plants (Bar et al., 2009a,b; Song and Kim, 2009) and microorganisms such as fungi, actinomycetes, bacteria, and algae (Ingle et al., 2008; Birla et al., 2009) have been used for the synthesis of metal NP. Studies have reported that biological synthesis of NP has numerous advantages like eco-friendly, biodegradable, cost-effective and provides high yields as compared to that of physical and chemical methods (Husen and Siddiqi, 2014; Siddiqi and Husen, 2016).

Tollens, polyphenols, polyoxometalate, polysaccharides, and biological reduction have been found useful for mitigating the toxic chemicals (Iravani and Zolfaghari, 2013; Sharma et al., 2009). Sintubin et al. (2009) reported that plants, bacteria, yeast, or fungi produce biosurfactants and extracts that are used as reducing and capping agents in the biological reduction. These are the substitutes of capping and reducing agents used in the process of chemical reduction (Sharma et al., 2009)

However, biomolecules are known for stabilizing and capping the NPs. The shape of biomolecules can be regulated by pH which is considered as vital factor in the biosynthesis of NPs (Verma and Mehata, 2016). A study has reported the synthesis of gold NPs by using *Verticillium luteoalbum* and the variations in their shape and size were observed at various pH levels (Gericke and Pinches, 2006). At pH 3, the shape of generated NPs was found spherical and smaller in size while at pH 7 and 9 the shape is irregular, not defined with the larger size as recorded in the results.

Several researches have revealed the effect of temperature on synthesis and size of NPs by varying the temperature. The impact of temperature (10°C, 27°C, and 40°C) on the size of NPs obtained by *Trichoderma viride* has been

Nanoscience in Biotechnology

recorded with UV–visible spectra. The results of the study has revealed that at 40°C wavelength of the lower region, that is, 405 nm was obtained while at 27°C and 10°C, the wavelength of higher regions at 420 and 451 nm was recorded, respectively. This suggested that higher the wavelength, higher the size of NPs. The studies have also confirmed with TEM analysis, that 2- and 4 nm were small-sized NPs obtained at high temperature of 40°C whereas at lower temperature, large-sized NPs were obtained (Mohammed Fayaz et al., 2009).

9.3.3.1 MODE OF BIOSYNTHESIS

There are two methods, that is, intracellular and extracellular by which biosynthesis of NPs can be achieved (Mukherjee et al., 2008; Shaligram et al., 2009).

Intracellular method—After isolation of microbes like bacteria, actinomycetes, and algae, inoculated microorganisms are incubated in the growth medium and cell biomass is harvested and added to a beaker containing metal salt solution. The beaker is kept in a shaker in dark at optimum temperature and pressure (Govindraju et al., 2008; Thakkar et al., 2010). In fungi, after centrifugation, the freeze–dried mycelium of fungi is mixed with the metal salt solution, and then incubated for 24 h (Mukherjee et al., 2001) in a shaker at ambient pressure and temperature (Chen et al., 2003).

Extracellular method—For bacterial synthesis, bacterial culture is centrifuged at 10,000 rpm and the supernatant is suspended in a metal salt solution (Ogi et al., 2010). However, in fungi, the filtrate of the mycelium is treated with metal salt solution and incubated for 24 h at optimum condition (Fayaz et al., 2009).

TABLE 9.1 List of Biosynthesized Nanoparticles from Various Microorganisms.

Microorganisms	Types of nanoparticles	Location	Size (nm)	References
Bacillus licheniformis	Silver	Extracellular	50	Kalishwaralal et al. (2008)
Enterobacteria	Silver	Extracellular	52.5	Shahverdi et al. (2007a,b)
Lactobacillus sporogenes	Zinc oxide	Intracellular	5–15 Hexagonal	Prasad and Jha (2009)

TABLE 9.1 *(Continued)*

Microorganisms	Types of nanoparticles	Location	Size (nm)	References
Verticillium luteoalbum	Gold	Intracellular, extracellular	100 Spherical, triangular, hexagonal	Gericke and Pinches (2006)
Salmonella typhimurium	Copper	Extracellular	40–60	Ghorbani et al. (2015)
Planomicrobium sp.	Titanium dioxide	Extracellular	8.89, spherical	Malarkodi et al. (2013)
Schizosaccharomyces pombe	Cadmium sulfide (CdS)	Intracellular	1–1.5, hexagonal lattice	Kowshik et al. (2002)
Rhodobacter sphaeroides	CdS	Intracellular	8, hexagonal lattice	Bai et al. (2006)
Desulfobacteraceae	CdS	Intracellular	2–5, hexagonal lattice	Labrenz et al. (2000)
Fusarium oxysporum	Strontium carbonate (SrCO$_3$)	Extracellular	10–50, needle like	Ahmad et al. (2002)
Lactobacillus crispatus	Titanium dioxide	Extracellular	70–114, spherical, oval	Salman et al. (2014)
Sulfate-reducing bacteria	Ferrous sulfide (FeS)	Extracellular	2, spherical	Watson et al. (1999)
Coriolus versicolor	CdS	Extracellular	100–200, spherical	Sanghi and Verma (2009)

Sources: Adapted from Xiangqian et al. (2011) and Fariq et al. (2017).

9.4 APPLICATIONS OF NANOSCIENCE IN BIOTECHNOLOGY

Since the last few decades, metal NPs have gained a serious focus of researchers due to their optical, catalytic, magnetic, chemical stability, and antimicrobial activities. Owing to these diverse properties, metal NPs appear to be the most favorable material for research in different field of science. Nanoscience has numerous applications in biology and its respective science, for example, drug delivery, biomedical engineering, tissue engineering, bioremediation, agriculture, food technology, and inter alia (Dasgupta et al., 2015) (Fig. 9.2).

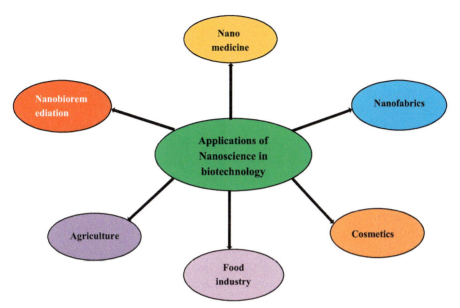

FIGURE 9.2 Applications of nanobiotechnology in various fields.

9.4.1 IN NANOMEDICINE

Nanobiotechnology has applications in various fields of nanomedicine like NP-based drug delivery, nano-vaccinology, and molecular nanotechnology. Nanomaterials are used for the control, prevention, monitoring, diagnosis, and treatment of various diseases (Tinkle et al., 2014). In medical applications, silver NPs are reported to possess anti-inflammatory, antibacterial, antifungal, and antiviral properties. These are also considered effective against diseases, infections, inflammation, and slow healing.

Silver NPs are used in bone cements, bandages, wound dressings, dental filling, sutures, female hygiene, catheters, surgical instruments, prosthesis, and contraceptive devices (Singh and Nalwa, 2011). Although nanomedicine has become an advance field of nanobiotechnology, it is in infancy stage (Silva, 2004; Kubik, 2005).

9.4.1.1 AS ANTITUMOR AND ANTICANCER AGENTS

In recent years, cancer has become a prominent reason for mortality around the world and cancer malaise is a complex disorder with many possible causes. It might be the result of the unregulated proliferation of different cell lines accompanied by the absence of apoptosis. Chemotherapeutics, surgery, and radiations are the conventional methods of treatment that may have several side effects too (Jabir et al., 2012). Therefore, there is a dire need of alternative replacements which will have less or negligible side effects.

Various drug delivery and tumor detection practices have been successfully developed with the help of nanotechnology (Sutradhar and Amin, 2014). Microbial-synthesized NPs possess various properties that seem to have prominent potential to facilitate cross biological barriers and molecular interactions without any harmful impact on viable cells.

In an experiment, platinum NP synthesized by using *Saccharomyces boulardii* showed anticancer activities against MCF-7 and A431 cell lines (Borse et al., 2015). The breast cancer malevolence is one of the main reasons for mortality in women across the world. Silver NPs synthesized from *Cryptococcus laurentii* have shown promising results against tumor and breast cancer cell lines (Ortega et al., 2015). The anticancerous properties of selenium nanorods synthesized from *Streptomyces bikiniensis* were found responsible for the death of MCF-7 and Hep-G2 human cancer cells (Ahmad et al., 2015). A recent study has suggested that gold NPs synthesized by using *Streptomyces cyaneus* were examined for anticancerous activities against MCF7 (breast carcinoma cells) and HEPG-2 (human liver cells) where they have exhibited promising results (El-Batal et al., 2015). Since several reports have determined the anticancerous activities of biosynthesized NPs. However, there is the necessity to explain the immune response and toxicity before applying them for therapeutic and diagnostic purposes.

9.4.1.2 IN DRUG-DELIVERY SYSTEMS

Considering the small size of NPs, the NP drug carriers can be transported to the desired target sites. They enhance the stability of different therapeutic agents, for instance, oligonucleotides and peptides (Emerich and Thanos, 2006). Conventional drug-delivery system has numerous demerits that have been replaced by NPs up to certain extent for safety, biocompatibility, biodistribution, and specific targeting (DeJong and Borm, 2008).

In a study, zinc oxide (ZnO) NPs synthesized from *Rhodococcus pyridinivorans* in combination with anthraquinone shows concentration-dependent cytotoxicity. Further, they have been used in HT-29 colon carcinoma cell in accordance with dose-dependence. Hereby, they can be utilized as anticancer drug delivery carriers (Kundu et al., 2014). A study revealed that thermophilic fungus *Humicola* sp. synthesized gadolinium oxide NPs conjugated with anticancer drug taxol can be used as a highly efficient drug delivery agent. These NPs can further be used against tumorigenic cells (Khan et al., 2014).

9.4.1.3   IN DIAGNOSTICS

Biochemical tests, microscopy, immunoassays, molecular diagnostics, and culturing of microorganisms are some of the methods used for diagnosis of various diseases. These techniques are laborious, time-consuming, and tedious (Tallury et al., 2010). In recent time, magnetic NPs, fluorescent NPs (dye-filled NPs and quantum dots), metallic NPs have been efficiently utilized to track, identify various pathogens. These are the alternative methods for various conventional methods that can be used for identification and diagnosis of various diseases. Nanomaterials facilitate rapid, easy-to-use, ultrasensitive, and accurate aid for diagnosis of certain pathogens.

In a study, Chauhan et al. (2011) demonstrated that potency of gold NPs synthesized by using *Candida albicans* for the diagnosis of liver cancer cells through their binding property with cancer cell surface-specific antibodies. However, the detection and diagnosis of diseases through nanomaterials is at its infancy.

9.4.1.4   NANOBIOSENSORS

Biosensors are the sensing instruments, manufactured by the combination of organic-sensing components and a transducer-based detector system. Biosensor is an effective diagnostic tool as compared to any conventional device, based on selectivity and sensitivity. Biosensing is a diverse field, has uncountable applications such as detection of pathogens in food, diagnosis of diseases, and detection of toxic pollutants, and enhances the crop production in agriculture. However, biosensors are costly and time-consuming. Thus, nanomaterial-based nanobiosensors introduced are capable of detection up to nanoscale with high accuracy and pace.

Studies have revealed that selenium NPs synthesized with the *Bacillus subtilis* (Wang et al., 2010) and modified electrode have shown promising applications in the detection of H_2O_2 in clinical, pharmaceutical, food, environmental, and industrial analyses (Zheng et al., 2010).

In medical science, biosensors have been used for the identification of antigens, carcinogens, and causal elements of various metabolic disorders. There are various other applications of biosensors in the field of medications such as diagnosis of cancer (Grimm et al., 2004), detection of HIV-AIDS (Alterman et al., 2001), detection of glucose in diabetic patients (Pickup et al., 2005), detection of urinary tract infections (Drummond et al., 2003). The association of nanomaterials with detecting enzymes facilitates recycling and reuses of these detecting enzymes that become more stable, sensitive, and cost-effective. With the implication of nanoscale technologies like NEMS (nanoelectromechanical systems) and MEMS (microelectromechanical systems), various biochips and microarray-based testing kits were manufactured, which are useful in diagnosis of more than one disease at a particular time (Haun et al., 2010).

9.4.1.5 ANTIMICROBIAL ACTIVITY

Increase and prevalence of antibiotic-resistant pathogenic microbial strains have led to the increase in a number of infectious diseases. Over the last few decades, due to consistent exposure to antibiotics, microbes acquire immunity and resistance against these drugs (Seil and Webster, 2012). Thus, biosynthesized NPs with strong bacteriostatic and bactericidal properties are reported as alternative antibiotics for the conventional ones. NPs bind to the cell membranes and then perforate inside and interfering with the replication process by interacting with DNA (Sunkar and Nachiyar, 2012). Furthermore, many reports recommended that using a mixture of conventional antibiotics with NPs leads to the improvement in antimicrobial competence. The silver NPs obtained by using *Brevibacterium frigoritolerans* exhibited antibiotic effects against pathogenic species of *Bacillus cereus, Salmonella enterica, Bacillus anthracis, Escherichia coli, C. albicans,* and *Vibrio parahaemolyticus.* The combination of biosynthesized silver NPs and antibiotics such as—novobiocin, penicillin G, rifampicin, vancomycin, oleandomycin, penicillin G, and lincomycin—exhibited improved synergistic effect (Singh et al., 2015). A study has revealed that titanium oxide NPs synthesized by using *Planomicrobium* sp. exhibited antimicrobial activity against

Aspergillus niger, B. subtilis, and *Klebsiella planticola* (Malarkodi et al., 2013).

9.4.2 IN NANO FABRICS

Numerous fabrics such as nylon, rayon, cotton, and aramid are being used across the world. These types of fabric may cause itching, rashes, foul body odor, and various bacterial and fungal infections due to sweating. Since nanomaterials like silver NPs are known to possess antimicrobial properties. Thus, in the modern time, nanobiotechnology has been introduced in the textile industry.

Silver NP-coated fabrics are used in the manufacturing of an extensive variety of products in clothing, socks, shoes, innerwear, bandages, surgical face masks, etc. Duran et al. (2007) reported that cotton fabrics modified with silver NPs show a remarkable antibacterial activity against *S. aureus*. Silver NPs (AgNP) in textile fibers regulate the secretion of antibacterial agents conferring sterile properties and expand the biocidal activities even after several washing cycles (Gao and Cranston, 2008). Silver NP-coated fabrics amended with hexadecyltrimethoxysilane yields superhydrophobic cotton fabrics. This superhydrophobic cotton fabric possesses anti-odor, antibacterial, antistatic properties, UV protection, and electrical conductivity (Xue et al., 2012).

Disposable surgical face masks are now trending in daily routine to forbid the inhalation of harmful gases. Human beings use N-95 masks to protect themselves from air pollution and some life-threatening pathogens like coronavirus. The progress in research is going on where scientists are trying to develop a face mask integrated with nanofilters that can give protection against several pathogens and other pollutants.

9.4.3 IN COSMETICS

Cosmetics like lotions, fairness creams, and sunscreen lotions are composed of different ingredients that contain toxic elements such as mercury, lead, and cadmium. Thus, high usage of cosmetics may cause harmful effects to the skin. These effects may be either acute or chronic and can cause several skin diseases like rashes, acne, itching, scars, etc. In the present scenario, nanotechnology is increasing its footprints in the field of cosmetics to overcome these harmful effects.

Nanobiotechnology has overspread by formulization of nanoemulsion-based sunscreen lotions by applying ultrasonication technique. The nanoemulsion system was attained by the combination of avocado oil that acts as an active ingredient and various nonionic surfactants. The physical and chemical sunscreen components such as titanium dioxide and octyl methoxycinnamate, respectively, are mixed until a lotion of stable composition with a total SPF (sun protection factor) 3 is obtained. The reports have revealed that the nanoemulsion system proficiently improves UV filter characteristics of octyl methoxycinnamate (Silva et al., 2013).

Study has revealed that nanoemulsion-based cosmetic gel has increased the efficiency of mangosteen extracts due to the hydrophobic nature of mangosteen extract. Further, this enhances the bioefficacy of the extract as natural anti-oxidant, antifungal, antibacterial, and antiviral agent (Mulia et al., 2018).

9.4.4 IN THE FOOD INDUSTRY

The lifestyle of human beings has changed due to urbanization and industrialization. The demand for the processed food has increased in modern days. Several kinds of preservatives and additives are added to preserve the taste, texture, appearance of food products. Nanotechnology has played the vital role in enhancing the efficacy of the preservative in food products. Silver-based nanomaterials are used as effective bactericidal agents in the food packaging and processing, food preservation, food storage, water purification and thereby increase the shelf life of foodstuffs (Duncan, 2011; Rhim et al., 2013). Applications of nanobiotechnology in designing "smart packaging" and to detect the variation of spoiling gases or pathogens before it becomes contaminated (Silvestre et al., 2011).

Nanobiosensors are widely used in the identification and diagnosis of harmful microorganisms. Nanotechnology-based biosensors interact with the top surface of food products and, thus, preserve the taste, flavor, and odor of foodstuffs (Ghaani et al., 2016).

9.4.4.1 FOOD PROCESS

One of the contributions of nanobiotechnology in food processing is that the nanocarrier delivery system and organic nanosized additives are extensively used for nutrients, foodstuff, animal feed, and supplements. Moreover,

Nanoscience in Biotechnology

the capsulated vitamins that have high efficacy are delivered into blood-streams via the digestive system (Berekaa, 2015). For instance, KD Pharma BEXBACH GMBH (Germany) furnishes encapsulated omega-3 fatty acids in two individual forms, that is, powder and suspension. This capsulation technique is used for yielding particles at the micro and nanoscale levels.

There are different nanotechniques used for encapsulation and delivery of functional ingredients.

- **Edible coating**—Gelatin-based edible coatings containing cellulose nanocrystal are used to maintain the texture of food (Fakhouri et al., 2014).
- **Hydrogels**—Protein hydrogels are used to protect capsules or drugs from an external environment (Qui and Park, 2001).
- **Nanoemulsion**—β-carotene-based nanoemulsion improves the stability of food for a long duration (Kong et al., 2011).

9.4.4.2 FOOD PACKAGING AND LABELING

In the food packaging and labeling, it is necessary to maintain the taste, freshness, quality, safety, etc. of the foodstuffs to a satisfying level. In supply chain management, it is necessary for the producers to label and package their food items. Nanobiosensors are used as indicators of the food texture and color change in packed food products. This process has found worthwhile applications in the packaging of meat and milk (Bumbudsanpharoke and Ko, 2015).

In the process of food packaging, plastic polymer–coated nanomaterials are used frequently that help to improve functional or mechanical properties of food products (Berekaa, 2015). These nanomaterial serves as antimicrobial barriers that preserve the food products from any source of contamination.

9.4.5 IN AGRICULTURE

With rising population, the demand for food has increased across the globe. To meet the increasing demand of food, the farmers have relied on the synthetic input of fertilizers and pesticides. The prolonged use of synthetic pesticides and fertilizers causes soil erosion, reduces N_2 content, changes the pH of the soil, increases salinity, and kills beneficial microorganisms as well. Thus, modern technologies are adopted to make the agriculture system sustainable. Biomass conversion technology is one of the techniques used in the field of agriculture. Improvement in farming practices, monitoring of water quality, control of nutrients, and quality of fertilizer and pesticides can be made feasible with the application of nanotechnology (Mukhopadhyay, 2014; Prasad et al., 2014). Report has stated that germination rate in *Lycopersicum esculentum* (tomato) seeds was increased with SiO_2NPs (silicon dioxide NPs) (Siddiqui and AlWhaibi, 2014).

Further, in the agriculture sector, applications of biosensor-controlled delivery systems, nanofiltration, nanotubes, fullerenes, etc. have been reported to improve the soil health and crop productivity (Sabir et al., 2014). Nanobiosensors are also used to detect heavy metals, herbicides, and residues of pesticides present in the soil.

9.4.5.1 NANOFERTILIZERS

In agriculture system, chemical fertilizers are known to reduce carbon, phosphorus, and nitrogen content of the soil, which decline soil fertility, soil quality and cause land degradation. Although synthetic fertilizers have improved the crop production and yield, these fertilizers are costly and have become burden on the farmers (DeRosa, 2010). Thus, keeping the negative effect of these chemical fertilizers into consideration, the farmers have shifted to nonconventional ways of agriculture. In nanofertilizers, nanomaterials nurture themselves by intake of nutrients and act as the nutrient carrier and mitigate the loss of essential mobile nutrients and thereby improve the crop production (Kah et al., 2018) and plant metabolism. They have also found applications in the absorption of mobile nutrients through nanometric pores (Rico et al., 2011). Therefore, nanotechnology can serve as an alternative to sustainable development in agriculture system.

Further, nanostructured materials like polyacrylic acid, clay minerals, chitosan, zeolite, hydroxyapatite, etc., are used to manufacture fertilizers. Urea-based hydroxyapatite NPs are efficient than various chemical fertilizers like urea, ammonium nitrate, etc. as the former one releases N_2 for minimum 60 days whereas the latter one releases N_2 for 30 days for the plant growth (Kottegoda et al., 2011).

9.4.5.2 NANOPESTICIDES

The use of pesticides in agriculture system to protect the crops from pests and insects is a regular practice. Out of total applied pesticides, only 0.1%

Nanoscience in Biotechnology 257

reaches the pest and the remaining 99.9% affects nearby surroundings, including crops (Carriger et al., 2006). It can have negative impacts on human and animal health. In order to find the solution of this problem, we have shifted our focus on biopesticides. Biopesticides are eco-friendly, but their applications are limited as they are environment-dependent and slow against pathogens. To overcome these limitations, nanopesticides have manifested feasible potential and can serve as alternative to these biopesticides. Nano-material can regulate the secretion of active ingredients and slow degradation of pesticides. Nanomaterials offer nanocapsules, inorganic-engineered NPs, and nanoemulsions that are used to enhance the quality of food and soil fertility. Thus, nanopesticides provide an efficient and long-term pest control system (Chhipa, 2017).

A study had revealed that chlorfenapyr-modified silica NPs possessed twofold higher insecticidal activity against cotton bollworm in comparison to chlorfenapyr-modified microparticles (Song et al., 2012b). This concluded that nanomaterial-based pesticides regulate the release of active ingredients to control various plant pathogens. The recent reports have demonstrated that the nano-silica has good potential to control insects and pests in stored grains (Gamal, 2018).

Aluminum (Stadler et al., 2012), silver (Kim et al., 2012), and copper (Gogos et al., 2012) NPs possess antimicrobial activities against various microbial pathogens. In a report, it was estimated that a pesticide composed of silica-silver NPs has been found efficient against various fungal species— *Colletotrichum gloeosporioides, Rhizoctonia solani*, and *Magnaporthe grisea*. These pathogenic fungi were removed within 3 days from infected leaves after spraying silica silver–based nanopesticides (Park et al., 2006).

9.4.6   IN NANOBIOREMEDIATION

In the last few decades, an exponential increase in urbanization and industrialization has enhanced the environmental pollution. Environmental pollution has become serious concern across the globe. Various kinds of noxious chemicals like dyes, radioactive compounds, heavy metals, pesticides are released from different industries. These are the slow degradable compounds and cannot be easily remediated from the contaminated sites. They enter the human body through the food chain and can cause various diseases (Lhomme et al., 2008). Nanobiotechnology can be applied for bioremediation of hydrocarbon, uranium, heavy metal, wastewater, groundwater, and solid waste. Nanosized dendrimers, engineered NPs, single enzyme NPs, carbon nanotubes, nanoiron, and its derivatives, etc. are some important nanomaterials used in nanobioremediation (Rizwan et al., 2014). Engineered polymeric NPs can be applied in soil remediation (Tungittiplakorn et al., 2004) and remediation of hydrophobic contaminants (Tungittiplakorn et al., 2005). Biogenic uraninite NPs can be applied for bioremediation of uranium (Bargar et al., 2008). NPs that are biosynthesized from various species like *Centaurea virgata, Gundelia tournefortii, Scariola orientalis, Elaeagnus angustifolia, Reseda lutea, Bacillus* sp., and *Noaea mucronata* have been found to mitigate heavy metals mainly Ni, Cu, Pb, and Zn (Ingle et al., 2014).

Wastewater treatment is necessary as the water sources are being exhausted day by day. Moreover, the polluted water can cause diseases like diarrhea, dysentery, typhoid, cholera, etc. In the conventional technologies, some disinfectants like ozone, chloramine, chlorine, and chlorine dioxide, etc. are added to treat the water used for drinking purposes (Zwiener et al., 2007). The water loses its actual odor and taste due to the addition of chemical substances. Its consumption for a long time might cause several diseases like nausea, vomiting, kidney or liver damage, insomnia, cancer, diarrhea, etc. (Khlifi and Hamza-Chaffai, 2010). In recent time, gold NP–mediated technologies are being used for the treatment of wastewater (Das et al., 2009). Bimetallic gold-palladium NPs act as an active catalyst to biodegrade TCE (trichloroethylene) (Wong et al., 2009). Further, gold NP–integrated water purification devices can efficiently eliminate halocarbon-based pesticides from drinking water (Pradeep and Anshup, 2009).

9.5 CONCLUSION AND FUTURE PROSPECTS

This chapter emphasizes on the different routes of synthesis of NPs and their efficiency in distinct fields of science. Several physical and chemical approaches have been adopted for their synthesis in the last few decades. However, they are expensive, highly reactive and cause detrimental effects on environmental and human health. Thus, biological methods, including synthesis by plants and microbial synthesis of NP, have been used to combat various types of problems due to their low cost, eco-friendly, and nontoxic approaches. But microbial synthesis is slightly a slow and challenging process due to optimizations of certain physicochemical parameters. Nanomaterials possess specific metallic, oxidant, optical, and biomedical properties. By the virtue of which they are used in various sectors such

Nanoscience in Biotechnology

as agriculture, bioremediation, textiles, biomedical, and food technology. Furthermore, several new applications in the distinct fields are expected in the near future. Moreover, from the recent researches, it is expected that nanoscience is an emerging technology and can also be used for the welfare of mankind and environment.

KEYWORDS

- **nanobiotechnology**
- **nanoscience**
- **nanoparticles**

REFERENCES

Ahmad, A.; Mukherjee, P.; Senapati, S.; Mandal, D.; Khan, M. I.; Kumar, R.; Sastry, M. Extracellular Biosynthesis of Silver Nanoparticles Using the Fungus *Fusarium oxysporum*. *Colloids Surf. B* **2003a**, *28*, 313–318.

Ahmad, A.; Senapati, S.; Khan, M. I.; Kumar, R.; Ramani, R.; Shrinivas, V.; Sastry, M. Intracellular Synthesis of Gold Nanoparticles by a Novel Alkalotolerant Actinomycete, *Rhodococcus* Species. *Nanotechnology* **2003b**, *14*, 824–828.

Ahmad, M. S.; Yasser, M. M.; Sholkamy, E. N.; Ali, A. M.; Mehanni, M. M. Anticancer Activity of Biostabilized Selenium Nanorods Synthesized by *Streptomyces bikiniensis* Strain Ess_am A-1. *Int. J. Nanomed.* **2015**, *10*, 3389.

Ahmad, P.; Mukherjee, D.; Mandal et al. Enzyme Mediated Extracellular Synthesis of CdS Nanoparticles by the Fungus, *Fusarium oxysporum*. *J. Am. Chem. Soc.* **2002**, *124* (41), 12108–12109.

Ahmed, S.; Ikram, S. Biosynthesis of Gold Nanoparticles: A Green Approach. *J. Photochem. Photobiol.* **2016**, *161*, 141–153.

Alterman, M.; Sjobom, H.; Safsten, P; et al. P1/P1 modified HIV Protease Inhibitors as Tools in Two New Sensitive Surface Plasmon Resonance Biosensor Screening Assays. *Eur. J. of Pharm. Sci.* **2001**, *13* (2), 203–212.

Asanithi, P.; Chaiyakun, S.; Limsuwan, P. Growth of Silver Nanoparticles by DC Magnetron Sputtering. *J. Nanomater.* **2012**, *2012*, 963609.

Bai, H. J.; Zhang, Z. M.; Gong, J. Biological Synthesis of Semiconductor Zinc Sulfide Nanoparticles by Immobilized *Rhodobacter sphaeroides*. *Biotechnol. Lett.* **2006**, *28* (14), 1135–1139.

Bansal, R. C.; Goyal, M. *Activated Carbon Adsorption*; CRC Press: Boca Raton, FL, 2005.

Bar, H.; Bhui, D. K.; Sahoo, G. P.; Sarkar, P.; De, S. P.; Misra, A. Green Synthesis of Silver Nanoparticles Using Latex of *Jatropha curcas*. *Colloids Surf. A: Physicochem. Eng. Asp.* **2009a**, *339*, 134–139.

Bar, H.; Bhui, D. K.; Sahoo, G. P.; Sarkar, P.; Pyne, S.; Misra, A. Green Synthesis of Silver Nanoparticles Using Seed Extract of Jatropha Curcas. *Colloids Surf. A: Physicochem. Eng. Asp.* **2009b**, *348*, 212–216.

Bargar, J. R.; Bernier-Latmani, R.; Giammar, D. E.; Tebo, B. M. Biogenic Uraninite Nanoparticles and Their Importance for Uranium Remediation. *Elements* **2008**, *4*, 407–412.

Berekaa, M. M. Nanotechnology in Food Industry; Advances in Food Processing, Packaging and Food Safety. *Int. J. Curr. Microbiol. App. Sci.* **2015**, *4*, 345–357.

Birla, S. S.; Tiwari, V. V.; Gade, A. K.; Ingle, A. P.; Yadav, A. P.; Rai, M. K. Fabrication of Silver Nanoparticles by Phoma Glomerata and Its Combined Effect against *Escherichia coli Pseudomonas aeruginosa* and *Staphylococcus aureus. Lett. Appl. Microbiol.* **2009**, *48*, 173–179.

Bodzek, M.; Konieczny, K. Membrane Techniques in the Removal of Inorganic Anionic Micro Pollutants Form Water Environment-State of the Art. *Arch. Environ. Protect.* **2011**, *37* (2), 15–29.

Borse, V.; Kaler, A.; Banerjee, U. C. Microbial Synthesis of Platinum Nanoparticles and Evaluation of Their Anticancer Activity. *Int. J. Emerg. Trends Electr. Electron.* **2015**, *11*.

Bumbudsanpharoke, N.; Ko, S. Nano-Food Packaging: An Overview of Market, Migration Research, and Safety Regulations. *J. Food Sci.* **2015**, *80*, R910–R923. doi: 10.1111/1750-3841.12861.

Buzea, C.; Pacheco, I. I.; Robbie, K. Nanomaterials and Nanoparticles: Sources and Toxicity. *Biointerphases* **2007**, *2*, MR17–MR71. doi: http://dx.doi.org/10.1116/1.2815690.

Cao, J.; Hu, X. Synthesis of gold nanoparticles using halloysites. *e-J Surf Sci Nanotechnol* **2009**, *7*, 813–815.

Carriger, J. F.; Rand, G. M.; Gardinali, P. R.; Perry, W. B.; Tompkins, M. S.; Fernandez, A. M. Pesticides of Potential Ecological Concern in Sediment from South Florida Canals: An Ecological Risk Prioritization for Aquatic Arthropods. *Soil Sediment Contam.* **2006**, *15*, 2145.

Chauhan, A.; Zubair, S.; Tufail, S.; Sherwani, A.; Sajid, M.; Raman, S. C.; Azam, A.; Owais, M. Fungus-Mediated Biological Synthesis of Gold Nanoparticles: Potential in Detection of Liver Cancer. *Int. J. Nanomed.* **2011**, *6*, 2305–2319.

Chen, C. Y.; Chiang, C. L. Preparation of Cotton Fibers with Antibacterial Silver Nanoparticles. *Mater. Lett.* **2008**, *62* (21–22), 3607–3609.

Chen, J. C.; Lin, Z. H.; Ma, X. X. Evidence of the Production of Silver Nanoparticles via Pretreatment of *Phoma* sp. 32883 with Silver Nitrate. *Lett. Appl. Microbiol.* **2003**, *37*, 105–108.

Chhipa, H. Nanofertilizers and Nanopesticides for Agriculture. *Environ. Chem. Lett.* **2017**, *15*, 15–22.

Das, S. K.; Das, A. R.; Guha, A. K. Gold Nanoparticles: Microbial Synthesis and Applications in Water Hygiene Management. *Langmuir* **2009**, *25* (14), 8192–8199.

Dasgupta, N.; Ranjan, S.; Mundekkad et al. Nanotechnology in Agro-Food: From field to Plate. *Food Res. Int.* **2015**, *69*, 381–400.

DeJong, W. H.; Borm, P. J. Drug Delivery and Nanoparticles: Applications and Hazards. *Int. J. Nanomedicine* **2008**, *3*, 133.

DeRosa, M. C. Nanotechnology in Fertilizers. *Nat. Nanotechnol.* **2010**, *5*, 91.

Drummond, T. G.; Hill, M. G.; Barton, J. J. Electrochemical DNAsensors. *Nat. Biotechnol.* **2003**, *21*, 1192–1199.

Duncan, T. V. Applications of Nanotechnology in Food Packaging and Food Safety: Barrier Materials, Antimicrobials and Sensors. *J. Colloid Interf. Sci.* **2011**, *363* (1), 1–24.

Ehud, G. Plenty of Room for Biology at the Bottom: An Introduction to Bionanotechnology. *Imperial College Press,* **2007**, *677*, 6. ISBN 978-1-86094.

El-Batal, A. l.; Mona, S. S.; Al-Tamie, M. S. S. Biosynthesis of Gold Nanoparticles Using Marine *Streptomyces cyaneus* and Their Antimicrobial, Antioxidant Antitumor (in Vitro) Activities. *J. Chem. Pharm. Res.* **2015**, *7*, 1020–1036.

Emerich, D. F.; Thanos, C. G. The Pinpoint Promise of Nanoparticle-Based Drug Delivery and Molecular Diagnosis. *Biomol. Eng.* **2006**, *23* (4), 171–184.

Fakhouri, F. M.; Casari, A. C. A.; Mariano, M.; Yamashita, F.; Mei, L. I.; Soldi, V. et al. Effect of a Gelatin-Based Edible Coating Containing Cellulose Nanocrystals (CNC) on the Quality and Nutrient Retention of Fresh Strawberries during Storage. In *Proceedings of the IOP Conference Series: Materials Science and Engineering, Conference 1 2nd International Conference on Structural Nano Composites (NANOSTRUC 2014), Vol. 64*; Madrid, 2014. doi: 10.1088/1757.

Fariq, A.; Khan, T.; Yasmin, A. Microbial Synthesis of Nanoparticles and Their Potential Applications in Biomedicine, *J. Appl. Biomed.* **2017**, *15*, 241–248.

Fayaz, A. M.; Balaji, K.; Girilal, M.; Kalaichelvan, P. T.; Venkatesan, R. Mycobased Synthesis of Silver Nanoparticles and Their Incorporation into Sodium Alginate films for Vegetable and Fruit Preservation. *J. Agric. Food Chem.* **2009**, *57*, 6246–6252.

Feynman, R. Lecture at the California Institute of Technology, December 29, 1959.

Gajbhiye, M.; Kesharwani, J.; Ingle, A.; Gade, A.; Rai, M. Fungus-Mediated Synthesis of Silver Nanoparticles and Their Activity against Pathogenic Fungi in Combination with Fluconazole. *Nanomed. Nanotechnol. Biol. Med.* **2009**, *5*, 382–386.

Gamal, Z. M. M.; Nano-Particles: A Recent Approach for Controlling Stored Grain Insect Pests. *Acad. J. Agric. Res.* **2018**, *6*, 088–094.

Gao, X.; Cui, Y.; Levenson, R. M.; Chung, L. W. K.; Nie, S. In Vivo Cancer Targeting and Imaging with Semiconductor Quantum Dots. *Nat. Biotechnol.* **2004**, *22* (8), 969–976.

Gao, Y. Cranston, R. Recent Advances in Antimicrobial Treatments of Textiles. *Text. Res. J.* **2008**, *78* (1), 60–72.

Gericke, M.; Pinches, A. Microbial Production of Gold Nanoparticles. *Gold Bull.* **2006**, *39*, 22–28.

Ghaani, M.; Cozzolino, C. A.; Castelli, G.; Farris, S. An Overview of the Intelligent Packaging Technologies in the Food Sector. *Trends Food Sci. Tech.* **2016**, *51*, 1–11. doi: 10.1016/j.tifs.2016.02.008.

Ghorbani, H. R.; Mehr, F. P.; Poor, A. K. Extracellular Synthesis of Copper Nanoparticles Using Culture Supernatants of *Salmonella typhimurium. Orient J. Chem.* **2015**, *31*, 527–529.

Gogos, A.; Knauer, K.; Bucheli, T. D. Nanomaterials in Plant Protection and Fertilization: Current State, Foreseen Applications, and Research Priorities. *J. Agri. Food Chem.* **2012**, *60*, 9781–9792.

Govindraju, K.; Basha, S. K.; Kumar, V. G. Singaravelu, G. Silver, Gold and Bimetallic Nanoparticles Production Using Single-Cell Protein (Spirulina Platensis) Geitler. *J. Mater. Sci.* **2008,** *43,* 5115–5122.

Grimm, J.; Perez, J. M.; Josephson, L.; Weissleder, R. Novel Nanosensors for Rapid Analysis of Telomerase Activity. *Cancer Res.* **2004,** *64* (2), 639–643.

Hamdy, A. S.; Butt, D. P. Novel Anti-Corrosion Nanosized Vanadia-Based Thin Films Prepared by Sol–Gel Method for Aluminum Alloys. *J. Mater. Process. Technol.* **2007,** *181* (1), 76–80.

Hanauer, M.; Lotz, A.; Pierrat, S.; Sonnichsen, C.; Zins, I. Separation of Nanoparticles by Gel Electrophoresis According to Size and Shape. *Nano Lett.* **2007,** *7* (9), 2881–2885.

Haun, J. B.; Yoon, T. J.; Lee, H.; Weissleder, R. Magnetic Nano Particle Biosensors. *Wiley Interdiscipl. Rev.* **2010,** *2* (3), 291–304.

Hosseini, M. R.; Sarvi, M. N. Recent Achievements in the Microbial Synthesis of Semiconductor Metal Sulfide Nanoparticles. *Mater. Sci. Semicond. Process* **2015,** *40,* 293–301.

Husen, A.; Siddiqi, K. S. Phytosynthesis of Nanoparticles: Concept, Controversy and Application. *Nano Res Lett* **2014,** 9, 229.

Husen, A.; Siddiqi, K. S. Plants and Microbes Assisted Selenium Nanoparticles: Characterization and Application. *J. Nanobiotechnol.* **2014,** *12,* 28.

Iijima, S. Carbon Nanotubes: Past, Present, and Future. *Phys. B: Condensed Matter.* **2002,** *323* (1), 1–5.

Ingle, A. P.; Seabra, A. B.; Duran, N.; Rai, M. Nanoremediation: A New and Emerging Technology for the Removal of Toxic Contaminant from Environment, Chapter 9. In Microbial Biodegradation and Bioremediation; Das, S., Ed., 1st edn.; Elsevier, 2014.

Ingle, A.; Gade, A.; Pierrat, S. Sonnichsen, C.; Rai, M. K. Mycosynthesis of Silver Nanoparticles Using the Fungus Fusarium Acuminatum and Its Activity against Some Human Pathogenic Bacteria. *Curr. Nanosci.* **2008,** *4,* 141–144.

Iravani, S.; Zolfaghari, B. Green Synthesis of Silver Nanoparticles Using Pinus Eldarica Bark Extract. *BioMed. Res. Int.* **2013,** Article ID 639725.

Jabir, N.; Shams, T. S.; Ashraf, G. M.; Shakil, S.; Damanhouri, G. A.; Kamal, M. A. Nanotechnology-Based Approaches in Anticancer Research. *Int. J. Nanomed.* **2012,** *7,* 4391–4408.

Jha, A. K.; Prasad, K.; Prasad, K.; Kulkarni, A. R. Plant System: Nature's Nanofactory. *Colloids Surf. B: Biointerf.* **2009,** *73,* 219–223.

Jung, J. H.; Oh, H. C.; Noh, H. S.; Ji, J. H.; Kim, S. S. Metal Nanoparticle Generation Using a Small Ceramic Heater with a Local Heating Area. *J. Aerosol Sci.* **2006,** *37* (12), 1662–1670.

Kah, M.; Kookana, R. S.; Gogos, A.; Bucheli, T. D. A Critical Evaluation of Nanopesticides and Nanofertilizers Against Their Conventional Analogues. *Nat. Nanotechnol.* **2018,** *13,* 677684.

Kalishwaralal, K.; Deepak, V.; Ramkumarpandian, S.; Nellaiah, H.; Sangiliyandi, G. Extracellular Biosynthesis of Silver Nanoparticles by the Culture Supernatant of *Bacillus licheniformis. Mater. Lett.* **2008,** *62,* 4411–4413.

Karnik, B.; Davies, S.; Baumann, M.; Masten, S. The Effects of Combined Ozonation and Filtration on Disinfection By-Product Formation. *Water Res.* **2005,** *39* (13), 2839–2850.

Kelly, K. L.; Coronado, E.; Zhao, L. L.; Schatz, G. C. Theoptical Properties of Metal Nanoparticles: The Influence of Size, Shape, and Dielectric Environment. *J. Phys. Chem. B,* **2003,** *107* (3), 668–677.

Khan, S. A.; Gambhir, S.; Ahmad, A. Extracellular Biosynthesis of Gadolinium Oxide (Gd 203) Nanoparticles, Their Biodistribution and Bioconjugation with the Chemically Modified Anticancer Drug Taxol. *Beilstein J. Nanotechnol.* **2014,** *5*, 249–257.

Khlifi, R.; Hamza Chaffai, A. Head and Neck Cancer Due to Heavy Metal Exposure via Tobacco Smoking and Professional Exposure: A Review. *Toxicol. Appl. Pharmacol.* **2010,** *248* (2), 71–88.

Kim, H. J.; Phenrat, T.; Tilton, R. D.; Lowry, G. V. Effect of Kaolinite, Silica Fines and pH on Transport of Polymer-Modified Zero Valent Iron Nano-Particles in Heterogeneous Porous Media. *J. Colloid Interf. Sci.* **2012,** *370*, 1–10.

Kim, J. S. Antibacterial Activity of Ag+ Ion-Containing Silver Nanoparticles Prepared Using the Alcohol Reduction Method. *J. Ind. Eng. Chem.* **2007,** *13* (4), 718–722.

Kong, M.; Chen, X. G.; Kweon, D. K.; Park, H. J. Investigations on Skin Permeation of Hyaluronic Acid Based Nanoemulsion as Transdermal Carrier. *Carbohydr. Polym.* **2011,** *86*, 837–843. doi: 10.1016/j.carbpol.2011.05.027.

Kottegoda, N.; Munaweera, I.; Madusanka, N.; Karunaratne, V. A Green Slow-Release Fertilizer Composition Based on Urea-Modified Hydroxyapatite Nanoparticles Encapsulated Wood. *Curr. Sci.* **2011,** *101*, 73–78.

Kowshik, M.; Deshmuke, N.; Vogal et al. Microbial Synthesis of Semiconductor CdS Nanoparticles, Their Characterization, and Their Use in the Fabrication of an Ideal Diode. *Biotechnol. Bioeng.* **2002,** *78* (5), 583–588.

Kubik, T.; Bogunia-Kubik, K.; Sugisaka, M. Nanotechnology on Duty in Medical Applications. *Curr. Pharm. Biotechnol.* **2005,** *6*, 17–33.

Kundu, D.; Hazra, C.; Chatterjee, A.; Chaudhari, A.; Mishra, S. Extracellular Biosynthesis of Zinc Oxide Nanoparticles Using *Rhodococcus pyridinivorans* NT2: Multifunctional Textile Finishing, Biosafety Evaluation and in Vitro Drug Delivery in Colon Carcinoma. *J. Photochem. Photobiol. B* **2014,** *140*, 194–204.

Labrenz, M.; Druschel, G. K.; Thomsen-Ebert et al. Formation of Sphalerite (ZnS) Deposits in Natural Biofilms of Sulfate-Reducing Bacteria. *Science* **2000,** *290* (5497), 1744–1747.

Lamni, R.; Sanjines, R.; Parlinska-Wojtan, M.; Karimi, A.; Levy, F. Microstructure and Nanohardness Properties of Zr –Al–N and Zr –Cr–N thin films. *J. Vacuum Sci. Technol. A: Vacuum, Surf.Films* **2005,** *23* (1–2), 593–598.

Lhomme, L.; Brosillon, S.; Wolbert, D. Photocatalytic Degradation of Pesticides in Pure Water and a Commercial Agricultural Solution on TiO_2 Coated Media. *Chemosphere* **2008,** *70*, 381–386.

Malarkodi, C.; Chitra, K.; Rajeshkumar, S.; Gnanajobitha, G.; Paulkumar, K.; Vanaja, M.; Annadurai, G. Novel Eco-Friendly Synthesis of Titanium Oxide Nanoparticles by Using *Planomicrobium* sp. and Its Antimicrobial Evaluation. *Der Pharmacia Sinica* **2013,** *4*, 59–66.

Maretti, L.; Billone, P. S.; Liu, Y.; Scaiano, J. C. Facile Photochemical Synthesis and Characterization of Highly Fluorescent Silver Nanoparticles. *J. Am. Chem. Soc.* **2009,** *131* (39), 13972–13980.

Miyake, S.; Kawasaki, S.; Yamazaki, S. Nanotribology Properties of Extremely Thin Diamond-Like Carbon Films at High Temperatures with and Without Vibration. *Wear*, **2013,** *300* (12), 189–199.

Mukherjee, P.; Ahmad, A.; Mandal, D.; Senapati, S.; Sainkar, S. R.; Khan, M. I.; Ramani et al. Bioreduction of $AuCl_4^-$ Ions by the Fungus *Verticillium* sp. and Surface Trapping of the Gold Nanoparticles Formed. *Angew Chem. Int. Ed.* **2001,** *40* (19), 3585–3588.

Mukherjee, P.; Roy, M.; Mandal, B. P.; Dey, G. K.; Mukherjee, P. K.; Ghatak, J.; Tyagi et al. Green synthesis of Highly Stabilized Nanocrystalline Silver Particles by a Non-Pathogenic and Agriculturally Important Fungus *T. asperellum*. *Nanotechnology* **2008**, *19*, 103–110.

Mukhopadhyay, S. S. Nanotechnology in Agriculture: Prospects and Constraints. *Nanotechnol. Sci. Appl.* **2014**, *7*, 63–71. doi: 10.2147/NSA.S39409.

Mulia, K.; Ramadhan, R. M.; Krisanti, E. A. Formulation and Characterization of Nanoemulgel Mangosteen Extract in Virgin Coconut Oil for Topical Formulation. In *MATEC Web of Conferences*, Vol. 156, 2018 (EDP Sciences), 01013 p.

Narayanan, K. B.; Sakthivel, N. Biological Synthesis of Metal Nanoparticles by Microbes. *Adv. Colloid Interf. Sci.* **2010**, *156*, 1–13.

Ogi, T.; Saitoh, N.; Nomura, T.; Konishi, Y. Room-Temperature Synthesis of Gold Nanoparticles and Nanoplates Using Shewanella Algae Cell Extract. *J. Nanopart Res.* **2010**.

Pal, S. L.; Jana, U.; Manna, P. K.; Mohanta, G. P.; Manavalan, R. Nanoparticle: An Overview of Preparation and Characterization. *J. Appl. Pharmaceut. Sci.* **2011**, *1*, 228–234.

Park, H. J.; Kim, S. H.; Kim, H. J.; Choi, S. H. A New Composition of Nanosized Silica-Silver for Control of Various Plant Diseases. *Plant Pathol. J.* **2006**, *22*, 295–302.

Pastoriza-Santos, I.; Liz-Marzan, L. M. Reduction of Silver Nanoparticles in DMF Formation of Monolayers and Stable Colloids. *Pure Appl. Chem.* **2000**, *72* (1–2), 83–90.

Pickup, J. C.; Hussain, F.; Evans, N. D.; Sachedina, N. In Vivo Glucose Monitoring: The Clinical Reality and the Promise. *Biosens. Bioelectron.* **2005**, *20* (10), 1897–1902.

Pradeep, T.; Anshup. Noble Metal Nanoparticles for Water Purification: A Critical Review. *Thin Solid Films* **2009**, *517* (24), 6441–6478.

Prasad, K.; Jha, A. K. Zno Nanoparticles: Synthesis and Adsorption Study. *Nat. Sci.* **2009**, *1*, 129.

Prasad, R. Synthesis of Silver Nanoparticles in Photosynthetic Plants. *J. Nanopart.* **2014**, 963961. doi: 10.1155/2014/963961.

Qui, Y.; Park, K. Environment Sensitive-Hydrogels for Drug Delivery. *Adv. Drug Devl. Rev.* **2001**, *53*, 321–339.

Rai, M.; Yadav, A.; Gade, A. Current Trends in Phytosynthesis of Metal Nanoparticles. *Crit. Rev. Biotechnol.* **2008**, *28* (4), 277–284.

Ranjan, S.; Dasgupta, N.; Chakraborty, A. R. et al. Nanoscience and Nanotechnologies in Food Industries: Opportunities and Research Trends. *J. Nanoparticle Res.* **2014**, *16*, 2464.

Rhim, J. W.; Park, H. M.; Ha, C. S. Bio-Nanocomposites for Food Packaging Applications. *Progress in Polymer Science* **2013**, *38* (10), 1629–1652.

Rico, C. M.; Majumdar, S.; Duarte-Gardea, M.; Peralta-Videa, J. R.; Gardea-Torresdey, J. L. Interaction of Nanoparticles with Edible Plants and Their Possible Implications in the Food Chain. *J. Agric. Food Chem.* **2011**, *59*, 3485–3498.

Rizwan, M.; Singh, M.; Mitra, C. K.; Morve, R. K. Ecofriendly Application of Nanomaterials: Nanobioremediation. *J. Nanoparticles* **2014**, 1–7.

Sabir, S.; Arshad, M.; Chaudhari, S. K. Zinc Oxide Nanoparticles for Revolutionizing Agriculture: Synthesis and Applications. *Sci. World J.* **2014**, 8. doi: 10.1155/2014/925494.

Salman, J. A. S.; Ibrahem, K. H.; Ali, F. A. Effect of Culture Media on Biosynthesis of Titanium Dioxide Nanoparticles Using *Lactobacillus crispatus*. *Int. J. Adv. Res* **2014**, *2*, 1014–1021.

Sanghi, R.; Verma, P. A Facile Green Extracellular Biosynthesis of CdS Nanoparticles by Immobilized Fungus. *Chem. Eng. J.* **2009,** *155* (3), 886–891.

Sato-Berru, R.; Redon, R.; Vazquez-Olmos, A.; Saniger, J. M. Silver Nanoparticles Synthesized by Direct Photoreduction of Metal Salts. Application in Surface-Enhanced Raman Spectroscopy. *J. Raman Spectrosc.* **2009,** *40* (4), 376–380.

Sau, T. K.; Rogach, A. L. Nonspherical Noble Metal Nanoparticles: Colloid-Chemical Synthesis and Morphology Control. *Adv. Mater.* **2010,** *22* (16), 1781–1804.

Scott, N.; Chan, H. *Nanoscale Science and Engineering for Agriculture and Food System Report*; National Planning Workshop: Washington, DC, 2002.

Seil, J. T.; Webster, T. J. Antimicrobial Applications of Nanotechnology: Methods and Literature. *Int. J. Nanomed.* **2012,** *7*, 2767.

Seshan, K. *Handbook of Thin-Film Deposition Processes and Techniques— Principles, Methods, Equipment and Applications*; William Andrew/Noyes: Norwich, NY, 2002.

Shahverdi, A. R.; Fakhimi, A.; Shahverdi, H. R.; Minaian, S. Synthesis and Effect of Silver Nanoparticles on the Antibacterial Activity of Different Antibiotics against *Staphylococcus aureus* and *Escherichia coli*. *Nanomedicine: NBM* **2007a,** *3*, 168–171.

Shahverdi, A. R.; Minaeian, S.; Shahverdi, H. R.; Jamalifar, H.; Nohi, A. A. Rapid Synthesis of Silver Nanoparticles Using Culture Supernatants of *Enterobacteria*: A Novel Biological Approach. *Process Biochem.* **2007b,** *42*, 919–923.

Shaligram, N. S.; Bule, M.; Bhambure, R. M.; Singhal, R. S.; Singh, S. K.; Szakacs, G.; Pandey, A. Biosynthesis of Silver Nanoparticles Using Aqueous Extract from the Compactin Producing Fungal Strain. *Process Biochem.* **2009,** *44*, 939–948.

Sharma, V. K.; Yngard, R. A.; Lin, Y. Silver Nanoparticles: Green Synthesis and Their Antimicrobial Activities. *Adv. Colloid Interf. Sci.* **2009,** *145* (1–2), 83–96.

Shi, W.; Wang, Z.; Qiucen, Z.; Zhang, Q. C.; Leong, C.; He, M. Q. et al. Superconductivity in Bundles of Double-Wall Carbon Nanotubes. *Sci. Rep.* **2012,** *2* Article number 625.

Siddiqi, K. S.; Husen, A. Green Synthesis, Characterization and Uses of Palladium/ Platinum Nanoparticles. *Nano Res. Lett.* **2016,** *11*, 482.

Siddiqi, K. S.; Husen, A.; Rao, R. A. K. A Review on Biosynthesis of Silver Nanoparticles and Their Biocidal Properties. *J. Nanobiotechnol.* **2018,** *16*, 14.

Siddiqui, M. H.; Al-Whaibi, M. H. Role of Nano-SiO$_2$ in Germination of Tomato (*Lycopersicum esculentum* Seeds Mill.). *Saudi J. Biol. Sci.* **2014,** *21*, 13–17.

Silva, F. F.; Ricci-Junior, E.; Mansur, C. R.; Nanoemulsions Containing Octyl Methoxycinnamate and Solid Particles of TiO$_2$: Preparation, Characterization and in Vitro Evaluation of the Solar Protection Factor. *Drug Dev. Ind. Pharm.* **2013,** *39*, 1378–1388.

Silva, G. A.; Introduction to Nanotechnology and Its Applications to Medicine. *Surg. Neurol.* **2004,** *61*, 216–220.

Silvestre, C.; Duraccio, D.; Cimmino, S. Food Packaging Based on Polymer Nanomaterials. *Progress Polym. Sci.* **2011,** *36* (12), 1766–1782.

Singaravelu, G.; Arockiamary, J. S.; Ganesh Kumar, V.; Govindraju, K. A Novel Extracellular Synthesis of Monodisperse Gold Nanoparticles Using Marine Alga, *Sargassum wightii* Greville. *Colloid Surf. B: Biointerf.* **2007,** *57*, 97–101.

Singh, P.; Kim, Y. J.; Singh, H.; Wang, C.; Hwang, K. H.; Farh, M. E. A.; Yang, D. C. Biosynthesis, Characterization and Antimicrobial Applications of Silver Nanoparticles. *Int. J. Nanomed.* **2015,** *10*, 2567.

Singh, P.; Kim, Y.; Singh, H.; Wang, C.; Hwang, K. H.; Farh, M. E.; Yang, D. C. Biosynthesis, Characterization, and Antimicrobial Applications of Silver Nanoparticles. *Int. J. Nanomed.* **2014,** *10,* 2567.

Singh, R.; Nalwa, H. S. Medical Applications of Nanoparticles in Biological Imaging, Cell Labeling, Antimicrobial Agents, and Anticancer Nanodrugs. *J. Biomed. Nanotechnol.* **2011,** *7* (4), 489–503.

Sintubin, L.; De Windt, W.; Dick, J.; Mast, J.; van der Ha, D.; Verstraete, W.; Boon, N. Lactic Acid Bacteria as Reducing and Capping Agent for the Fast and Efficient Production of Silver Nanoparticles. *Appl. Microbiol. Biotechnol.* **2009,** *84* (4), 741–749.

Song, J. Y.; Kim, B. S. Rapid Biological Synthesis of Silver Nanoparticles Using Plant Leaf Extracts. *Bioproc. Biosyst. Eng.* **2009,** *44,* 1133–1138.

Song, M. R.; Cui, S. M.; Gao, F.; Liu, Y. R.; Fan, C. L.; Lei, T. Q.; Liu, D. C. Dispersible Silica Nanoparticles as Carrier for Enhanced Bioactivity of Chlorfenapyr. *J. Pest. Sci.* **2012b,** *37,* 258–260.

Stadler, T.; Buteler, M.; Weaver, DK.; Sofie, S. Comparative Toxicity of Nanostructured Alumina and a Commercial Inert Dust for *Sitophilus oryzae* (L.) and *Rhyzopertha dominica* (F.) at Varying Ambient Humidity Levels. *J. Stored Prod. Res.* **2012,** *48,* 81–90.

Sunkar, S.; Nachiyar, C. V. Biogenesis of Antibacterial Silver Nanoparticles Using the Endophytic Bacterium *Bacillus cereus* Isolated from *Garcinia xanthochymus*. *Asian Pac. J. Trop. Biomed.* **2012,** *2,* 953–959.

Tallury, P.; Malhotra, A.; Byrne, L. M.; Santra, S. Nanobioimaging and Sensing Of Infectious Diseases **2010**.

Taniguchi, N. Proceedings of International Conference on Precision Engineering (ICPE), Tokyo, Japan, **1974**.

Thakkar, K. N.; Mhatre, S. S.; Parikh, R. Y. Biological Synthesis of Metallic Nanoparticles. *Nanomed.: Nanotechnol., Biol. Med.* **2010,** *6* (2), 257–262.

Tinkle, S.; McNeil, S. E.; Muhlebach, S.; Bawa, R.; Borchard, G.; Barenholz, Y. C. et al. Nanomedicines: Addressing the Scientific and Regulatory Gap. *Ann. NY Acad. Sci.* **2014,** *1313,* 35–36.

Tong, H.; Ouyang, S.; Bi, Y.; Umezawa, N.; Oshikiri, M.; Ye, J. Nanophotocatalytic Materials: Possibilities and Challenges. *Adv. Mater.* **2012,** *24* (2), 229–251.

Tungittiplakorn, W.; Cohen, C.; Lion, L. W. Engineered Polymeric Nanoparticles for Bioremediation of Hydrophobic Contaminants. *Environ Sci. Technol.* **2005,** *39,* 1354–1358.

Tungittiplakorn, W.; Lion, L. W.; Cohen, C.; Kim, J. Y. Engineered Polymeric Nanoparticles for Soil Remediation. *Environ. Sci. Technol.* **2004,** *38,* 1605–1610.

Vatta, L. L.; Sanderson, R. D.; Koch, K. R. Magnetic Nanoparticles: Properties and Potential Applications. *Pure Appl. Chem.* **2009,** *78* (9), 1793–1801.

Verma, A.; Mehata, M. S. Controllable Synthesis of Silver Nanoparticles Using Neem Leaves and Their Antimicrobial Activity. *J. Radiat. Res. Appl. Sci.* **2016,** *9,* 109–115.

Wang, T.; Yang, L.; Zhang, B.; Liu, J. Extracellular Biosynthesis and Transformation of Selenium Nanoparticles and Application in H_2O_2 Biosensor. *Colloids Surf. B* **2010,** *80* (1), 94–102.

Watson, J. H. P.; Ellwood, D. C.; Soper, A. K.; Charnock, J. Nano Sized Strongly-Magnetic Bacterially-Produced Iron Sulfide Materials. *J. Magnet. Magnet. Mater.* **1999,** *203* (1–3), 69–72.

Wong, M. S.; Alvarez, P. J. J.; Fang, Y. L.; Akcin, N.; Nutt, M. O.; Miller, J. T.; Heck, K. N. Cleaner Water Using Bimetallic Nanoparticle Catalysts. *J. Chem. Technol. Biotechnol.* **2009,** *84,* 158–166.

Xiangqian, L.; Huizhong, X.; Zhe-Sheng, C.; Guofang, C. Biosynthesis of Nanoparticles by Microorganisms and Their Applications. *J. Nanomater.* **2011.**

Xue, C. H.; Chen, J.; Yin, W.; Jia, S. T.; Ma, J. Z. Superhydrophobic Conductive Textiles with Antibacterial Property by Coating Fibers with Silver Nanoparticles. *Appl. Surf. Sci.* **2012,** *258* (7), 2468–2472.

Zhang, W.; Qiao, X.; Chen, J. Synthesis of Silver Nanoparticles—Effects of Concerned Parameters in Water/Oil Microemulsion. *Mater. Sci. Eng. B* **2007,** *142,* 1–15.

Zhang, X.; Minear, R. A.; Guo, Y.; Hwang, C. J.; Barrett, S. E.; Ikeda, K.; Shimizu, Y.; Matsui, S. An Electrospray Ionizationtandem Mass Spectrometry Method for Identifying Chlorinated Drinking Water Disinfection Byproducts. *Water Res.* **2004,** *38* (18), 3920–3930.

Zheng, D.; Hu, C.; Gan, T.; Dang, X.; Hu, S. Preparation and Application of a Novel Vanillin Sensor Based on Biosynthesis of Au-Ag Alloy Nanoparticles. *Sens. Actuat. B: Chem.* **2010,** *148* (1), 247–252.

Zwiener, C.; Richardson, S. D.; De Marini, D. M.; Grummt, T.; Glauner, T.; Frimmel, F. H. Drowning in Disinfection Byproducts? Assessing Swimming Pool Water. *Environ. Sci. Technol.* **2007,** *41* (2), 363–372.

CHAPTER 10

PLANT PRODUCT-BASED NANOMEDICINE FOR MALIGNANCIES: TYPES AND THERAPEUTIC EFFECTS

ZUHA IMTIYAZ[1], TABISH MEHRAJ[2], ANDLEEB KHAN[3],
MIR TAHIR MAQBOOL[4], RUKHSANA AKHTER[5], MUFEED IMTIYAZ[6],
WAJHUL QAMAR[7], AZHER ARAFAH[8], and MUNEEB U. REHMAN[8*]

[1]*Clinical Drug Development of Herbal Medicine, College of Pharmacy, Taipei Medical University, Taipei, Taiwan*

[2]*Department of Pharmaceutical Sciences, School of Pharmacy, University of Mississippi, University, MS, United States*

3*Department of Pharmacology and Toxicology, College of Pharmacy, Jazan University, Jazan, Saudi Arabia*

[4]*National Center for Natural Products Research, School of Pharmacy, University of Mississippi, University, MS, United States*

[5]*Department of Clinical Biochemistry, Govt. Degree College (Baramulla), Khawaja Bagh, Baramulla, Jammu and Kashmir, India*

[6]*Khwaja Yunis Ali Medical College, Enayetpur Sharif Sirajgonj, Bangladesh*

[7]*Department of Clinical Pharmacology & Toxicology and Research Centre, College of Pharmacy, King Saud University, Riyadh, Saudi Arabia*

[8]*Department of Clinical Pharmacy, College of Pharmacy, King Saud University, Riyadh, Saudi Arabia*

Corresponding author. E-mail: muneebjh@gmail.com

ABSTRACT

The field of nanomedicine is immensely diverse and expansive. It has been extensively explored for delivering various drugs and as a field of therapeutic strategy. In recent years, research has shown that nanoparticles make it possible to deliver the drugs that are either metabolized on the way to the target or have an abrupt release which interferes with the bioavailability as well as the efficacy. Therefore, making the controlled release, higher bioavailability, and specificity achievable. The nanoparticles can be designed to meet specific needs. This can be done by modifying the charge, functional group or also by creating various hybrids. Nanomedicine also makes it possible to co-deliver multiple drugs that help in impeding the major problem such as drug resistance. Combing the science of phytomedicine and nanomedicine also helps circumvent the challenges that are usually encountered when designing and developing novel drugs from phytochemicals. This chapter revolves around discussing various therapeutic benefits of nanomedicine and dissecting its application in clinical drug development.

10.1   INTRODUCTION TO NANOMEDICINE

Nanomedicine is the science of nanotechnology dealing with nanoscale materials and systems, which display unique and better biological, physical, and chemical characteristics, used for the identification, control, and treatment of diseases (Soares et al., 2018). The growth in the field of nanomedicine technology can improve the treatment conditions and diagnosis by utilizing the unique characteristics of engineered nanoparticles to recognize the specific markers of the diseases and to deliver the drugs to their respective target sites (Kim et al., 2010; Weiss, 2009; Davis et al., 2010). Using nanotechnology in the field of medicine is advantageous from drug delivery to their efficacy. Numerous properties of nanoparticles, which include the electrical, optical, and magnetic, could be controlled through altering their shape, size, physicochemical properties, and surface characterization, etc. Due to their flexible tunable properties, it allows the researchers and scientists around the world to create enormous amounts of materials with distinct properties and advantages to tailor the engineering of biomedical, diagnostic, and electronic devices (Zhang et al., 2016).

The field of nanomedicine has matured drastically during the last few decades. Many nanomedicines have been commercialized and many of them are in clinical trials (Chan, 2017; Etheridge et al., 2013). There has been an acceptance of the risk and benefit of the cost trade-offs for utilizing nanomedicine for toxicological and pharmacokinetic/pharmacodynamic studies. Several reports are published that have characterized nanoscale materials for their side effects (Ragelle et al., 2017; Chen et al., 2016). Similarly, for budding technologies in the area of risk investigation, there is a necessity of flexibility. Risk screening approaches such as Nano Guidance for Risk Informed Deployment (NanoGRID) outline, which was formed by the US Army Engineering R & D Center (Paluri et al., 2017), and the European Union's Sustainable Nanotechnologies Project (Collier et al., 2015) are the two models of multistakeholder works to form an all-inclusive assessment of nanomaterial risk for various applications which includes therapeutics and medical devices (Subramanian et al., 2014).

Tremendous research has been carried out for anticancer compounds in combination with nanoparticles (NPs) as combinational therapy. Various health benefits, including drug resistance, minimization of therapeutic dose, synergism, potentiation, and efficacies, have been observed by adopting combinational therapy by entrapping anticancer drugs with chemo-sensitizing agents. For example, in the case of curcumin, from the last few years, nanoformulations are being formulated and evaluated for their efficacy against different cancer models (both in vivo and in vitro). These formulations of curcumin have also been screened in drug-resistant models of cancer. Despite the advancement of the combinatorial models, the shift from preclinical experiments to clinical trials is always challenging. Currently, 20 drugs containing curcumin in combination with other drugs are undergoing clinical trials, out of which only one-tenths are nanoformulations. The bulk manufacturing of nanoformulations, based on their pharmacokinetic and physicochemical limitations, is a major challenge. The main challenges while formulating the nanoformulation are the nanosized dose having poor bioavailability, undefined ability to bind the target, and curcumin instability issues. There is much more emphasis on the alternation of naturally occurring forms of curcumin and its use in formulations such as the prodrug approach or covering the surface of curcumin to combat the problems with stability and to improve the pharmaceutical and therapeutic ability of its nanoformulations (Rycroft et al., 2018).

Nanoparticles (NPs) are the principal constituents of nanomedicine, and presently, numerous types of NPs exist, but there is no uniform nomenclature in the literature; however, researchers use names, for example, nonbiological complex drugs (NBCDs), engineered nanomaterials, nanomedicals/nanomedicines. Certain nanomaterials can function like that of globular biological macromolecules (Batra et al., 2019). The examples are lipid micelles (Torchilin, 2017), different polymeric nanostructures (Zhang et al., 2014), protein constructs (Lee et al., 2007), ribonucleic acid (RNA) (Guo, 2010), carbon dots (Baker and Baker, 2010), carbon nanotubes (CNTs), nanodiamonds (NDs), graphene (Yu et al., 2005), and inorganic materials such as mesoporous silica NP (MSNP) (Baughman et al., 2002). It is necessary to establish safety profile of NPs while administering them into the body in addition to their uses in diagnosis and therapeutics.

10.2 ADVANTAGES OF NANOMEDICINE OVER CONVENTIONAL METHODS

There are many restrictions related to the traditional disease treatments with utilizing the accessible medications, such as short half-lives in blood, low bioavailability and solubility, and nonspecificity. These limitations associated with conventional treatments result in the need for high drug dosage requirements. The high dosage administrations are one of the factors responsible for the adverse effect associated with conventional treatments of drugs, when used at therapeutic dose. While addressing the problems of the conventional treatment of the diseases, the applications of nano-based drug-delivery systems such as NPs, nanocapsule dendrimers, and micelles have been developed which are effective in a way for the delivery recombinant proteins of conventional drugs, vaccines, and genetic materials (Moghimi et al., 2005).

In one of the recent research conducted by Xiao et al., where they developed nanomedicine with sustained release of the drug of choice and targeted drug delivery, for the treatment of rheumatic arthritis, and achieved success in rheumatic arthritis animal models. Presently, antirheumatic drugs require high drug dosage and frequent administration, leading to serious side effect and challenging in terms of patient compliance. Nanomedicines have the potential to pave the way for drug solubilization, particularly for targeting tissues or cells, envisaging the manner of drugs in vivo and so on (Xiao et al., 2019). Therefore, the development of nanomedicine makes it quite possible to overcome the limitations of the presently used rheumatic arthritis therapeutic drugs. Multifunctional NPs are of a greater advantage in terms of their capability of cutting-edge nanotechnologies. Multifunctional NPs have a synergetic effect in detecting the abnormal biological situation and obtaining a therapeutic efficacy for rheumatic arthritis (Kesharwani et al., 2019) (Fig. 10.1).

Plant Product-Based Nanomedicine for Malignancies

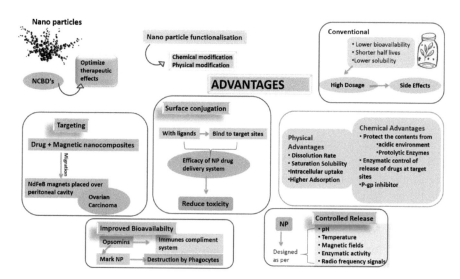

FIGURE 10.1 Advantages of nano-based formulations in terms of improved bioavailability, higher efficacy, and reduced drug load, thereby reducing the requirement of drug dosage hence minimizing toxicity.

10.3 FUNCTIONALIZATION OF NANOPARTICLES

Functionalization of NPs is a remarkable characteristic of NPs, which provides an opportunity to engineer them as per required specifications (Galvin et al., 2012; Nam and Luong, 2019). Based on physical and chemical conjugations, one can modify or functionalize the surface of the NPs to get our required specification or targeting properties. The process of functionalization can provide a directed delivery of drugs; enhance the efficacy, sustained release, prolong the effect of the drug; enable the targeted drug delivery, detect of tumor or any infected tissue; provide feedback concerning drug efficacy; and reduce the sheer effects of blood flow (Shakeri et al., 2020).

10.3.1 TARGETING

Numerous perspectives must be considered while defining focused on NPs, which incorporates focusing on capacity and shirking for reticuloendothelial system (RES) framework. Numerous elements should be viewed as when building focused on NPs, including size, biocompatibility, shirking of the RES, and its solidness in the blood, just as the capacity of the NPs to empower controlled medication discharge (Ventola, 2012). NPs with attractive polymer nanocomposites or magnetoliposomes installed with sedate atoms have incredible potential for focused medication conveyance.

Many aspects have to be taken into account while formulating targeted NPs, which include targeting ability and avoidance for RES. Numerous elements should be taken into account when formulating NPs with targeted drug delivery, which includes biocompatibility, size and eliminating opsonization, and drug stability in the blood, and its controlled release (Ventola, 2012). NPs with magnetic polymer nanocomposites or magnetoliposomes are efficient for targeted drug delivery. In one study, MRI established that the NPs with magnetic property were transferred toward NdFeB magnets, which were placed external to the peritoneal cavity, over the human ovarian carcinoma grafts (Yuksel et al., 2018). Contrasting agents can also be filled in the NPs for imaging purposes (Naseri et al., 2018).

10.3.2 SURFACE CONJUGATION

Some candidate biomolecules such as the peptides, which penetrate the cells (CPPs), are attached to the surfaces of nanoparticles that enhance their intracellular delivery. Surface biomolecule–nanoparticle conjugates include means for gene therapy such as small inhibitory RNA (siRNA) and fluorescent imaging dyes (Hong and Nam, 2014). In order to enhance the drug efficacy, surface conjugation with ligands would reduce the toxicity associated with conventional drugs at high doses while binding to their respective target sites (Yoo et al., 2019). Similarly, in cancer treatment strategy involving biomolecule–nanoparticle conjugates, surface conjugation can be conducted with a biomarker or a molecule such as protein, peptide, or nucleic which can bind to tumor cell receptors (Spicer et al., 2018)

10.3.3 IMPROVED BIOAVAILABILITY

Most of the phytochemicals are being consumed as plant foods, including vegetables, fruits, spices, and beverages. The prevalence of phytochemicals in food makes them easily available with least apparent toxicity attributed to their role in influencing human health and has beneficial impact on several diseases. In view of their potential health benefits, numerous studies are being conducted to determine the critical factors affecting their bioavailability. As far as phytochemical bioavailability is concerned, phytochemicals could be grouped into lipid-soluble (carotenoids, tocochromanols, curcuminoids) and water-soluble (polyphenols and phenolics). The main factors that influence bioavailability of phytochemicals include (1) biological aspects such as ADME (absorption, distribution, metabolism, and excretion) and (2) identification of food factors such as micro- and macro-composition, physical form, and phytochemical concentration (Manach et al., 2004). Poor absorption and rapid metabolism and excretion are the most challenging issues faced by researchers while implementing the use of phytochemicals against various diseases like cancer, diabetes, etc. It has been observed that nanocarrier drug-delivery systems are beneficial for increasing the bioavailability of phytochemicals. For example, it has been reported that encapsulation of anthocyanins in nanocarriers improves their anti-oxidant potential by increasing their scavenging of the free radicals. In this, anthocyanins were entrapped with 60% efficacy in nanocarrier based on polylactide-co-glycolide (Amin et al., 2017).

Opsonization is a process by which the NPs are usually removed from circulatory system by the proteins present in the immune system known as opsonins. These opsonins mask the NPs for engulfment by phagocytes and macrophages. Charged NPs are difficult to clear from the circulation than neutral NPs, while as the hydrophobic NPs are cleared faster than hydrophilic ones. Therefore, researchers are taking interests in designing of neutral NPs by conjugating it with a hydrophilic polymer of interest such as polyethylene glycol (PEG) for prolonged circulation into the body. Pegylation can be done for the enhancement of bioavailiblalty of the nanomedicine. Lipid coatings on the surface can also aid in the biocompatibility (Suk et al., 2016).

10.3.4 CONTROLLED RELEASE

For the targeting- or site-specific delivery of the nanocarrier, it can be designed in a way so that it activates the drug release when triggered. The characteristics of NPs that can be utilized to functionalize NPs are magnetic field, pH, enzymatic activity, temperature, and other factors like light, radio frequency, etc. (Karimi et al., 2016). NPs designed with pH-sensitive substances can be made to release the drug for the pH sensitive target sites, such as the mild acidic environment present in tumor tissues (pH 6.8), cellular vesicles, lysosomes (pH 4.5–5.0), endosomes (pH 5.5–6.0) (Cao et al., 2019). The thermal sensitive linkers consist of nucleic acids, peptides lipids, carbohydrates or proteins, and polymers can also be utilized for the attachment of one or more agents for controlled release of the drug from the NPs. The drug release from NPs can also be achieved by incorporating certain bonds that get degraded in specific conditions at target site (Mackay and Chilkoti, 2008; Vaddiraju et al., 2010; Spicer et al., 2018)

10.3.5 TUNABILITY

NPs have a variety of tunable, optical, magnetic, biologic, mechanical, and electric properties that vary significantly from the same materials in bulk pertaining to improved quantum mechanics which occurs close to the nano-sized scale (Jeevanandam et al., 2018).

10.4 POTENTIAL ADVANTAGES AND APPLICATIONS OF NANOMEDICINE

Nanomedicine is believed to have a revolutionary effect on the health-care system in near future. It has a potentiality for the improvement of the drug safety and efficacy of traditionally marketed therapeutic drugs (Ventola, 2012). The greater surface area of the NPs provides an opportunity for enhancing the saturation, solubility, dissolution rate, and intracellular uptake of drugs, thereby enhancing their in vivo efficacies (Patra et al., 2018). Properties like entrapment efficiency, release drug characteristics, and surface functionalization can improve therapeutic targeting and bioavailability (Patel and Agrawal, 2011). It can also enhance the efficacy of NP formulations greatly. Nanotechnologies have already reformed analytical techniques of biologicals by designing devices that inspect biomarkers. These assessments are much efficient in terms of their reliability, cost-effectiveness, and sensitivity. Nanotechnologies have the ability to discover early indications of metabolic disorders that could aid in the control of disorders such as obesity and diabetes. Even though nanomedicine is still emerging, various drugs that make use of nano-based technology are approved and commercialized (Fornaguera and García-Celma, 2017).

10.4.1 PHYSICAL AND BIOCHEMICAL ADVANTAGES OF ORAL NANOMEDICINE

The oral drug delivery is globally preferred method for consuming drugs because of advantages of being noninvasive, pain free, and convenient to

Plant Product-Based Nanomedicine for Malignancies 277

manage. However, the oral administration is also associated with certain drawbacks like bioavailability, pharmacokinetics, and pharmacodynamics. Bioavailability is the major challenge for orally administered drugs. Before reaching to its target site, a drug has to overcome multiple compartments of the body or multiple barriers, including intestinal mucosa, epithelial, or endothelial layers, where a drug comes under different cellular conditions like pH and temperature variations or enzymatic catalysis. To overcome such challenges, encapsulation of drug into nanocarrier would protect the drug from environmental variations in the human body thereby considerably enhancing the bioavailability and efficacy of oral dosage forms (Reinholz et al., 2018).

NPs offer a greater surface area because of their nanosize that can allow a greater capacity of adsorption for the loading of the drugs (Ramesan and Sharma, 2009). Muco-adhesion can be enhanced because of the interaction between the mucous coating of the GI tract and the drug, which results in the enhancement of the adsorption of the drugs, nucleotides, and proteins. Various hypotheses of enhanced muco-adhesion have been followed such as the interactions being electrostatic between the endothelial layer and positively charged NP, or by Van der Waals forces of attraction. Eudragit poly (acrylic acid), sodium alginate, and chitosan are the muco-adhesive polymers (Kawashima et al., 2000). NPs entrap and protect the drug and the excipients from the acidic environment and enzymes, which are proteolytic in nature (Ensign et al., 2012; Luo et al., 2015). The release of the drug can be controlled by these enzymes, which are present in the GI tract. In the pH-responsive polymers such as hydroxypropyl methylcellulose phthalate, and hydroxypropyl methyl acetate, Eudragit (L100-55, L100, and S100) dispenses the drugs to desired area. Various reports suggest that NPs could be utilized as a P-gp inhibitor. There are polymers such as polyalkylcyano-acrylates and coating agents like polysorbate 80 that can overcome P-gp-mediated drug efflux (Niazi et al., 2016; Kapse-Mistry et al., 2014).

10.5 APPLICATIONS OF NANOMEDICINE: NANOTECHNOLOGY AND PHYTOCHEMICALS (NANOTECHNOLOGY-BASED DELIVERY OF PHYTOCHEMICALS)

Plant-based naturally derived products have been utilized by humankind for centuries for several ailments (Veeresham, 2012). The natural compounds in the plant known as the phytochemicals have been linked to their cura-tive properties. There are various population studies suggesting the use of phytochemicals being effective against neurodegeneration diseases, cardiovascular diseases. There is enough evidence to confirm many advantages of phytochemicals present in plants with respect to health (Hosseini and Ghorbani, 2015). Generally, all of the biological processes that occur within the body, which includes cancer initiation to progression, are known to take place at the nanoscale level (Jeevanandam et al., 2018). Because of this, the important role played by nanotechnology in the area of medicine has succeeded. Nanotechnology is one of the progressive dynamic technologies that particularly involve the nanoscale level (10-9 m) investigation (Bayda et al., 2019) (Fig. 10.2; Table 10.1).

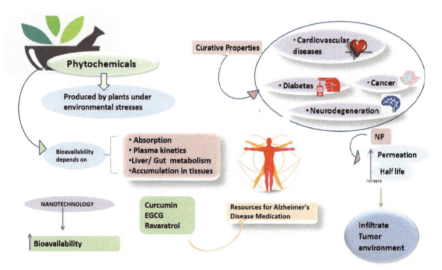

FIGURE 10.2 Factors affecting efficacy of phytochemicals and application of nanotechnology in improving their therapeutic efficacy.

10.6 NANOFORMULATION-BASED PHYTOCHEMICALS IN DISEASES

It has been found that several phytochemicals are very effective against Alzheimer's disease by acting on specific targets, influencing pathogenesis of Alzheimer's disease. In vivo and in vitro neurodegenerative disease models have shown beneficial effect of numerous phytochemicals, including curcumin, EGCG, and resveratrol. Phytochemicals are proposed to be used as promising resources for the growth of new drugs aiming Alzheimer's disease in addition to their beneficial effects against several neurodegenerative diseases, including Parkinson's disease and Alzheimer's disease. Phytochemicals are apparently least toxic, cost–effective, and easily available. Due

Plant Product-Based Nanomedicine for Malignancies 279

TABLE 10.1 Examples of Different Phytochemicals, Sources, and Their Applications in Nanomedicine.

Phytochemical	Source	Chemical structure	Application in nanomedicine	References
Apigenin	Plants belonging to Artemisia, Achillea, Matricaria, and Tanacetum. Parsley, celery, celeriac, chamomile tea		In vivo hepatocellular carcinoma. Therapeutic approach of Apigenin-loaded liposomes in patients with lung cancer.	Salehi et al. (2019), Bhattacharya et al. (2018), Jin et al. (2017)
Berberine	Rhizomes, stems, bark of *Berberis vulgaris*, and roots.		Nanoformulations have shown a potential in treating the metabolic syndrome. Berberine nanoparticles show a positive antibacterial effect.	Imenshahidi and Hosseinzadeh (2019), Taghipour et al. (2019), Sahibzada et al. (2018)
Combretastatin A-4	South African tree *Combretum caffrum* (bark)		Antitumor effect, antineovasculature, antiproliferation, and apoptosis induction, active targeting. Polymeric micelles increase the effectiveness against primary murine breast tumors after IV drug delivery, sustained drug release of the nanocarrier, increase in the residence time.	Mustafa et al. (2019); Yang et al. (2012), Wakaskar et al. (2015)
Curcumin	Turmeric or *Curcuma longa*		Curcumin-loaded NPs have been shown to alleviate the disorders related to central nervous system (CNS), Alzheimer's disease, Huntington disease, epilepsy, and amyotrophic Lateral Sclerosis Multiple sclerosis. Parkinson's disease	Hewlings and Kalman (2017); Yavarpour-Bali et al. (2019)

TABLE 10.1 (Continued)

Phytochemical	Source	Chemical structure	Application in nanomedicine	References
Ellagic acid	Pomegranate, grape, raspberries, strawberries, and blackberries.		NPs containing Ellagic acid with PCL Poly(ε-caprolactone) enhance the bioavailability in terms of oral drug delivery ellagic acid being a bioactive substance in cancer therapy.	Arab et al. (2019), Madyand Shaker (2017)
Emodin	Traditional Chinese herb: *Polygonum multiflorum*		A stable nanoformulation of emodin loaded with TPGS, which is a pegylating agent showed enhanced therapeutics.	Zheng et al. (2019), Wang et al. (2012)
Epigallocatechin gallate	*Camellia sinensis* (green tea), plums, onions, pecans and hazelnuts, carob powder apple skin.		EGCG gold nanoparticles (AuNPs) possess anti-osteoclastogenic efficacy than free EGCG both in vitro and in vivo, in antibone resorption treatment.	Casanova et al. (2019), Zhu et al. (2019)
Ferulic acid (FA)	Vegetables, fruits, cereals, and some traditional medicinal plants like *Angelica sinensis* (Oliv.) Diels Coffee, apple, artichoke, peanut, female ginseng, flaxseed. Orange		FA—water-insoluble natural antioxidant with anticancer activity, when incorporated into nanosponges with cyclodextrin as a cross-linking agent increased the solubility of the drug up to 15-folds.	Zhao et al. (2015), Rezaei et al. (2019); Tao et al. (2019)

TABLE 10.1 (Continued)

Phytochemical	Source	Chemical structure	Application in nanomedicine	References
Gallic acid	Gallnut, witch hazel, Indian gooseberry North American white oak, sundew, hot chocolate, green tea, raspberry blackberry		Antibacterial activity of gallic acid was enhanced by conjugation with AuNPs, which has a potential in biomedical sciences.	Subramanian et al. (2016)
Gambogic acid (GA)	*Garcinia* species, including *G. hanburyi* and *G. morella*		GA is a novel antitumor drug, but because of its low water solubility, it is combined with poly(lactic-*co*-glycolic acid) (PLGA) with red-blood-cell NPs, which shows it can retain and enhance efficacy in colorectal cancer treatment as compared with free GA.	Wang and Chen (2012)
Honokiol	*Magnolia officinalis*		Honokiol NPs show a greater solubility profile and better` bioavailability offering a better therapeutic response.	Jiraviriyakul et al. (2019); Wu et al. (2018)
β-Lapachone	Lapacho tree (*Tabebuia avellanedae*) in South America.		β-Lapachone compromising poly(ethylene glycol)-block-poly(D,L-lactide) (PEG–PLA) polymer micelles for the treatment of NQO1(quinine oxidoreductase) overexpressing tumors.	Woo et al. (2006), Blanco et al. (2007)
Luteolin	Pollen, celery, broccoli, green pepper, , thyme, ragweed parsley ,chamomile tea, olive, carrots oil, peppermint, navel oranges		Luteolin folic acid–modified poly(ethylene glycol)-poly(e-caprolectone) nanomicelles used to prevent glioblastoma multiforme given as an intravenous formulation.	Miean and Mohamed (2001), Wu et al. (2019)

TABLE 10.1 (Continued)

Phytochemical	Source	Chemical structure	Application in nanomedicine	References
Noscapine	Plants of the poppy family (*Papaver somniferum*)		Noscapine loaded with human serum albumin NPs used to treat breast cancer and showed a better efficacy.	Yanran et al. (2018) Sebak et al. (2010)
Nobiletin	Citrus fruits		Nobiletin poly(ethylene glycol)-block-poly(e-caprolactone) micelles showed a sustained release of NOB and extended circulation. The NOB–PEG–PCL delivery system could be a promising treatment for osteoporosis.	Huang et al. (2016), Wang et al. (2019)
Resveratrol	Peanuts and grapes, grapevines, pines, and legumes.		The liposomal nanomedicines of resveratrol were found to be effective against glioblastomas and promising for active and passive targeting.	Lu et al. (2019); Langcake and Pryce (1976), Soleas et al. (1997), Jhaveri et al. (2018)
Silibinin	*Silybum marianum*		Silibinin-loaded NPs with chitosan resulted in the increased drug load, efficacy, and sustained release of silibinin, showing enhanced anticancer activity.	Cheung et al. (2010), Kuen et al. (2017)
Thymoquinone	*Nigella sativa*		Nanoformulations of thymoquinone enhance the bioavailability of the drug thus reducing the dosing.	Ijaz et al. (2017), El-Far et al. (2018)

TABLE 10.1 (Continued)

Phytochemical	Source	Chemical structure	Application in nanomedicine	References
Triptolide	*Tripterygium wilfordii* herb Hook F, known as the "thunder god vine"		Triptolide nanomedicine was developed for chemo-resistant pancreatic cancer cells and to avoid contact to healthy tissues for Triptolide by nucleolin-specific aptamer (AS1411)-mediated polymeric nanocarrier.	Chen et al. (2018), Wang et al. (2016)
Ursolic acid	(Rosemary, marjoram, lavender, thyme, and organum) leaves, fruits (apple fruit peel), berries, flowers.		Nanoformulation composed of chitosan, ursolic acid, and folate increases the drug-loading concentrations at the target site and reducing the side effects. In vivo results demonstrated that FA-CS-UA-NPs could lessen breast cancer burden in MCF-7 xenograft mouse model.	Jäger et al. (2009), Seo et al. (2018), Jin et al. (2016)
Zerumbone	*Zingiber zerumbet*		Zerumbone nanostructured lipid carriers are promising in terms of sustained release, for the treatment of leukemia.	Kalantari et al. (2017), Rahman et al. (2013)

to the complex mechanistic pathology of Alzheimer's diseases, it becomes a challenge for scientists around the world to identify the particular mechanism by which phytochemicals exhibit their effects. The nanoformulation-based phytochemicals exert their potential beneficial effects in Alzheimer's disease by targeting major points for the pathophysiology of Alzheimer's disease, including anti-amyloid beta immunotherapy, neurotransmitter models, metal stress reducers, and inflammatory inhibitors (Ovais et al., 2018). For example, Tiwari et al. have demonstrated that *Curcumin*-entrapped PLGA NPs (Cur-PLGA-NPs) induce in vitro proliferation of neural stem cells (NSCs) and In vivo differentiation in the hippocampus and subventricular zones of rats as compared to conventional curcumin (Tiwari et al., 2014). Quercetin is a bioactive phytochemical that is found in many dietary compounds. Nam and colleagues described its application in oncotherapy pertaining to its anticancer properties. Quercetin is found to have a regulatory effect on cell migration, growth, and cell apoptosis via various signaling pathways. Even though quercetin has great pharmaceutical significance, its use, as a potential therapeutic drug, is limited in terms of its lower bioavailability and poor aqueous solubility, and rapid digestion of quercetin is a major drawback. Therefore, in order to overcome these limitations, quercetin-based nanoformulations have been developed. The in vivo and in vitro studies have shown potential antiproliferative effect of quercetin-loaded NPs on different types of cancer cells (Nam et al., 2016). Salehi et al. suggested that curcumin has shown potential benefits on certain diseases that are growing increasingly every year and is presenting a global burden. The clinical reports have established the effectiveness of curcumin in cardiovascular diseases through its anti-atherosclerotic and its antihypercholesterolemic properties and its protective features against reperfusion and cardiac ischemia. However, the curcumin therapy has certain limitations with respect to its bioavailability. Curcumin nanomedicine formulations aim to overcome the limitation through the enhancement of the curcumin targeting, efficacy and pharmacokinetics, and cellular uptake. Curcumin nanomedicine is a potential therapeutic alternative in a new discovery phase (Salehi et al., 2020).

10.7 TYPES OF NANOPARTICLES AND THEIR ROLE IN CHEMOTHERAPEUTICS

NPs have been characterized into various categories and subcategories based on their molecular buildup. In this study, we have taken few broad categories and discussed their characteristic features and their role as drug-delivery agents and their effect on chemotherapeutic property of the natural products Figure 10.3.

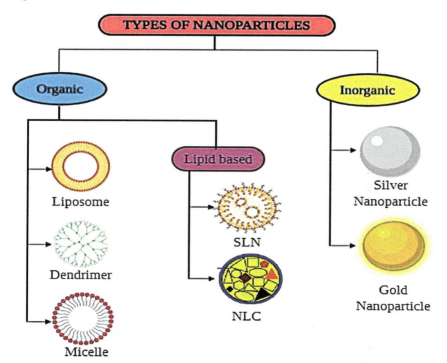

FIGURE 10.3 Schematic representation of broad categorization of nanoparticles.

10.7.1 ORGANIC NANOPARTICLES

10.7.1.1 LIPOSOMES

Liposomes, these are lipid carriers, they have mono or multiple layers of phospholipid structures, containing one hydrophilic head and two hydrophobic tails. And the choice of these phospholipids ranges between phosphatidylcholine to PEG. These are mostly used as nanocarriers (Sharma et al., 2018). Factors such as lipid-to-drug ratio in the encapsulation can influence the efficacy and the stability of the liposomal carrier as the presence of lipids makes it easy for them to get absorbed in liver instantaneously and then they are taken up by macrophages (He et al., 2019). To overcome this, ligands such as monosialoganglioside are coated on the surface of the liposome and incorporation of compounds like cholesterol is also done. It not only increases the physical properties of the liposome but also warrants an increased circulation time in the body. On loading the phytochemicals to the liposome, their solubility, stability, and circulation time increases. The advantage of using liposomes is that the hydrophobic phytochemicals are entrapped in the membrane and hydrophilic phytochemicals in their water compartments (Chen and Huang, 2019; Zong et al., 2016).

Resveratrol and curcumin are widely known and studied phytochemicals, they have been extensively explored for their anticancer properties. However, their poor bioavailability, selectivity, and absorption are the shortcoming limiting their clinical applications. Therefore, human epidermal growth factor receptor (HER)2-targetted immune-liposomes coupled with trastuzumab were used to carry drugs (resveratrol and curcumin), and they displayed exemplary antiproliferative effect in two breast cancer cell lines, MCF7 and JIMT1 (Catania et al., 2013). Liposomal curcumin was also found effective in two prostate cancer cell lines wherein it was able to inhibit the proliferation of two prostate cancer cell lines, LNCaP and C4-2B (Thangapazham et al., 2008). Artemisinin is a famous phytochemical known for its antimalarial effect, and recent studies have reported its anticancer property. In order to develop new carrier strategies for better delivery and activation inside the cancer tissue, it was loaded with transferrin-conjugated liposomes. Results showed that conjugation of transferrin resulted in better delivery of artemisinin and also it was observed the cytotoxic effect of artemisinin on colon cancer cells, HCT8 (Leto et al., 2016). Honokiol, one more anticancer phytochemical that was formulated into PEGylated liposomal honokiol to have better solubility also enhanced its anticancer effect (Wang et al., 2011).

10.7.1.2 DENDRIMER

Dendrimers are globular shape, with branched three-dimensional structure, and these are macromolecules with extensively branched structure containing numerous peripheral groups. The globular structure and the presence of branching serve as an advantage for the hydrophobic drug encapsulation. Structurally they are divided into a central core, branching units, and the terminally present functional groups (Castro et al., 2018; Wilczewska et al., 2012). Thus, resulting in the presence of nanocavities and the environment of these cavities depend on the core. The production of these dendrimers can take place in two ways, one is from the outside to in (convergent) and other is from outside to the central core. The systemic properties of the dendrimers are mainly influenced by the functional groups (Kambhampati and Kannan, 2013; Wilczewska et al., 2012). Dendrimers can be designed to attain various sizes and shapes, it properties can also be controlled like the arrangement of polar and no-polar layers alternatively, including the drug solution. The addition of the hydrophilic species to the external extensions can help make them water soluble which is not very likely when working with other polymers. Their unique properties make them highly suitable for intracellular drug-delivery system specifically against cancer (Baker, 2009).

Paclitaxel, a taxane found to have anticancer properties when formulated into dendrimers with polyamidoamine (PAMAM) carrier in conjugation trastuzumab (PAMAM-ptx-trastuzumab) was found to induce cell-cycle arrest in G1 and S phase in SKBR-3 cells. In both breast cancer cells, MCF-7 and SKBR-3, this conjugate was found to induce significant level of cell death through necrosis and apoptosis (Marcinkowska et al., 2019). Gallic acid, a polyphenol when formulated as dendrimers with PAMAM, showed higher toxicity in MCF-7 cells as compared to the free gallic acid (Sharma et al., 2011). Capsaicin, one more phytochemical reported to possess anticancer effect, was prepared into dendrosomal nanoformulation. On analyzing its effect on cancer cells, MCF-7 and HEp 2, results showed that the dendrimer of capsaicin was significantly cytotoxic (Carlin, 2015).

10.7.1.3   MICELLES (POLYMERIC MICELLES)

Micelles are a nonstructured collection of amphiphilic block copolymer that aggregate on their own to form a spherical vesical in aqueous solution. The inner core of the polymeric micelles contains the tail and the outer shell is the head. In a micelle, the tail is hydrophobic containing hydrocarbons of the fatty acid chains, thus making the core of the polymeric micelle hydrophobic, which can be used to load the hydrophobic drugs (Kalhapure and Renukuntla, 2018). The tail of a micelle is hydrophilic which makes external shell the same and helps in water solubility and therefore stabilizes the core. Preparation of the polymeric micelle is simple but the complicated part is to control the rate of release of the drug, therefore the modification of the shell is done (Alai et al., 2015). The molecular weight of these polymeric structures is high which leads to maximum storage inside the tissue of solid cancer and the entry of these polymeric micelles is easier in cancer tissue as compared to other tissues. The modified chemical bonds on the shell guarantee the release of drug in the tumor environment, thus preventing the release of drug in the bloodstream (Wakaskar, 2018).

Polymeric micelles of luteolin were formulated and when analyzed for the effect on C-26 colon cancer cells, it was observed that the encapsulation of the compound increases the levels of bioavailability and cytotoxicity. When analyzed in vivo to analyze the rate of bioavailability, results showed that luteolin encapsulated in monomethoxy poly(ethylene glycol)-poly(Ɛ-caprolactone) (MPEG-PCL) micelles had significantly higher bioavailability as compared to free luteolin (Qiu et al., 2013). Honokiol, a phytochemical with multiple medicinal properties, was encapsulated in a polymeric micelle using MPEG and PCL micelles, and its effect was analyzed in CT-26, a colorectal carcinoma cell line. Results showed that the presence of micelles enhanced the dose-related cytotoxicity in CT-26 (Dong et al., 2010). Another study formulated the micelles of honokiol using poly(Ɛ-caprolactone)-poly (ethylene glycol)-poly(Ɛ-caprolactone) copolymer (PCEC). They analyzed the effect of these honokiol polymer micelles on A549, a lung cancer cell line and results showed that they had a significant antiproliferative effect (Wei et al., 2009). A recent study, reported the encapsulation of curcumin using methoxy poly(ethylene glycol)-poly(D/L-lactide) (mPEG-PLA) and their results showed a significant increase in the cellular uptake of the polymeric micelles of curcumin. It was also observed that the polymeric micelles of curcumin were more cytotoxic as compared to free curcumin (Kumari et al., 2016).

10.7.2   INORGANIC NANOPARTICLES

10.7.2.1   GOLD NANOPARTICLES

God being a multifunctional material is essential for its biomedical applications. Gold NPs (AuNPs) are the suspended nanosized gold particles, also referred to as colloidal gold. The wide use of gold in nanobiotechnology is due to its properties such as bacteriostatic, anti-oxidative, and anticorrosive effect as well as multisurface functionalities (Connor and Broome, 2018). In order to functionalize the core of the AuNP, a monolayer of thiol group moieties is added. The convenience of functionalizing the AuNP provides a vast range of nanobiological assemblies. Some key attributes that make AuNPs highly useful in diagnostics and drug delivery are its size, shape, low toxicity, high biocompatibility, and large surface-to-volume ratio. One essential physical property of AuNPs is surface plasmon resonance (SPR), the ability to quench fluorescence. Due to localized SPR (LSPR), wherein gold particles can absorb light at specific wavelengths which results in

Plant Product-Based Nanomedicine for Malignancies

photothermal and photoacoustic properties making them useful for the treatment of hyperthermic cancers (Dube et al., 2017; Liu et al., 2018).

Kaempferol was conjugated with AuNP and it was observed that they induced apoptosis in MCF-7 cells. It also inhibited vascular endothelial growth factor (VEGF)-induced angiogenesis (Srinivas Raghavan et al., 2015). Green synthesis of the AuNP in conjugation with the plant extracts is also beneficial for its anticancer properties. *Catharanthus roseus* and *Carica papaya and their active compounds are well known for their anticancer properties. The extracts from the leaves of these plants were used for the synthesis of AuNP, and it was observed that their effect on HepG2 liver cancer cells and MCF-7 breast cancer cells was enhanced (Muthukumar et al., 2016). Licochalcone A*, isolated from *Glycyrrhiza inflate, is a potent anticancer compound, it was loaded on hollow AuNP by ultrasonic method to overcome its poor bioavailability. NIR irradiation was used to overcome its low solubility and it was reported that the method was successful (Sun et al., 2017). Codelivery is also an exciting method to functionalize the drugs and improve their pharmacological effects by synergetic interaction. One such study was done using curcumin and sulforaphane, they possess anticancer effect but the outcome is limited due to low water solubility and poor bioavailability. Combined delivery using iron oxide (Fe_3O_4) and AuNPs was performed and a dramatic increase in the biological activity of curcumin and sulforaphane was observed. Results showed that they had a significant effect on MCF-7 cells by the induction of apoptosis and necrosis, wherein they also inhibited the cell migration (Danafar et al., 2018).*

10.7.2.2 SILVER NANOPARTICLES

Silver NPs (AgNPs), also known as colloidal silver, are widely used due to unique their chemical, physical, and biological properties such as catalytic activity, chemical stability, high conductivity, and LSPR. It has been reported that the ROS released on the surface of AgNPs can cause cell death in any microbial or mammalian cell death (Siddiqi et al., 2018). This peculiar phenomenon guarantees the bactericidal and fungicidal properties to AgNPs. Due to their properties, AgNPs are used as antibacterial agents' in industrial, health care-related products, and it is also used in medical device coatings, optical sensors, drug delivery, pharmaceuticals, and also diagnostics (Markowska et al., 2013). They are also significantly used in the form of prostheses and possess huge potential in preventing wound inflammation and promoting wound healing in form of topical administration. However, the biological activity of AgNPs is controlled by the factors like particle morphology, particle composition, dissolution rate, efficiency of ion release, surface chemistry, and cell type (Abbasi et al., 2016). The choice of reducing agent used also plays a key role in determining the toxicity of AgNP. The cytotoxic effect of anticancer agents loaded in AgNP reported in human cancer cell lines is due to the cellular damage caused by loss of membrane integrity, apoptosis, and oxidative stress (Zhang et al., 2016).

Recently, the inorganic NPs are being synthesized using green production strategy, in which a plant extract of the plants possessing medicinal properties is employed for the synthesis. Similarly, *Cucumis prophetarum* was used to synthesize AgNP and it was found to possess antimicrobial and anticancer activity. It was observed that when analyzed on MCF-7 cells, these AgNPs inhibited proliferation (Hemlata et al., 2020). Extract of plant *Caesalpinia pulcherrima* has been traditionally used for treating cancer, microbial infections, and many more conditions. Deepika et al synthesized AgNPs using *C. pulcherrima* and analyzed its effect on HCT116, colon cancer cells, their results showed a significant level of cytotoxicity (Deepika et al., 2020).

10.7.3   LIPID BASED

10.7.3.1   SOLID LIPID NANOPARTICLES/CARRIER (SLN/SLC)

SLNs are the colloidal drug transporters constituted of hydrophobic solid lipids consisting of triglycerides, fatty acids, and waxes, and the hydrophilic shell is made up of phospholipids and surfactants. The use of emulsifiers is done to control the dispersion of lipids in water. SLNs are commonly produced by micro-emulsification and high-pressure homogenization (Tapeinos et al., 2017). SLNs are advantageous due to high lipophilic lipid matrix for the dispersion of drugs that results into the encapsulation of various molecules such as proteins, antigens, drugs, and nucleotides. Additionally, promoting the delivery of molecules in different specific cells and tissues, SLNs are advantageous over conventional drug carriers as they can protect themselves from the chemical degradation (Mishra et al., 2018). Sustained and targeted effect of SLNs enhances the bioavailability of the drugs. The advantages of SLNs are stability, biocompatibility, tolerability, and protection of labile drugs. However, the new generation of these nanocarriers is developed to overcome the limitations of SLNs such as low loading capacity and the release of drugs during storage (da Silva Santos et al., 2019).

Silibinin, an active constituent of silymarin was loaded into lipid NPs and they evaluated the effect in MDA-MB-23, a breast cancer cell line. It was observed that cellular uptake of silibinin-loaded SLN was enhanced as compared to free silibinin. Cell toxicity of the silibinin-loaded SLN was also significantly higher than free silibinin (Xu et al., 2013). Study showed that when breast cancer mice model was treated with paclitaxel-loaded SLN, the tumor size reduced remarkably and also the lower degree of inhibition was observed (Serpe et al., 2004). A study showed that hyaluronic acid (HA)-coated SLN loaded with paclitaxel; in this study, the HA was used because cancer stem cells express CD44 that binds to HA. Their results showed that the cellular uptake of HA-SLN-paclitaxel was significantly increased, and also the higher rate of apoptosis was induced in CD44[+] cells. Results also showed that the low dose of paclitaxel in the conjugate was able to exhibit significant antitumor effects in comparison to free paclitaxel (Shen et al., 2015)

10.7.3.2 NANOSTRUCTURED LIPID CARRIERS (NLC)

It is a new type of nanocarrier, it is an improvement over solid lipid NPs/carriers (SLN/SLC). The composition of NLC is the solid lipids and the liquid lipids, including an aqueous phase containing surfactants resulting in an unstructured solid lipid matrix (Fang et al., 2013). They are made up of solid lipid core, which are further stabilized in aqueous solutions by emulsifiers resulting in the encapsulation of large variety of drugs and also decreasing the drug expulsion rate during storage. Varieties of active ingredients have been incorporated to improve the bioavailability, water solubility, absorption rate, and circulation time (Khosa et al., 2018). These make NLC and improvement over SLN by their high stability, enhanced drug-loading capacity, and controlled rate of drug release. NLCs are extensively used for topical application in products for skin healing. In recent years, NLCs have been widely used as drug carriers for anticancer therapy (Uner, 2006).

Quercetin is a common dietary flavonoid found to possess properties such as anti-oxidant activity, protection against tissue injury, and many more (Anand David et al., 2016). It has also been reported to have anticancer activity; however, its low solubility in water and poor stability has limited the use as a chemotherapeutic agent. In order to overcome this, Sun et al synthesized the quercetin-NLC (Q-NLC) and on analyzing the effect on MCF-7 and MDA-MB-231 breast cancer cells, they observed that Q-NLC increased the apoptosis and it had higher cytotoxicity. They also reported that a significant increase in the intracellular content of quercetin (Sun et al., 2014). Another phytochemical curcumin has been for long known for its anticancer properties; however, poor bioavailability and rapid metabolism in vivo limit its therapeutic use. On delivering curcumin using NLCs (cur-NLC) for brain cancer treatment, Chen et al. observed that in A172 brain cancer cells, cur-NLC induced cytotoxicity, apoptosis, and ROS significantly higher than free curcumin. They also reported that in mice, the blood level of curcumin when delivered as cur-NLC increased 6.4-fold as compared to free curcumin (Chen et al., 2016).

10.8 CONCLUSION

This chapter focused on the plant product-based nanoformulations for the cure of malignancies of different types. The green synthesis of nanoformulations is a major breakthrough in the present scenario of the medicinal importance of the plants. Plants as we already know constitute the rich source of various conventional drugs for the treatment of a wide variety of diseases, including cancers. The active compounds from the plants are used as an alternative therapy along with the conventional treatment and have reported better results than the conventional treatments alone. The majority of the plant products have low solubility and scare bioavailability in humans and animals which makes them less effective. Nanotechnology is a blooming field that has led to combat the stated limitations of the free compound forms. NPs loaded with plant products provide greater potential than their parent compounds against a wide variety of cancers and malignancies. Withal, the maximum studies on nanoformulated medicinal plant products are in vitro based which has to be translated in vivo and clinically to recognize them as effective drugs. Further research to elucidate the mechanism of action of these nanoformulations is required for advancement in the field of cancer prevention.

ACKNOWLEDGMENT

The special thanks are due to Research Centre, College of Pharmacy, King Saud University, Riyadh and Deanship of Scientific Research, King Saud University, Kingdom of Saudi Arabia.

KEYWORDS

- **nanomedicine**
- **drug delivery**
- **combinational therapy**
- **bioavailability**
- **EPR effect**
- **phytochemicals**

REFERENCES

Abbasi, E.; Milani, M.; Fekri Aval, S.; Kouhi, M.; Akbarzadeh, A.; Tayefi Nasrabadi, H.; Nikasa, P.; Joo, S. W.; Hanifehpour, Y.; Nejati-Koshki, K.; Samiei, M. Silver Nanoparticles: Synthesis Methods, Bio-applications and Properties. *Crit. Rev. Microbiol.* **2016**, *42* (2), 173–180.

Alai, M. S.; Lin, W. J.; Pingale, S. S. Application of Polymeric Nanoparticles and Micelles in Insulin Oral Delivery. *J. Food Drug Analys.* **2015**, *23* (3), 351–358.

Amin, F. U.; Shah, S. A.; Badshah, H.; Khan, M.; Kim, M. O. Anthocyanins Encapsulated by PLGA@PEG Nanoparticles Potentially Improved Its Free Radical Scavenging Capabilities via p38/JNK Pathway against Aβ1-42-induced Oxidative Stress. *J. Nanobiotechnol.* **2017**, *15* (1), 12. doi:10.1186/s12951-016-0227-4

Anand David, A. V.; Arulmoli, R.; Parasuraman, S. Overviews of Biological Importance of Quercetin: A Bioactive Flavonoid. *Pharmacogn. Rev.* **2016**, *10* (20), 84–89.

Arab, H. H.; Gad, A. M.; Fikry, E. M.; Eid, A. H. Ellagic Acid Attenuates Testicular Disruption in Rheumatoid Arthritis via Targeting Inflammatory Signals, Oxidative Perturbations and Apoptosis. *Life Sci.* **2019**, *239*, 117012. doi:10.1016/j.lfs.2019.117012

Baker, J. R., Jr. Dendrimer-Based Nanoparticles for Cancer Therapy. *Hematol. Am. Soc. Hematol. Educ. Program* **2009**, 708–719.

Baker, S. N.; Baker, G. A. Luminescent Carbon Nanodots: Emergent Nanolights. *Angew Chem. Int. Ed. Engl.* **2010**, *49* (38), 6726–6744. doi:10.1002/anie.200906623

Batra, H.; Pawar, S.; Bahl, D. Curcumin in Combination with Anti-cancer Drugs: A Nanomedicine Review. *Pharmacol. Res.* **2019**, *139*, 91–105. doi:10.1016/j.phrs.2018.11.005

Baughman, R. H.; Zakhidov, A. A.; de Heer, W. A. Carbon Nanotubes—the Route Toward Applications. *Science* **2002**, *297* (5582), 787–792. doi:10.1126/science.1060928

Bayda, S.; Adeel, M.; Tuccinardi, T.; Cordani, M.; Rizzolio, F. The History of Nanoscience and Nanotechnology: From Chemical-Physical Applications to Nanomedicine. *Molecules* **2019**, *25* (1), 112. doi:10.3390/molecules25010112

Bhattacharya, S.; Mondal, L.; Mukherjee, B.; Dutta, L.; Ehsan, I.; Debnath, M. C.; Gaonkar, R. H.; Pal, M. M.; Majumdar, S. Apigenin Loaded Nanoparticle Delayed Development of Hepatocellular Carcinoma in Rats. *Nanomedicine* **2018**, *14* (6), 1905–1917. doi:10.1016/j.nano.2018.05.011

Blanco, E.; Bey, E. A.; Dong, Y.; Weinberg, B. D.; Sutton, D. M.; Boothman, D. A.; Gao, J. Beta-lapachone-containing PEG-PLA Polymer Micelles as Novel Nanotherapeutics against NQO1-Overexpressing Tumor Cells. *J. Control. Rel. Off. J. Control. Rel. Soc.* **2007,** *122* (3), 365–374. https://doi.org/10.1016/j.jconrel.2007.04.014

Cao, Z.; Li, W.; Liu, R. et al. pH- and Enzyme-Triggered Drug Release as an Important Process in the Design of Anti-tumor Drug Delivery Systems. *Biomed. Pharmacother.* **2019,** *118,* 109340. doi:10.1016/j.biopha.2019.109340

Carlin, G. M. Dendrosomal Capsaicin Nanoformulation for the in Vitro Anticancer Effect on Hep 2 And Mcf-7 Cell Lines. *Int. J. Appl. Bioeng.* **2015,** *9* (2).

Casanova, E.; Salvadó, J.; Crescenti, A.; Gibert-Ramos, A. Epigallocatechin Gallate Modulates Muscle Homeostasis in Type 2 Diabetes and Obesity by Targeting Energetic and Redox Pathways: A Narrative Review. *Int. J. Mol. Sci.* **2019,** *20* (3), 532. doi:10.3390/ijms20030532

Castro, R. I.; Forero-Doria, O.; Guzmán, L. Perspectives of Dendrimer-based Nanoparticles in Cancer Therapy. *Anais da Academia Brasileira de Ciencias* **2018,** *90* (2 suppl 1), 2331–2346.

Catania, A.; Barrajón-Catalán, E.; Nicolosi, S.; Cicirata, F.; Micol, V. Immunoliposome Encapsulation Increases Cytotoxic Activity and Selectivity of Curcumin and Resveratrol against HER2 Overexpressing Human Breast Cancer Cells. *Breast Cancer Res. Treat.* **2013,** *141* (1), 55–65.

Chan, W. C. W. Nanomedicine 2.0. *Acc. Chem. Res.* **2017,** *50* (3), 627–632. doi:10.1021/acs.accounts.6b00629

Chen, F.; Huang, G. Application of Glycosylation in Targeted Drug Delivery. *Eur. J. Med. Chem.* **2019,** *182,* 111612.

Chen, G.; Roy, I.; Yang, C.; Prasad, P. N. Nanochemistry and Nanomedicine for Nanoparticle-based Diagnostics and Therapy. *Chem. Rev.* **2016,** *116* (5), 2826–2885. doi:10.1021/acs.chemrev.5b00148

Chen, S. R.; Dai, Y.; Zhao, J.; Lin, L.; Wang, Y.; Wang, Y. A Mechanistic Overview of Triptolide and Celastrol, Natural Products from Tripterygium wilfordii Hook F. *Front. Pharmacol.* **2018,** *9,* 104. https://doi.org/10.3389/fphar.2018.00104

Chen, Y.; Pan, L.; Jiang, M.; Li, D.; Jin, L. Nanostructured Lipid Carriers Enhance the Bioavailability and Brain Cancer Inhibitory Efficacy of Curcumin Both in Vitro and in Vivo. *Drug Deliv.* **2016,** *23* (4), 1383–1392.

Cheung, C. W. Y.; Gibbons, N.; Johnson, D. W.; Nicol, D. L. Silibinin—A Promising New Treatment for Cancer. *Anti-Cancer Agents Med. Chem.* **2010,** *10* (3), 186–195.

Collier, Z. A.; Kennedy, A. J.; Poda, A. R.; Cuddy, M. F.; Moser, R. D.; MacCuspie, R. I.; Harmon, A.; Plourde, K.; Haines, C. D.; Steevens, J. A. Tiered Guidance for Risk-Informed Environmental Health and Safety Testing of Nanotechnologies. *J. Nanopart. Res.* **2015,** *17* (3), 155. doi:10.1007/s11051-015-2943-3

Connor, D. M.; Broome, A. M. Gold Nanoparticles for the Delivery of Cancer Therapeutics. *Adv. Cancer Res.* **2018,** *139,* 163–184.

da Silva Santos, V.; Badan Ribeiro, A. P.; Andrade Santana, M. H. Solid Lipid Nanoparticles as Carriers for Lipophilic Compounds for Applications in Foods. *Food Res. Int.* (Ottawa, Ont.) **2019,** *122,* 610–626.

Danafar, H.; Sharafi, A.; Kheiri, S.; Kheiri Manjili, H. Co-delivery of Sulforaphane and Curcumin with PEGylated Iron Oxide-Gold Core Shell Nanoparticles for Delivery to Breast Cancer Cell Line. *Iran. J. Pharm. Res.* **2018,** *17* (2), 480–494.

Davis, M. E.; Zuckerman, J. E.; Choi, C. H. J.; Seligson, D.; Tolcher, A.; Alabi, C. A.; Yen, Y.; Heidel, J. D.; Ribas, A. Evidence of RNAi in Humans from Systemically Administered siRNA via Targeted Nanoparticles. *Nature* **2010,** *464,* 1067–1070.

Deepika, S.; Selvaraj, C. I.; Roopan, S. M.; Screening Bioactivities of *Caesalpinia pulcherrima* L. Swartz and Cytotoxicity of Extract Synthesized Silver Nanoparticles on HCT116 Cell Line. *Mater. Sci. Eng. C, Mater. Biol. App.* **2020,** *106,* 110279.

Dong, P.; Wang, X.; Gu, Y.; Wang, Y.; Wang, Y.; Gong, C.; Luo, F.; Guo, G.; Zhao, X.; Wei, Y.; Qian, Z. Self-Assembled Biodegradable Micelles Based on Star-Shaped PCL-b-PEG Copolymers for Chemotherapeutic Drug Delivery. *Colloids Surf. A: Physicochem. Eng. Aspects* **2010,** *358* (1), 128–134.

Dube, T.; Mandal, S.; Panda, J. J. Nanoparticles Generated from a Tryptophan Derivative: Physical Characterization and Anti-Cancer Drug Delivery. *Amino Acids* **2017,** *49* (5), 975–993.

El-Far, A. H.; Jaouni, S. K.; Li, W.; Mousa, S. A. Protective Roles of Thymoquinone Nanoformulations: Potential Nanonutraceuticals in Human Diseases. *Nutrients* **2018,** *10* (10), 1369. doi:10.3390/nu10101369

Ensign, L. M.; Cone, R.; Hanes, J. Oral Drug Delivery with Polymeric Nanoparticles: The Gastrointestinal Mucus Barriers. *Adv. Drug Deliv. Rev.* **2012,** *64* (6), 557–570. doi:10.1016/j.addr.2011.12.009

Etheridge, M. L.; Campbell, S. A.; Erdman, A. G.; Haynes, C. L.; Wolf, S. M.; McCullough, J. The Big Picture on Nanomedicine: The State of Investigational and Approved Nanomedicine Products. *Nanomedicine* **2013,** *9* (1), 1–14. doi:10.1016/j.nano.2012.05.013

Fang, C. L.; Al-Suwayeh, S. A.; Fang, J. Y. Nanostructured Lipid Carriers (NLCs) for Drug Delivery and Targeting. *Recent Patents Nanotechnol.* **2013,** *7* (1), 41–55.

Fornaguera, C.; García-Celma, M. J. Personalized Nanomedicine: A Revolution at the Nanoscale. *J. Pers. Med.* **2017,** *7* (4), 12. doi:10.3390/jpm7040012.

Galvin, P.; Thompson, D.; Ryan, K. B.; et al. Nanoparticle-Based Drug Delivery: Case Studies for Cancer and Cardiovascular Applications. *Cell Mol. Life Sci.* **2012,** *69* (3), 389–404. doi:10.1007/s00018-011-0856-6

Guo, P. The Emerging Field of RNA Nanotechnology. *Nat. Nanotechnol.* **2010,** *5* (12), 833–842. https://doi.org/10.1038/nnano.2010.231

He, H.; Lu, Y.; Qi, J.; Zhu, Q.; Chen, Z.; Wu, W. Adapting Liposomes for Oral Drug Delivery. *Acta Pharm. Sinica B.* **2019,** *9* (1), 36–48.

Hemlata, Meena, P. R.; Singh, A. P.; Tejavath, K. K. Biosynthesis of Silver Nanoparticles Using *Cucumis prophetarum* Aqueous Leaf Extract and Their Antibacterial and Antiproliferative Activity Against Cancer Cell Lines. *ACS Omega* **2020,** *5* (10), 5520–5528.

Hewlings, S. J.; Kalman, D. S. Curcumin: A Review of Its' Effects on Human Health. *Foods* **2017,** *6* (10), 92. doi:10.3390/foods6100092

Hong, C. A.; Nam, Y. S. Functional Nanostructures for Effective Delivery of Small Interfering RNA Therapeutics. *Theranostics* **2014,** *4* (12), 1211–1232. Published 2014 Sep 19. doi:10.7150/thno.8491

Hosseini, A.; Ghorbani, A. Cancer Therapy with Phytochemicals: Evidence from Clinical Studies. *Avicenna J. Phytomed.* **2015,** *5* (2), 84–97.

Huang, H.; Li, L.; Shi, W.; Liu, H.; Yang, J.; Yuan, X.; Wu, L. The Multifunctional Effects of Nobiletin and Its Metabolites In Vivo and In Vitro. *Evidence-based Complement. Altern. Med.: eCAM* **2016,** 2918796. https://doi.org/10.1155/2016/2918796

Ijaz, H.; Tulain, U. R.; Qureshi, J.; et al. Review: Nigella sativa (Prophetic Medicine): A Review. *Pak. J. Pharm. Sci.* **2017,** *30* (1), 229–234.

Imenshahidi, M.; Hosseinzadeh, H. Berberine and Barberry (*Berberis vulgaris*): A Clinical Review. *Phytother. Res.* **2019,** *33* (3), 504–523. doi:10.1002/ptr.6252

Jäger, S.; Trojan, H.; Kopp, T.; Laszczyk, M. N.; Scheffler, A. Pentacyclic Triterpene Distribution in Various Plants—Rich Sources for a New Group of Multi-Potent Plant Extracts. *Molecules (Basel, Switzerland).* **2009,** *14* (6), 2016–2031. https://doi.org/10.3390/molecules14062016

Jeevanandam, J.; Barhoum, A.; Chan, Y. S.; Dufresne, A.; Danquah, M. K. Review on Nanoparticles and Nanostructured Materials: History, Sources, Toxicity and Regulations. *Beilstein J. Nanotechnol.* **2018,** *9*, 1050–1074. doi:10.3762/bjnano.9.98

Jhaveri, A.; Deshpande, P.; Pattni, B.; Torchilin, V. Transferrin-Targeted, Resveratrol-Loaded Liposomes for the Treatment of Glioblastoma. *J. Control. Release.* **2018,** *277*, 89–101. doi:10.1016/j.jconrel.2018.03.006

Jin, H.; Pi, J.; Yang, F.; et al. Folate-Chitosan Nanoparticles Loaded with Ursolic Acid Confer Anti-Breast Cancer Activities in Vitro and in Vivo. *Sci. Rep.* **2016,** *6*, 30782. doi:10.1038/srep30782

Jin, X.; Yang, Q.; Zhang Y. Synergistic Apoptotic Effects of Apigenin TPGS Liposomes and Tyroservatide: Implications for Effective Treatment of Lung Cancer. *Int. J. Nanomed.* **2017,** *12*, 5109–5118. doi:10.2147/IJN.S140096

Jiraviriyakul, A.; Songjang, W.; Kaewthet, P.; Tanawatkitichai, P.; Bayan, P.; Pongcharoen, S. Honokiol-Enhanced Cytotoxic T Lymphocyte Activity against Cholangiocarcinoma Cells Mediated by Dendritic Cells Pulsed with Damage-Associated Molecular Patterns. *World J. Gastroenterol.* **2019,** *25* (29), 3941–3955. doi:10.3748/wjg.v25.i29.3941

Kalantari, K.; Moniri, M.; Boroumand Moghaddam, A.; Abdul Rahim, R.; Bin Ariff, A.; Izadiyan, Z.; Mohamad, R. A Review of the Biomedical Applications of Zerumbone and the Techniques for Its Extraction from Ginger Rhizomes. *Molecules (Basel, Switzerland).* **2017,** *22* (10), 1645. https://doi.org/10.3390/molecules22101645

Kalhapure, R. S.; Renukuntla, J. Thermo- and pH Dual Responsive Polymeric Micelles and Nanoparticles. *Chemico-biol. Interact.* **2018,** *295*, 20–37.

Kambhampati, S. P.; Kannan, R. M. Dendrimer Nanoparticles for Ocular Drug Delivery. *J. Ocular Pharm. Therap. Off. J. Assoc. Ocular Pharmacol. Therap.* **2013,** *29* (2), 151–165.

Kapse-Mistry, S.; Govender, T.; Srivastava, R.; Yergeri, M. Nanodrug Delivery in Reversing Multidrug Resistance in Cancer Cells. *Front. Pharmacol.* **2014,** *5*, 159. doi:10.3389/fphar.2014.00159

Karimi, M.; Ghasemi, A.; Sahandi Zangabad, P.; et al. Smart Micro/Nanoparticles in Stimulus-Responsive Drug/Gene Delivery Systems. *Chem. Soc. Rev.* **2016,** *45* (5), 1457–1501. doi:10.1039/c5cs00798d

Kawashima, Y.; Yamamoto, H.; Takeuchi, H., Kuno, Y. Mucoadhesive DL-Lactide/Glycolide Copolymer Nanospheres Coated with Chitosan to Improve Oral Delivery of Elcatonin. *Pharm. Dev. Technol.* **2000,** *5* (1), 77–85. doi:10.1081/pdt-100100522

Kesharwani, D.; Paliwal, R.; Satapathy, T.; Das Paul, S. Rheumatiod Arthritis: An Updated Overview of Latest Therapy and Drug Delivery. *J. Pharmacopuncture* **2019,** *22* (4), 210–224. doi:10.3831/KPI.2019.22.029

Khosa, A.; Reddi, S.; Saha, R. N. Nanostructured Lipid Carriers for Site-Specific Drug Delivery. *Biomed. Pharmacotherap.* **2018,** *103*, 598–613.

Plant Product-Based Nanomedicine for Malignancies 297

Kim, B. Y. S.; Rutka, J. T.; Chan, W. C. W. Nanomedicine. N. *Engl. J. Med.* **2010,** *363,* 2434–2443.

Kuen, C. Y.; Fakurazi, S.; Othman, S. S.; Masarudin, M. J. Increased Loading, Efficacy and Sustained Release of Silibinin, a Poorly Soluble Drug Using Hydrophobically-Modified Chitosan Nanoparticles for Enhanced Delivery of Anticancer Drug Delivery Systems. *Nanomaterials (Basel)* **2017,** *7* (11), 379. Published 2017 Nov 8. doi:10.3390/nano7110379

Kumari, P.; Swami, M. O.; Nadipalli, S. K.; Myneni, S.; Ghosh, B.; Biswas, S. Curcumin Delivery by Poly (Lactide)-based Co-polymeric Micelles: An in Vitro Anticancer Study. *Pharm Res.* **2016,** *33* (4), 826–841.

Langcake, P.; Pryce, R. J. The Production of Resveratrol by Vitis vinifera and Other Members of the Vitaceae as a Response to Infection or Injury. *Physiol. Plant Pathol.* **1976,** *9,* 77–86.

Lee, Y., Fukushima, S., Bae, Y., Hiki, S., Ishii, T., Kataoka, K. A protein nanocarrier from charge-conversion polymer in response to endosomal pH. *J. Am. Chem. Soc.* **2007,** *129* (17), 5362–5363. https://doi.org/10.1021/ja071090b

Leto, I.; Coronnello, M.; Righeschi, C.; Bergonzi, M. C.; Mini, E.; Bilia, A. R. Enhanced Efficacy of Artemisinin Loaded in Transferrin-Conjugated Liposomes versus Stealth Liposomes against HCT-8 Colon Cancer Cells. *ChemMedChem.* **2016,** *11* (16), 1745–1751.

Liu, Y.; Crawford, B. M.; Vo-Dinh, T. Gold Nanoparticles-Mediated Photothermal Therapy and Immunotherapy. *Immunotherapy.* **2018,** *10* (13), 1175–1188.

Lu, J.; Yang, J.; Zheng, Y.; Fang, S.; Chen, X. Resveratrol Reduces Store-Operated Ca2+ Entry and Enhances the Apoptosis of Fibroblast-Like Synoviocytes in Adjuvant Arthritis Rats Model via Targeting ORAI1-STIM1 Complex. *Biol. Res.* **2019,** *52* (1), 45. https://doi.org/10.1186/s40659-019-0250-7

Luo, D.; Carter, K. A.; Lovell, J. F. Nanomedical Engineering: Shaping Future Nanomedicines. *Wiley Interdiscip. Rev. Nanomed. Nanobiotechnol.* **2015,** *7* (2), 169–188. doi:10.1002/wnan.1315

Mackay, J. A.; Chilkoti, A. Temperature Sensitive Peptides: Engineering Hyperthermia-Directed Therapeutics. *Int. J. Hyperthermia* **2008,** *24* (6), 483–495. doi:10.1080/02656730802149570

Mady, F.M; Shaker, M. A. Enhanced Anticancer Activity and Oral Bioavailability of Ellagic Acid through Encapsulation in Biodegradable Polymeric Nanoparticles. *Int. J. Nanomed.* **2017,** *12,* 7405–7417. doi:10.2147/IJN.S14774

Manach, C.; Scalbert, A.; Morand, C.; Rémésy, C.; Jiménez, L. Polyphenols: Food Sources and Bioavailability. *Am. J. Clin. Nutr.* **2004,** *79* (5), 727–747. doi:10.1093/ajcn/79.5.727

Marcinkowska, M.; Stanczyk, M.; Janaszewska, A.; Gajek, A.; Ksiezak, M.; Dzialak, P.; Klajnert-Maculewicz, B. Molecular Mechanisms of Antitumor Activity of PAMAM Dendrimer Conjugates with Anticancer Drugs and a Monoclonal Antibody. *Polymers* **2019,** *11* (9), 1422.

Markowska, K.; Grudniak, A. M.; Wolska, K. I. Silver Nanoparticles as an Alternative Strategy against Bacterial Biofilms. *Acta Biochim. Polonica.* **2013,** *60* (4), 523–530.

Miean, K. H.; Mohamed, S. Flavonoid (myricetin, quercetin, kaempferol, luteolin, and apigenin) Content of Edible Tropical Plants. *J. Agric. Food Chem.* **2001,** *49* (6), 3106–3112. doi:10.1021/jf000892m

Mishra, V.; Bansal, K. K.; Verma, A.; Yadav, N.; Thakur, S.; Sudhakar, K.; Rosenholm, J. M. Solid Lipid Nanoparticles: Emerging Colloidal Nano Drug Delivery Systems. *Pharmaceutics* **2018,** *10* (4).

Moghimi, S. M.; Hunter, A. C.; Murray, J. C. Nanomedicine: Current Status and Future Prospects. *FASEB J.* **2005,** *19* (3), 311–330. doi:10.1096/fj.04-2747rev

Mustafa, M.; Anwar, S.; Elgamal, F.; Ahmed, E. R.; Aly, O. M. Potent Combretastatin A-4 Analogs Containing 1,2,4-Triazole: Synthesis, Antiproliferative, Anti-Tubulin Activity, and Docking Study. *Eur. J. Med. Chem.* **2019**, *183*, 111697. doi:10.1016/j.ejmech.2019.111697

Muthukumar, T.; Sudhakumari, Sambandam, B.; Aravinthan, A.; Sastry, T. P.; Kim, J.-H. Green Synthesis of Gold Nanoparticles and Their Enhanced Synergistic Antitumor Activity Using HepG2 and MCF7 Cells and Its Antibacterial Effects. *Process Biochem.* **2016**, *51* (3), 384–391.

Nam, J. S.; Sharma, A. R.; Nguyen, L. T.; Chakraborty, C.; Sharma, G.; Lee, S. S. Application of Bioactive Quercetin in Oncotherapy: From Nutrition to Nanomedicine. *Molecules* **2016**, *21* (1), E108. doi:10.3390/molecules21010108

Nam, N. H.; Luong, N. H. Nanoparticles: Synthesis and Applications. *Mater. Biomed. Eng.* **2019**, 211–240. doi:10.1016/B978-0-08-102814-8.00008-1

Naseri, N.; Ajorlou, E.; Asghari, F.; Pilehvar-Soltanahmadi, Y. An Update on Nanoparticle-Based Contrast Agents in Medical Imaging, Artificial Cells. *Nanomed. Biotechnol.* **2018**, *46* (6), 1111–1121. doi:10.1080/21691401.2017.1379014

Niazi, M.; Zakeri-Milani, P.; Najafi Hajivar, S.; Soleymani Goloujeh, M.; Ghobakhlou, N.; Shahbazi Mojarrad, J.; Valizadeh, H. Nano-Based Strategies to Overcome *p*-glycoprotein-Mediated Drug Resistance. *Expert Opin. Drug Metab. Toxicol.* **2016**, *12* (9), 1021–1033. doi:10.1080/17425255.2016.1196186

Ovais, M.; Zia, N.; Ahmad, I.; Khalil, A,T.; Raza, A.; Ayaz, M.; Sadiq, A.; Ullah, F.; Shinwari, Z. K. Phyto-Therapeutic and Nanomedicinal Approaches to Cure Alzheimer's Disease: Present Status and Future Opportunities. *Front. Aging Neurosci.* **2018**, *10*, 284. doi:10.3389/fnagi.2018.00284

Paluri, S. L. A.; Ryan, J. D.; Lam, N. H.; Nepal, D.; Sizemore, I. E. Analytical-Based Methodologies for Examining the In Vitro Absorption, Distribution, Metabolism, and Elimination (ADME) of Silver Nanoparticles. *Small.* **2017**, *13* (23). doi:10.1002/smll.201603093

Patel, V. R.; Agrawal, Y. K. Nanosuspension: An Approach to Enhance Solubility of Drugs. *J. Adv. Pharm. Technol. Res.* **2011**, *2* (2), 81–87. doi:10.4103/2231-4040.82950

Patra, J. K.; Das, G.; Fraceto, L. F.; Campos, E.; Rodriguez-Torres, M.; Acosta-Torres, L. S.; Diaz-Torres, L. A.; Grillo, R.; Swamy, M. K.; Sharma, S.; Habtemariam, S.; Shin, H. S. Nano Based Drug Delivery Systems: Recent Developments and Future Prospects. *J. Nanobiotechnol.* **2018**, *16* (1), 71. doi:10.1186/s12951-018-0392-8

Qiu, J.-F.; Gao, X.; Wang, B.-L.; Wei, X.-W.; Gou, M.-L.; Men, K.; Liu, X.-Y.; Guo, G.; Qian, Z.-Y.; Huang, M.-J. Preparation and Characterization of Monomethoxy Poly(ethylene glycol)-Poly(ε-caprolactone) Micelles for the Solubilization and in Vivo Delivery of Luteolin. *Int. J. Nanomed.* **2013**, *8*, 3061–3069.

Ragelle, H.; Danhier, F.; Préat, V.; Langer, R.; Anderson, D. G. Nanoparticle-Based Drug Delivery Systems: A Commercial and Regulatory Outlook as the Field Matures. *Expert Opin. Drug Deliv.* **2017**, *14* (7), 851–864. doi:10.1080/17425247.2016.1244187

Rahman, H. S.; Rasedee, A.; How, C. W.; Abdul, A. B.; Zeenathul, N. A.; Othman, H. H.; Saeed, M. I.; Yeap, S. K. Zerumbone-Loaded Nanostructured Lipid Carriers: Preparation, Characterization, and Antileukemic Effect. *Int. J. Nanomed.* **2013**, *8*, 2769–2781. https://doi.org/10.2147/IJN.S45313

Ramesan, R. M.; Sharma, C. P. Challenges and Advances in Nanoparticle-Based Oral Insulin Delivery. *Expert Rev. Med. Devices* **2009**, *6* (6), 665–676. doi:10.1586/erd.09.43

Reinholz, J.; Landfester, K.; Mailänder, V. The Challenges of Oral Drug Delivery via Nanocarriers. *Drug Deliv.* **2018,** *25* (1), 1694–1705. doi:10.1080/10717544.2018.1501119

Rezaei, A.; Varshosaz, J.; Fesharaki, M.; Farhang, A.; Jafari, S. M. Improving the Solubility and in Vitro Cytotoxicity (Anticancer Activity) of Ferulic Acid by Loading it Into Cyclodextrin Nanosponges. *Int. J. Nanomed.* **2019,** *14*, 4589–4599. doi:10.2147/IJN.S206350

Rycroft, T.; Trump, B.; Poinsatte-Jones, K.; et al. Nanotoxicology and Nanomedicine: Making Development Decisions in an Evolving Governance Environment. *J. Nanopart Res.* **2018,** *20*, 52. https://doi.org/10.1007/s11051-018-4160-3

Sahibzada, M.; Sadiq, A.; Faidah, H. S.; Khurram, M.; Amin, M. U.; Haseeb, A.; Kakar, M. Berberine Nanoparticles with Enhanced in Vitro Bioavailability: Characterization and Antimicrobial Activity. *Drug Design Dev. Therap.* **2018,** *12*, 303–312. https://doi.org/10.2147/DDDT.S156123

Salehi, B.; Del Prado-Audelo, M. L.; Cortés, H.; et al. Therapeutic Applications of Curcumin Nanomedicine Formulations in Cardiovascular Diseases. *J. Clin. Med.* **2020,** *9* (3), 746. doi:10.3390/jcm9030746

Salehi, B.; Venditti, A.; Sharifi-Rad, M.; et al. The Therapeutic Potential of Apigenin. *Int. J. Mol. Sci.* **2019,** *20* (6), 1305. doi:10.3390/ijms20061305

Sebak, S.; Mirzaei, M.; Malhotra, M.; Kulamarva, A.; Prakash, S. Human serum albumin nanoparticles as an efficient noscapine drug delivery system for potential use in breast cancer: preparation and in vitro analysis. *Int. J. Nanomed.* **2010,** 5, 525–532. doi:10.2147/ijn.s10443

Seo, D. Y.; Lee, S. R.; Heo, J. W.; No, M. H.; Rhee, B. D.; Ko, K. S.; Kwak, H. B.; Han, J. Ursolic acid in health and disease. *The Korean J. Physiol. Pharmacol Off. J. Korean Physiol. Soc. Korean Soc. Pharmacol.* **2018,** *22* (3), 235–248. https://doi.org/10.4196/kjpp.2018.22.3.235

Serpe, L.; Catalano, M. G.; Cavalli, R.; Ugazio, E.; Bosco, O.; Canaparo, R.; Muntoni, E.; Frairia, R.; Gasco, M. R.; Eandi, M. Cytotoxicity of Anticancer Drugs Incorporated in Solid Lipid Nanoparticles on HT-29 Colorectal Cancer Cell Line. *Eur. J. Pharm. Biopharm.* **2004,** *58* (3), 673–680.

Shakeri, S.; Ashrafizadeh, M.; Zarrabi, A.; Roghanian, R.; Afshar, E. G.; Pardakhty, A.; Mohammadinejad, R., Kumar, A., Thakur, V. K. Multifunctional Polymeric Nanoplatforms for Brain Diseases Diagnosis, Therapy and Theranostics. *Biomedicines* **2020,** *8* (1), 13. doi:10.3390/biomedicines8010013

Sharma, A.; Gautam, S. P.; Gupta, A. K. Surface Modified Dendrimers: Synthesis and Characterization for Cancer Targeted Drug Delivery. *Bioorg. Med. Chem.* **2011,** *19* (11), 3341–3346.

Sharma, D.; Ali, A. A. E.; Trivedi, L. R. An Updated Review on: Liposomes as Drug Delivery System. *PharmaTutor.* **2018,** *6* (2), 50–62.

Shen, H.; Shi, S.; Zhang, Z.; Gong, T.; Sun, X. Coating Solid Lipid Nanoparticles with Hyaluronic Acid Enhances Antitumor Activity against Melanoma Stem-like Cells. *Theranostics* **2015,** *5* (7), 755–771.

Siddiqi, K. S.; Husen, A.; Rao, R. A. K. A Review on Biosynthesis of Silver Nanoparticles and Their Biocidal Properties. *J. Nanobiotechnol.* **2018,** *16* (1), 14.

Soares, S.; Sousa, J.; Pais, A.; Vitorino, C. Nanomedicine: Principles, Properties, and Regulatory Issues. *Front. Chem.* **2018,** *6*, 360. doi: 10.3389/fchem.2018.00360

Soleas, G. J.; Diamandis, E. P.; Goldberg, D. M. Resveratrol: A Molecule Whose Time Has Come and Gone? *Clin. Biochem.* **1997,** *30*, 91–113.

Spicer, C. D.; Jumeaux, C.; Gupta, B.; Stevens, M. M. Peptide and Protein Nanoparticle Conjugates: Versatile Platforms for Biomedical Applications. *Chem. Soc. Rev.* **2018**, *47* (10), 3574–3620. doi:10.1039/c7cs00877e

Srinivas Raghavan, B.; Kondath, S.; Anantanarayanan, R.; Rajaram, R.. Kaempferol Mediated Synthesis of Gold Nanoparticles and Their Cytotoxic Effects on MCF-7 Cancer Cell Line. *Process Biochem.* **2015**, *50* (11), 1966–1976.

Subramanian, A. P.; Jaganathan, S. K.; Manikandan, A.; Pandiaraj, K,N.; Gomathi, N.; Supriyanto, E. Recent Trends in Nano-Based Drug Delivery Systems for Efficient Delivery of Phytochemicals in Chemotherapy. *RSC Adv.* **2016**, *6*, 48294–48314.

Subramanian, V.; Semenzin, E.; Hristozov, D.; Marcomini, A.; Linkov, I. Sustainable Nanotechnology: Defining, Measuring and Teaching. *Nano Today* **2014**, *9* (1), 6–9.

Suk, J. S.; Xu, Q.; Kim, N.; Hanes, J.; Ensign, L. M. PEGylation as a Strategy for Improving Nanoparticle-Based Drug and Gene Delivery. *Adv. Drug Deliv. Rev.* **2016**, *99* (Pt A), 28–51. doi:10.1016/j.addr.2015.09.012

Sun, M.; Nie, S.; Pan, X.; Zhang, R.; Fan, Z.; Wang, S. Quercetin-Nanostructured Lipid Carriers: Characteristics and Anti-Breast Cancer Activities in Vitro. *Colloids Surf. B, Biointerf.* **2014**, *113*, 15–24.

Sun, Y.-W.; Wang, L.-H.; Meng, D.-L.; Che, X. A Green and Facile Preparation Approach, Licochalcone A Capped on Hollow Gold Nanoparticles, for Improving the Solubility and Dissolution of Anticancer Natural Product. *Oncotarget* **2017**, *8* (62), 105673–105681.

Taghipour, Y. D.; Hajialyani, M.; Naseri, R.; Hesari, M.; Mohammadi, P.; Stefanucci, A.; Mollica, A.; Farzaei, M. H.; Abdollahi, M. Nanoformulations of Natural Products for Management of Metabolic Syndrome. *Int. J. Nanomed.* **2019**, *14*, 5303–5321. https://doi.org/10.2147/IJN.S213831

Tao, Z. H.; Li, C.; Xu, X. F.; Pan, Y. J. Scavenging Activity and Mechanism Study of Ferulic Acid against Reactive Carbonyl Species Acrolein. *J. Zhejiang Univ. Sci. B.* **2019**, *20* (11), 868–876. doi:10.1631/jzus.B1900211

Tapeinos, C.; Battaglini, M.; Ciofani, G. Advances in the Design of Solid Lipid Nanoparticles and Nanostructured Lipid Carriers for Targeting Brain Diseases. *J. Control. Release Off. J. Control. Release Soc.* **2017**, *264*, 306–332.

Thangapazham, R. L.; Puri, A.; Tele, S.; Blumenthal, R.; Maheshwari, R. K. Evaluation of a Nanotechnology-Based Carrier for Delivery of Curcumin in Prostate Cancer Cells. *Int. J. Oncol.* **2008**, *32* (5), 1119–1123.

Tiwari, S. K.; Agarwal, S.; Seth, B.; et al. Curcumin-Loaded Nanoparticles Potently Induce Adult Neurogenesis and Reverse Cognitive Deficits in Alzheimer's Disease Model via Canonical Wnt/β-Catenin Pathway. *ACS Nano.* **2014**, *8* (1), 76–103. doi:10.1021/nn405077y

Torchilin, V. P. Micellar Nanocarriers: Pharmaceutical Perspectives. *Pharm Res.* **2007**, *24* (1), 1–16. doi:10.1007/s11095-006-9132-0

Uner, M. Preparation, Characterization and Physico-Chemical Properties of Solid Lipid Nanoparticles (SLN) and Nanostructured Lipid Carriers (NLC): Their Benefits as Colloidal Drug Carrier Systems. *Die Pharmazie* **2006**, *61* (5), 375–386.

Vaddiraju, S.; Tomazos, I.; Burgess, D. J.; Jain, F. C.; Papadimitrakopoulos F. Emerging Synergy between Nanotechnology and Implantable Biosensors: A Review. *Biosens. Bioelectron.* **2010**, *25* (7), 1553–1565. doi:10.1016/j.bios.2009.12.001

Veeresham C. Natural Products Derived from Plants as a Source of Drugs. *J. Adv. Pharm. Technol. Res.* **2012**, *3* (4), 200–201. doi:10.4103/2231-4040.104709

Ventola CL. The Nanomedicine Revolution: Part 1: Emerging Concepts. *P T.* **2012**, *37* (9), 512–525.

Wakaskar, R. R. General Overview of Lipid–Polymer Hybrid Nanoparticles, Dendrimers, Micelles, Liposomes, Spongosomes and Cubosomes. *J. Drug Target.* **2018**, *26* (4), 311–318.

Wakaskar, R. R.; Bathena, S. P.; Tallapaka, S. B.; Ambardekar, V. V.; Gautam, N.; Thakare, R.; Simet, S. M.; Curran, S. M.; Singh, R. K.; Dong, Y.; Vetro, J. A. Peripherally Cross-Linking the Shell of Core-Shell Polymer Micelles Decreases Premature Release of Physically Loaded Combretastatin A4 in Whole Blood and Increases Its Mean Residence Time and Subsequent Potency against Primary Murine Breast Tumors after IV Administration. *Pharm. Res.* **2015**, *32* (3), 1028–1044. https://doi.org/10.1007/s11095-014-1515-z

Wang, C.; Liu, B.; Xu, X.; et al. Toward Targeted Therapy in Chemotherapy-Resistant Pancreatic Cancer with a Smart Triptolide Nanomedicine. *Oncotarget* **2016**, *7* (7), 8360–8372. doi:10.18632/oncotarget.7073

Wang, T.; Yin, X.; Lu, Y.; Shan, W.; Xiong, S. Formulation, Antileukemia Mechanism, Pharmacokinetics, and Biodistribution of a Novel Liposomal Emodin. *Int. J. Nanomed.* **2012**, *7*, 2325–2337. doi:10.2147/IJN.S31029

Wang, X. H.; Cai, L. L.; Zhang, X. Y.; Deng, L. Y.; Zheng, H.; Deng, C. Y.; Wen, J. L.; Zhao, X.; Wei, Y. Q.; Chen, L. J. Improved Solubility and Pharmacokinetics of PEGylated Liposomal Honokiol and Human Plasma Protein Binding Ability of Honokiol. *Int. J. Pharm.* **2011**, *410* (1–2), 169–174.

Wang, X.; Chen, W. Gambogic Acid Is a Novel Anti-Cancer Agent That Inhibits Cell Proliferation, Angiogenesis and Metastasis. *Anticancer Agents Med. Chem.* **2012**, *12* (8), 994–1000.

Wang, Y.; Xie, J.; Ai, Z.; Su, J. Nobiletin-Loaded Micelles Reduce Ovariectomy-Induced Bone Loss by Suppressing Osteoclastogenesis. *Int. J. Nanomed.* **2019**, *14*, 7839–7849. doi:10.2147/IJN.S213724

Wei, X.; Gong, C.; Shi, S.; Fu, S.; Men, K.; Zeng, S.; Zheng, X.; Gou, M.; Chen, L.; Qiu, L.; Qian, Z. Self-Assembled Honokiol-Loaded Micelles Based on Poly(ε-caprolactone)-Poly(Ethylene Glycol)-poly(ε-Caprolactone) Copolymer. *Int. J. Pharm.* **2009**, *369* (1), 170–175.

Weiss, P. A Conversation with Prof. Chad Mirkin: Nanomaterials Architect. *ACS Nano* **2009**, *3*, 1310−1317.

Wilczewska, A. Z.; Niemirowicz, K.; Markiewicz, K. H.; Car, H. Nanoparticles as Drug Delivery Systems. *Pharmacol. Reports PR* **2012**, *64* (5), 1020–1037.

Woo, H. J.; Park, K. Y.; Rhu, C. H.; et al. Beta-Lapachone, a Quinone Isolated from *Tabebuia avellanedae*, Induces Apoptosis in HepG2 Hepatoma Cell Line Through Induction of Bax and Activation of Caspase. *J. Med. Food* **2006**, *9* (2), 161–168. doi:10.1089/jmf.2006.9.161

Wu, C.; Xu, Q.; Chen, X.; Liu, J. Delivery Luteolin with Folacin-Modified Nanoparticle for Glioma Therapy. *Int. J. Nanomed.* **2019**, *14*, 7515–7531. doi:10.2147/IJN.S214585

Wu, W.; Wang, L.; Wang, L.; Zu, Y.; Wang, S.; Liu, P.; Zhao, X. Preparation of Honokiol Nanoparticles by Liquid Antisolvent Precipitation Technique, Characterization, Pharmacokinetics, and Evaluation of Inhibitory Effect on HepG2 Cells. *Int. J. Nanomed.* **2018**, *13*, 5469–5483. https://doi.org/10.2147/IJN.S178416

Xiao, S.; Tang, Y.; Lv, Z.; Lin, Y.; Chen, L. Nanomedicine—advantages for Their Use in Rheumatoid Arthritis Theranostics. *J. Control. Release.* **2019**, *316*, 302–316. doi:10.1016/j.jconrel.2019.11.008

Xu, P.; Yin, Q.; Shen, J.; Chen, L.; Yu, H.; Zhang, Z.; Li, Y. Synergistic Inhibition of Breast Cancer Metastasis by Silibinin-Loaded Lipid Nanoparticles Containing TPGS. *Int. J. Pharm.* **2013,** *454* (1), 21–30.

Yang, T.; Wang, Y.; Li, Z.; Dai, W.; Yin, J.; Liang, L.; Ying, X.; Zhou, S.; Wang, J.; Zhang, X.; Zhang, Q. Targeted Delivery of a Combination Therapy Consisting of Combretastatin A4 and Low-Dose Doxorubicin against Tumor Neovasculature. *Nanomed. Nanotechnol. Biol. Med.* **2012,** *8* (1), 81–92. https://doi.org/10.1016/j.nano.2011.05.003

Yanran, L.; Sijin, L.; Kate, T.; Isis, T.; Aaron Cravens, C. D. Smolke. Complete biosynthesis of noscapine and halogenated alkaloids in yeast. *Proc. Natl. Acad. Sci.* **2018,** *115* (17), E3922–E3931. doi:10.1073/pnas.1721469115

Yavarpour-Bali, H.; Ghasemi-Kasman, M.; Pirzadeh, M. Curcumin-Loaded Nanoparticles: A Novel Therapeutic Strategy in Treatment of Central Nervous System Disorders. *Int. J. Nanomed.* **2019,** *14*, 4449–4460. doi:10.2147/IJN.S208332

Yoo, J.; Park, C.; Yi, G.; Lee, D.; Koo, H. Active Targeting Strategies Using Biological Ligands for Nanoparticle Drug Delivery Systems. *Cancers (Basel).* **2019,** *11* (5), 640. doi:10.3390/cancers11050640

Yu, S. J., Kang, M. W.; Chang, H. C.; Chen, K. M.; Yu, Y. C. Bright Fluorescent Nanodiamonds: No Photobleaching and Low Cytotoxicity. *J. Am. Chem. Soc.* **2005,** *127* (50), 17604–17605. doi:10.1021/ja0567081

Yuksel, C.; Ankarali, S.; Yuksel, N. A. The Use of Neodymium Magnets in Healthcare and Their Effects on Health. *North Clin. Istanb.* **2018,** *5* (3), 268–273. doi:10.14744/nci.2017.00483

Zhang, R.; Yang, J.; Sima, M.; Zhou, Y.; Kopeček, J. Sequential Combination Therapy of Ovarian Cancer with Degradable *N*-(2-hydroxypropyl)Methacrylamide Copolymer Paclitaxel and Gemcitabine Conjugates. *Proc. Natl. Acad. Sci. USA* **2014,** *111* (33), 12181–12186. doi:10.1073/pnas.1406233111

Zhang, X. F.; Liu, Z. G.; Shen, W.; Gurunathan, S. Silver Nanoparticles: Synthesis, Characterization, Properties, Applications, and Therapeutic Approaches. *Int. J. Mol. Sci.* **2016a,** *17* (9).

Zhang, Y.; Yu, J.; Bomba, H. N.; Zhu, Y.; Gu, Z. Mechanical Force-Triggered Drug Delivery. *Chem Rev.* **2016b,** *116* (19), 12536–12563. doi:10.1021/acs.chemrev.6b00369

Zhao, Y.; Wu, Y.; Wang, M. Bioactive Substances of Plant Origin. In *Handbook of Food Chemistry*; Cheung, P., Mehta, B., Eds.; Springer: Berlin, Heidelberg, 2015.

Zheng, X. Y.; Yang, S. M.; Zhang, R.; Wang, S. M.; Li, G. B.; Zhou, S. W. Emodin-Induced Autophagy against Cell Apoptosis through the PI3K/AKT/mTOR Pathway in Human Hepatocytes. *Drug Des. Del. Therap.* **2019,** *13*, 3171–3180. doi:10.2147/DDDT.S204958

Zhu, S.; Zhu, L.; Yu, J.; Wang, Y.; Peng, B. Anti-Osteoclastogenic Effect of Epigallocatechin Gallate-Functionalized Gold Nanoparticles in Vitro and in Vivo. *Int. J. Nanomed.* **2019,** *14*, 5017–5032. doi:10.2147/IJN.S204628

Zong, W.; Hu, Y.; Su, Y.; Luo, N.; Zhang, X.; Li, Q.; Han, X. Polydopamine-Coated Liposomes as pH-Sensitive Anticancer Drug Carriers. *J. Microencapsulation* **2016,** *33* (3), 257–262.

CHAPTER 11

TOXICITY OF NANOPARTICLES IN PLANTS

SYED ALI ZULQADAR[1], MOHAMMAD ALI KHARAL[2], UMAIR RIAZ[1*],
RASHID IQBAL[3], BEHZAD MURTAZA[4], GHULAM MURTAZA[5], and
MUHAMMAD AKRAM QAZI[6]

[1]*Soil and Water Testing Laboratory for Research, Bahawalpur, Pakistan*

[2]*Department of Agriculture Extension, Khairpur Tamewali,
Government of Punjab, Pakistan*

[3]*Department of Agronomy, Faculty of Agriculture and Environmental
Sciences, The Islamia University of Bahawalpur, Bahawalpur, Pakistan*

[4]*Department of Environmental Sciences, COMSAT Vehari Campus,
Islamabad, Pakistan*

[5]*Institute of Soil & Environmental Sciences, University of Agriculture,
Faisalabad, Pakistan*

[6]*Rapid Soil Fertility Survey and Soil Testing Institute, Lahore, Punjab,
Pakistan*

*Corresponding author: E-mail: umairbwp3@gmail.com,
umair.riaz@uaf.edu.pk*

ABSTRACT

Nanotechnology leads to a huge scope of novel application in agricultural
sector, because nanoparticles have unique physicochemical properties,
that is, high surface area, tunable pore-size, high reactivity, and particle
morphology. It has been noticed from various studies that nanoparticles'
toxicity has exerted detrimental effects on plant growth and development.
Moreover, nanoscience contributes new ideas leading us to understand the
toxicity of nanoparticles in plants. The realistic elucidation of biochemical,

304 Diverse Applications of Nanotechnology in the Biological Sciences

physiological, and molecular mechanism of nanoparticles and their excess in plant lead to several abnormalities. To identify the differences among the adverse and beneficial effects of nanoparticles in plants, one should be well aware of the underlying principles of nanoparticles affiliations to the plants. This chapter concludes the recent knowledge on the toxicity of nanoparticles in plants by describing its mechanism of phytotoxicity and certain risks. Nanoparticles are being utilized in agriculture to (1) attain agricultural produce more vigorously and have maximum productivity, which, in response, reduces the utilization of energy and water and (2) generate lesser waste. This chapter overviews the nanoparticles phytotoxicity and explains the selection of respective studies for metal/metal-oxide nanoparticles and for carbon-based particles. Efforts are made to describe the uptake mechanisms, biotransformation, and translocation that are crucial to the exact interpretation of potential risks and exposure due to nanoparticles. More knowledge is required on the second-generation nanoparticles acute and critical toxicity, as well as the implications of molecular end points for the assessment of toxicity.

11.1 INTRODUCTION

Ongoing inventions in nanotechnology have affected businesses comprising fabrication, biomedical implications, gadgets/broadcast communications, agribusiness, sustainable power source, among others (Mama et al., 2015). Nanoparticles are extensively characterized as particles having measurement somewhere in the range of 1 and 100 nm in diameter (Auffan et al., 2009). In view of their special novel and properties highlights, nanoparticles had been broadly utilized in numerous parts of everyday life and vitality creation, remembering for impetuses, semiconductors, beautifying agents, tranquilize transporters and ecological vitality (Nel et al., 2006). The huge scope and unhindered utilization of nanoparticles have driven analysts to think about the issues, difficulties, and results of their ecological effect (Gottschalk et al., 2015; Tolaymat et al., 2015). Until this point, the convergence of nanoparticles in the earth is a lot lesser than the poisonous fixation (Batley et al., 2013). The capacitive well-being and natural impacts of nanoparticles should be completely assessed before they are generally marketed. At the point when nanoparticles enter the soil through cultivation, environmental testimony, downpour disintegration, surface spillover, or different pathways, the nanoparticles gather in the dirt as time passes in view of their less relocation capacity in soil.

Presentation demonstrating additionally showed that the groupings of soil nanoparticles are greater than those present in water or air, suggesting that dirt may be the principle wellspring of nanoparticles discharged into nature (Gottschalk et al., 2009). As essential makers, plants are important factor for any network to work as they are answerable for changing over sun-based vitality into natural issue that could be utilized by few other trophic gatherings (McKee and Filser, 2016). Plants fill in as a productive pathway for the movement of nanoparticles (Rico et al., 2011). Through the natural pecking order, nanoparticles could be collected in high trophical-level bags (Zhu et al., 2008). Living beings in the biological system could experience the ill effects of oxidative pressure incited by nanoparticles (Hong et al., 2014). As of late, explore around there has been centered around the collaboration among nanoparticles and plants, and the impacts of nanoparticles at biology, an evolved way of life and human wellbeing; assessing the upsides and downsides of nanoparticles needs interdisciplinary information (Tolaymat et al., 2015).

11.2 MECHANISM OF PHYTOTOXICITY

The speedy improvement of nanotechnology and far-reaching business occupations of organized nanomaterials raised the danger of appearance of manufactured nanoparticles (E-nanoparticles) onto the earth and particularly the soil-plant system (Guzman et al., 2006; Scheringer, 2008; Wiesner et al., 2009; Batley et al. 2013). Out of the blue, there are two fundamental courses by which E-nanoparticles could penetrate into the soil-plant structure. First is done by creating the usage of waste flood water that dependably contains SiO_2, ZnO, TiO_2, and Ag nanoparticles (Pradas del Real et al., 2016; Pan and Xing, 2012; Gottschalk and Nowack, 2011; Gottschalk et al., 2009). The other course is with the use of nano-agro-based synthetics, for instance, nanofertilizers, nano-pesticides, and nano-alterations, achieving the short zone of SiO_2, TiO_2, ZnO/Zn, FeOx/Fe, CuO/Cu/$CuOH_2$, CeO_2, and Ag nanoparticles into the soil plant structure (Chhipa, 2017; Kah, 2015; Kah et al., 2018).

In the wake of entrance into the soil-plant structure, nanoparticles will unavoidably interface with plants and in this way conceivably impact plant food security and perhaps physiology. Nanophytotoxicity is the most generally considered part of research identified with the communications among E-nanoparticles and plants (more than 430 papers have been distributed in the previous decade), and both destructive and gainful consequences for the physiological processes of plants, innate plus biochemical elevations had been resulted (Ma et al., 2010; Ghosh et al., 2010; Pakrashi et al., 2014; Wang et al., 2011; El-Temsah and Joner, 2012; Dietz and Herth, 2011; Zhao et al., 2012, 2016a, 2016b, 2017). The take-up, translocation by putting away of E-nanoparticles in plants acknowledge essential livelihoods in the affirmation of E-nanoparticles phytotoxicity and may in addition likewise influence human food security. Plant take-up of nanoparticles is ordinarily assembled and mix in with its phyto-noxiousness, and matters concentrating on phytotoxicity, take-up, gathering and translocation of E-nanoparticles in plants have been investigated from various aspects (Dietz and Herth, 2011; Tripathi et al., 2017; Schwab et al., 2016; Miralles et al., 2012; Gardea-Torresdey et al., 2014). Honestly, present information taking hush-money up and translocation instruments of E-nanoparticles in plants remains obliged and isn't exact. Schwab et al. have totally looked into the various physiological squares to the take-up and transport of E-nanoparticles in plants (Schwab et al., 2016). Notwithstanding, they are dealing with the impact of take-up of E-nanoparticles and plant physiology on plant transport. Another basic perspective is that E-nanoparticles are essentially shaky and abiotic or biotic changes of E-nanoparticles, for example, redox responses, variety, and crumbling of E-nanoparticles may happen in the rhizosphere or within the plants, and it would fabulously modify the toxicity, bioavailability, and destiny of E-nanoparticles (Lowry et al., 2012; Lin et al., 2010; Zhang et al., 2012; Ma et al., 2011; Wang et al., 2013; Lv et al., 2015). A favorable and consolidated study of ongoing data accepting bribes up, translocation, assembling, and change of E-nanoparticles in the soil plant structure is along these lines required. Besides, perception of the take-up, change of nanoparticles, and translocation in plants or diverse living things is especially dependent on the progression of explicit sensible techniques. In the prior decade, unmistakable progressed enlightening systems had been utilized to see the space and speciation of nanoparticles in animals at cell, tissue, and subcellular points, yet magnificent targets present here especially for numeric assessment and in situ district of nanoparticles in complicated structures, for instance, plants, soils, and different living things (Castillo-Michel et al., 2017). Here, frontline redesigns in the available and potential structures for the appraisal of relationship among plants and nanoparticles are investigated, and central troubles and future assessments are being studied.

Toxicity of Nanoparticles in Plants

11.3 TRANSLOCATION AND UPTAKE ROUTES OF NANOPARTICLES WITHIN THE PLANTS

11.3.1 UPTAKE AND FOLIAR EXPOSURE TO NANOPARTICLES

11.3.1.1 FOLIAR TAKE-UP AND TRANSLOCATION OF NANOPARTICLES IN PLANTS

There have been two diverse colleague procedures for plants with nanoparticles, to express foliar introduction and root presentation. The fingernail skin is viewed as the foremost regular limit against nanoparticles going. In surface tissue, the leaves of most plants are guaranteed by the uncontrolled catastrophe of water and the replacement of other solutions by oak bark (Pollard et al., 2008). There are two courses for taking solutions on the nail skin (cuticle pathway), diffusion and invasion (lipophilic pathways), and polar fluid areas (hydrophilic course) through the system of asymmetric nonmolar solvents through the technique of polar solvents assessed productive size running degree between 0.6 and 4.8 nm (Eichert and Goldbach, 2008; Popp et al., 2005; Eichert et al., 2008). Likewise, nanoparticles underneath 4.8 nm available for use may enter through the fingernail skin through the cuticular pathway clearly, while different appraisals have pronounced foliar take-up and assembling of nanoparticles more noteworthy than 5 nm. In any case, the course by which these nanoparticles are up taken is as of recently scattered. Along these lines, explicit thought might be given to the issues of whether nanoparticles, expressly to those more noteworthy more than 5 nm, can be taken through the leaves through the skin passage. The life cycle of the skin, weather conditions, and the implantation of leaf fingernails with insects and debility are additionally extreme; one good method at a time can be removed by critically examining the skin of the nails against the nanoparticles. In spite of the cuticular pathway, analyses have shown the take-up of hydrophilic substances by means of stomatal passages (Fig. 11.1). The physical appearance of stomatal gaps is around 25-µm long but 3–10 µm in width (Eichert et al., 2008). Be that as it would, considering the outstanding geometric turn of events and physiological restriction of stomata; the real limit of size evading of stomatal opening for nanoparticles entrance is so far a long way from reality. Eichert and Goldbach, using a wicked figuring approach, evaluated the proportionate pore length of this course to be more prominent than 20 nm (Eichert and Goldbach, 2008). This stomatal course is the standard demanded course of foliar take-up of nanoparticles to within tissues from the leaf surface. Different investigations fortify this pathway of NP uptake, containing the perspective on different nanoparticles or on the other hand their totals in stomatal leaf and the more important tissues of different plant species including *Citrullus lanatus, Cucurbita pepo, Allium porrum, Arabidopsis thaliana*, and *Lactuca sativa* using µ-XRF or TEM, CLSM (Eichert et al., 2008; Uzu et al., 2010; Corredor et al., 2009; Larue et al., 2014; Kurepa et al., 2010). Eichert et al. looked into the limit of size evading and even heterogenesity of the foliar stomatal take-up passage for fluorescent polystyrene of water-suspended nanoparticles (Eichert et al., 2008). The take-up of 43-nm nanoparticles by means of the stomatal passage was seen with CLSM, while zero take-up of 1.1-µm materials was seen. They also found that the vehicle of fluorescent nanoparticles in the apoplast of leaf when they got penetrated the substomatal trademark unfilled. The starter results show that the opening away from the stomatal passage assessed by Eichert and Goldbach was a fundamentally nothing. Kim et al. resulted that nano-zero-valent iron instigated the start of plasma H+-ATPase layer of development and progressed stomatal hole openings (Kim et al., 2015). In improvement, plant along with various stomatal size, thickness morphology, and leaf are required to have explicit cutoff points concerning foliar take-up of nanoparticle (Uzu et al., 2010). In a coherent report, CuO nanoparticles (20–100 nm) leaves were introduced into the collard greens of cubes, vegetable lettuce, and single-bit mass spectrometry/coupled plasma (SP-ICP) MS (Keller et al., 2018) at the end. Different nanoparticles and their aggregates have been shown to be washable with H_2O, and distant regions of nanoparticles are more prone to leaf tissue for their hydrophobicity and surface roughness (Keller et al., 2018). Furthermore, examinations are urged to effectively for investigation of the stomatal pathway of NP take-up by assembled plant types.

Nanoparticles could encounter critical hole transport by methods for the vascular framework in the wake of permitting the apoplast of leaf over the stomatal pathway. Generally photosynthetic, large-scale particles and sugars in the leaf, containing little proteins and RNA, are able to trade plunging by methods for the phloem structure to shoots and roots (Lough and Lucas, 2006) when everything is said in done, the huge separation transport of fluids in higher plants happens by strategies for the vascular framework, which is made out of the phloem and xylem coordinating tissues. In the jellium structure, the head of the stream is from the bottom to the top (from the root-shoot); however, the stream bearing is completely (from the shoot-root) in the phloem framework (Fig. 11.1). This morphology of plant vessels does not circulate as it normally would, indicating that the nanomaterials that slide

Toxicity of Nanoparticles in Plants

into the phloem do not return to their abnormal positions through the jellyfish vessels (Batley and Kirby, 2013). So, the phloem structure is a guide for the transfer of nanoparticles between the leaf and the root. Although many wonders consider nano-cell uptake, no tests have really provided evidence to alter the path of phloem change in the subtrail of nano-cells in plants. Uptake (foil) of four metal oxide nanoparticles of 24–47 nm size was detected by watermelon. They observed that following the path of the stomach using TEM, small nanoparticles could be transported to watermelon leaves, and that metal parts in the roots and shoots were identified, which stimulated the nano-con. Phloem went to the roots by means of cipher tubes (Wang et al., 2013). Take decomposed leaves of CeO_2 nanoparticles of key size 8 and 1 nm by *Cucumis sativus*. CC in most layers of CC was observed using TEM (Hong et al., 2014) in cells of ICP-OES and CO_2 nanoparticle treated plants. Determined by ICP-MS, Zhao et al. revealed that 97%–99% of Cu was deposited in the leaves and only 1%–3% of the Q was deposited in the root tissue after secondary to QTO with lettuce plants. And they recommended to restore the isolation of $CuOH_2$ nanoparticles enter the epidermal tissue, understand the path of Cu cells and transfer to another (Kurepa et al., 2010; Zhao et al. 2016c). The first urgent concern is that TEM does not prioritize the main evaluation observed between nanoparticles or individual metal particles, either between the source or the seed or between the metallic material observed in the material. This is a unique class of standard trademark evaluation to follow the NP progress of plants. Using larger scales, Wang et al. (2012), split/root evaluation and TEM diagnosis, found that 20–40 nm CuO nanoparticles between the corn roots and shoots were transorchestrated and burned by the gill and after a while the roots return through the shoot. Between these translations, CuO nanoparticles can be reduced from Cu II to Cu I. Ours and others. Did you see it accidentally xylem-phloem-based vehicle of 25-nm CeO_2 nanoparticles in cucumber using split roots tests? They further found, using μ-XRF and μ-XANES, that CeO_2 nanoparticles were moved toward shoots from roots by the xylem vessels and 15% of the nanoparticles were decreased to, while just CeO_2 nanoparticles were moved back to roots from shoots through the phloem vessel (Ma et al., 2017). These results brace the probability that nanoparticles can go on upward through the xylem and falling through the phloem, at any rate how the transport of nanoparticles between jelly and phloem is cloudy. The nanoparticles and their modified material can be orchestrated at the roots, so that they can sink into the rhizosphere and the transmasterminded leaves will probably migrate to the phylo-circle, which is in the rhizome or micro. It affects the composition of microorganisms.

11.3.1.2 PHYLLO-SPHERE FACTORS INFLUENCING FOLIAR TAKE-UP OF NANOPARTICLES VIA PLANTS

Indigenous elemental regular units contain useful or pathogenic microorganisms mixed in phylo-regions (Batley and Kirby, 2013; Kah et al., 2018). Philo-circle microorganisms can release extracellular material or surprise organized materials that detect signaling molecules. They can shape the mucous membrane to further absorb metal particles through nano-cells and plant leaves or to ensure leaf surfaces against nanoparticle lay-up (Batley and Kirby, 2013). Philo-circle microbial deciduous rocks on how nanoparticles are formed by plants or whether nanoparticles predict the growth of phyllospiric microorganisms. In addition, organ enhancement levels for plants are characterized by extensive growth at different stages of life and affect the growth of nanoparticles leaves. For example, the absence of fingernail skin at the onset of late-forming leaves and the weakening of the skin of old leaf toenails in general give the nanoparticles a high chance of leaf penetration (Honour et al., 2009). Suddenly, some standard leaves, for example, leaf structure, flax, chlorosis, and debris can strike the leaf's natural tissues, such as nails, epidermis, and thermophile (Sankaran et al. 2010). Leaf counters against nanoparticles become irritated or deformed when injured. The plant leaves. In particular, a pair of nanoparticles, for example, Cu(OH)2, (Zhao et al., 2016c, 2017) TiO_2 (Kah and Hofmann, 2014), and Ag/SiO_2 (Park et al., 2006), are used as nano-pesticides, which rain on leaf surfaces. Thus, developing NP foliar take-up risks. Vehicles sooner or later to different tissues.

11.3.1.3 ABILITY OF UTILIZING PLANT LEAVES AS CAPTORS OF CLIMATIC NANOPARTICLES

Different examinations have revealed that plants, particularly trees and supports, perceive tremendous occupations in the catch of airborne particles and filtration, for example, PM10 and PM2.5, and along these lines improved quality of air (Nguyen et al., 2015; Dzierzanowski et al. 2011). Pure and simple, nanoscale ultrafine cells are more likely to grow than cells other than airborne cells, because they can enter the human body by inhalation for all purposes and purposes (Gottschalk and Nowack, 2011; Seaton et al., 2010). For any situation, the evaluation is no different and plants of any degree can channel and contain natural nanoparticles. As researched above, the evaluation center halfway through the investigation shows that the leaves of plants can receive nanoparticles. The earth is twice as wide. These are the encounters of the surface (Vorholt, 2012). Filter rationality of conventional

Toxicity of Nanoparticles in Plants 311

nanoparticles suggests that it is entirely subject to plant species (Pradas del Real et al., 2016; Dzierzanowski et al., 2011), so that the effects of plants on airflow and especially the prevention of weather NP captors can be reviewed. Tests are required, including a production extension and an on-field survey of the barometer nanoparticles.

11.3.2 TRANSLOCATION AND ROOT UPTAKE OF NANOPARTICLES IN PLANTS

11.3.2.1 COMPONENTS IMPACTING ROOT TAKE-UP OF NANOPARTICLES BY PLANTS

There are determinedly dispersed reports on root colleague of plants with nanoparticles than on foliar presentation. In any case, some clashing outcomes have been gotten and the subject of plant root take-up of nanoparticles is as of recently scrappy. One fundamental explanation is that plant take-up of nanoparticles is affected by different segments, for instance, molecule size, surface functionalization, morphology, presentation conditions, plant species, plant improvement stage, root suffering quality (mischief or ailment), and rhizosphere structures. We extracted current data from standard nanoparticles in plants, including NP characteristics, plant species, performance conditions, explicit morphology, and, moreover, root u, qi, and its significant outcomes. One of the most focal properties affecting the root-up of a nano-cell plant is particle size. Plant-wise there are arguments for the shape of nanoparticles, although the union does collide. Sabo-Attwood et al. (2012) N-XRF and TEM use was observed by tobacco at the root take-up of 3.5-nm Au nano-cells, although 18-nm Au nano-cells were extracted from the surfaces to the outer surface (Pradas del Real et al., 2016; Auffan et al., 2009). For any condition, Taylor et al. (2014) the use of TEM has been found to be undisputed by u nanoparticles (7–108 nm) Arabidopsis thaliana sources. In addition, it was observed that SiO_2 nanoparticles were set to move up to 200 nm using Slomberg and Scoenfish TEM. However, through the installations of Arabidopsis thaliana it was found that the smaller particles had a molecular size (14, 50, 200 nm) (Slomberg and Schoenfisch, 2012) large-sized subordinate take-up and transport revealed. In TEM_2 nanoparticle brown (*Triticum beautivum*) using TEM and μ-XRF, nano-heads with a head width of less than 36 nm and without the need to disconnect or replace the entire plant are required, while those collected within the 36–140 nm brown root parenchyma—With significant divisions—not yet visible in the stall and not translated into shoots and the nanoparticles exceed 140 nm. Recognized. Collected in wheat

312 Diverse Applications of Nanotechnology in the Biological Sciences

roots. Making observations for sources by unreliable NP take-up is so far from the current estimate. One factor is that SEL is undeniable for different plant species and growth stages. Another important point is that the degree of nano-cells in the rhizosphere is completely harmless to this type of thing and is particularly excellent and widely accepted. Regardless, the molecular size used in the various evaluations is established on the general shape of the head nanoparticles. Of course, taking nanoparticles through plants is risky, however, not a nanoparticle habit.

Surface charge is another factor that affects the root take-up and translation of nanoparticles in plants. Around, the top of the base of the plant is guaranteed by a layer of edge cell stick, which contains the root flood which is not properly collected. Avalon et al. Using the X-shaft of nano-tomography (nano-CT) and hyperepic imaging microscopy, *Arabidopsis thaliana* was found to emit nanoparticles (strictly 12 nm) of unevenly charged cements. Nano-cells in stem tissues. The nanoparticles (12 nm) were in contrast, did not absorb the glue and decided to translate the roots into apoplast. Coelmel et al. Laser transmission using independent binomial-mass spectrometry (LA-ICPMS), surface activity was found to inadvertently reduce the translocation of u nano-cells (focus diameter 2 nm) in root take-up and rice (*Oryza sativa* L). A group of people gathered at the sources followed the business: u nano-particle (+)> u nano-particle (0)> u nano-particle (n), although the opposite game plan was guaranteed to shoot, in which the contrast was shown unevenly. Nanoparticles charged by the convection system have been shown to have comparable effects on (Koelmel et al., 2013) wheat (Triticum beautivum). After examining the improvement of the nano-crown, the roots of the recharged elements were firmly turned on the surfaces of Si.

11.3.2.2 COORDINATED ROOT TAKE-UP AND TRANSLOCATION PATHWAYS OF NANOPARTICLES IN PLANTS

As the plant progresses toward the roots, the nanoparticles move away from the surface-absorbing bat. The roots of the plant have a large surface area depending on the proximity of the root hairs that can transmit pastes or small particles, for example, trademark acid, and if all else fails the root surface is negatively charged. These highlights cause the nanoparticles to become dangerous due to the positive surface charge and deposit on the surface of the roots (Wiesner et al., 2009). For lay-up and translocation to take place, nanoparticles must cross the progression of the physical source limit from

Toxicity of Nanoparticles in Plants

the surface. In the final vehicle, including the gillum vessels, root surface nail skin, epidermis, cortex, and endodermis, it burns to the gill through the Casperian bandage system (Fig. 11.1). The root surrounds the action and process of the skin of the surface nails, which is not actually defined by the skin of the leaf surface nails. Although the nanoparticles were able to dig into the base surface, the skin of the nails eventually began to become dull. However, the skin of the nails forms the root hairs and root tips of the root and accessory roots, as such nanoparticles may actually be familiar with the epidermis in these areas (Scheringer, 2008).

When the root is pushed toward the epidermis, most plants have two important pathways for nano-cell root-up and transport. Most evaluations have proposed the apoplastic pathway (Fig. 11.1), in which the nanoparticles enter the cell division pores directly from the bat and after a short time pass into the space between the cell divider and the plasma film or intracellular. Experience the place. Goliath number studies looked at nanoparticles or their aggregates using TEM or CLSM in the root apoplastic space, thus proposing the area of the apoptoplastic pathway for nanoparticles in plant roots. For the model, 20-nm ZnO nanoparticles in Rigrass roots (Lin and Xing, 2008), 12-nm u nano-cells in Arabidopsis thaliana roots (Avellan et al., 2017), 20- to 80-nm Ag nano cells in Arabidopsis forage roots (Geisler-Lee et al., 2013), 43-nm CuO cells in Allsholtzia roots of cucumber. In 76- and 22-nm La_2O_3 nano-cells, an irreversible problem is that the displacement in plant cell dividers is estimated to be in the range of 5–20 nm. Naturally, nanoparticles can only shrink by more than 20 nm. However, more than 20 nano-significant nanoparticles have been observed at intercellular locations. Nanoparticles can start hitting phone dividers and expand the size of the holes (Gottschalk and Nowack, 2011). Another potential spec is that nanoparticles may enter intercellular locations or cause bodily injury through sources contaminated with gel or underground herbicides. Mechanical injuries, for example, are unlucky during transplantation (Dietz and Herth, 2011). Nanoparticles as a whole, through the apoplastic pathway, are found in the epidermis, cortex, and endodermis, although the Caspian stripe, asymmetric cell dividers are affected by the belt of the material. The approach to the cells of the endodermis around the vascular structure is resolved by lipophilic hydrocarbons, to the macro- and nano-cell compartment in the vascular cylinder. Distance from Caspian stripes by root apoplast, for example, an area through the root tip where the Casparian strip has not yet formed (Lv et al., 2015; Schymura et al., 2017), or the level root crossing point cut by the Casparian strip (Dietz and Herth, 2011; Lv et al., 2015; McCully, 1995).

KEYWORDS

- **nanoparticles**
- **plants**
- **toxicity**
- **carbon-based materials**
- **metals**

REFERENCES

Amarie, S.; Zaslansky, P.; Kajihara, Y.; Griesshaber, E.; Schmahl, W. W.; Keilmann, F. Nano-FTIR Chemical Mapping of Minerals in Biological Materials. *Beilstein J. Nanotechnol.* **2012**, *3*, 312–323.

Aubert, T.; Burel, A.; Esnault, M. A.; Cordier, S.; Grasset, F.; Cabello-Hurtado, F. Root Uptake and Phytotoxicity of Nanosized Molybdenum Octahedral Clusters. *J. Hazard. Mater.* **2012**, *219*, 111–118.

Auffan, M.; Rose, J.; Bottero, J. Y.; Lowry, G. V.; Jolivet, J. P.; Wiesner, M. R. Towards a Definition of Inorganic Nanoparticles from an Environmental, Health and Safety Perspective. *Nat. Nanotechnol.* **2009**, *4*, 634–641.

Avellan, A.; Schwab, F.; Masion, A.; Chaurand, P.; Borschneck, D.; Vidal, V.; Rose, J.; Santaella, C.; Levard, C. Nanoparticle Uptake in Plants: Gold Nanomaterial Localized in Roots of *Arabidopsis thaliana* by X-ray Computed Nanotomography and Hyperspectral Imaging. *Environ. Sci. Technol.* **2017**, *51*, 8682–8691.

Bao, D. P.; Oh, Z. G.; Chen, Z. Characterization of Silver Nanoparticles Internalized by Arabidopsis Plants Using Single Particle ICP-MS Analysis. *Front. Plant Sci.* **2016**, *7*, 32.

Batley, G. E.; Kirby J. K.; McLaughlin, M. J. Fate and Risks of Nanomaterials in Aquatic and Terrestrial Environments. *Acc. Chem. Res.* **2013**, *46*, 854–862.

Beattie, I. R.; Haverkamp, R. G. Silver and Gold Nanoparticles in Plants: Sites for the Reduction to Metal. *Metallomics* **2011**, *3*, 628–632.

Becker, L. Fullerenes in the 1.85-Billion-Year-Old Sudbury Impact Structure. *Science* **1994**, *265*, 1644–1644.

Bernard, S.; Beyssac, O.; Benzerara, K. Raman Mapping Using Advanced Line-Scanning Systems: Geological Applications. *Appl. Spectrosc.* **2008**, *62*, 1180–1188.

Birbaum, K.; Brogioli, R.; Schellenberg, M.; Martinoia, E.; Stark, W. J.; Gunther, D.; Limbach, L. K. No Evidence for Cerium Dioxide Nanoparticle Translocation in Maize Plants. *Environ. Sci. Technol.* **2010**, *44*, 8718–8723.

Bone, A. J.; Colman, B. P.; Gondikas, A. P.; Newton, K. M.; Harrold, K. H.; Cory, R. M.; Unrine, J. M.; Klaine, S. J.; Matson, C. W.; Di Giulio, R. T. Biotic and Abiotic Interactions in Aquatic Microcosms Determine Fate and Toxicity of Ag Nanoparticles: Part 2-Toxicity and Ag Speciation, *Environ. Sci. Technol.* **2012**, *46*, 6925–6933.

Castillo-Michel, H. A.; Larue, C.; Pradas del Real, A. E.; Cotte, M.; Sarret, G. Practical Review on the Use of Synchrotron Based Micro- and Nano- X-ray Fluorescence Mapping

Toxicity of Nanoparticles in Plants

and X-ray Absorption Spectroscopy to Investigate the Interactions between Plants and Engineered Nanomaterials. *Plant. Physiol. Biochem.* **2017**, *110*, 13–32.

Chhipa, H. Nanofertilizers and Nanopesticides for Agriculture. *Environ. Chem. Lett.* **2017**, *15*, 15–22.

Corredor, E.; Testillano, P. S.; Coronado, M. J.; Gonzalez- Melendi, P.; Fernandez-Pacheco, R.; Marquina, C.; Ibarra, M. R.; de la Fuente, J. M.; Rubiales, D.; Perez-De-Luque, A.; Risueno, M. C. Nanoparticle Penetration and Transport in Living Pumpkin Plants: In situ Subcellular Identification. *BMC Plant Biol.* **2009**, *9*, 45–53.

Coutris, C.; Joner, E. J.; Oughton, D. H. Aging and Soil Organic Matter Content Affect the Fate of Silver Nanoparticles in Soil. *Sci. Total Environ.* **2012**, *420*, 327–333.

Cui, D.; Zhang, P.; Ma, Y. H.; He, X.; Li, Y. Y.; Zhang, J.; Zhao, Y. C.; Zhang, Z. Y. Effect of Cerium Oxide Nanoparticles on Asparagus Lettuce Cultured in an Agar Medium. *Environ. Sci. Nano.* **2014**, *1*, 459–465.

Dan, Y. B.; Zhang, W. L.; Xue, R. M.; Ma, X. M.; Stephan, C.; Shi, H. L. Characterization of Gold Nanoparticle Uptake by Tomato Plants Using Enzymatic Extraction Followed by Single-Particle Inductively Coupled Plasma-Mass Spectrometry Analysis. *Environ. Sci. Technol.* **2015**, *49*, 3007–3014.

Dan, Y.; Ma, X.; Zhang, W.; Liu, K.; Stephan, C.; Shi, H. Single Particle ICP-MS Method Development for the Determination of Plant Uptake and Accumulation of CeO_2 Nanoparticles. *Anal. Bioanal. Chem.* **2016**, *408*, 5157–5167.

Deng, Y.; Petersen, E. J.; Challis, K. E.; Rabb, S. A.; Holbrook, R. D.; Ranville, J. F.; Nelson, B. C.; Xing, B. Multiple Method Analysis of TiO_2 Nanoparticle Uptake in Rice (*Oryza sativa* L.) Plants. *Environ. Sci. Technol.* **2017**, *51*, 10615–10623.

Dietz, K. J.; Herth, S. Plant Nanotoxicology. *Trends Plant Sci.* **2011**, *16*, 582–589.

Dimkpa, C. O.; Latta, D. E.; McLean, J. E.; Britt, D. W.; Boyanov, M. I.; Anderson, A. J. Fate of CuO and ZnO Nano and Micro Particles in the Plant Environment. *Environ. Sci. Technol.* **2013**, *47*, 4734–4742.

Dimkpa, C. O.; McLean, J. E.; Latta, D. E.; Manangon, E.; Britt, D. W.; Johnson, W. P.; Boyanov, M. I.; Anderson, A. J. CuO and ZnO nanoparticles: Phytotoxicity, Metal Speciation, and Induction of Oxidative Stress in Sand-Grown Wheat. *J. Nanopart. Res.* **2012**, *14*, 1125.

Dzierzanowski, K.; Popek, R.; Gawronska, H.; Saebo, A.; Gawronski, S. W. Deposition of Particulate Matter of Different Size Fractions on Leaf Surfaces and in Waxes of Urban Forest Species. *Int. J. Phytorem.* **2011**, *13*, 1037–1046.

Eder, S. H. K.; Gigler, A. M.; Hanzlik, M.; Winklhofer, M. Sub-Micrometer-Scale Mapping of Magnetite Crystals and Sulfur Globules in Magnetotactic Bacteria Using Confocal Raman Micro-Spectrometry. *PLoS One* **2014**, *9*, e107356.

Eichert, T.; Goldbach, H. E. Equivalent Pore Radii of Hydrophilic Foliar Uptake Routes Instomatous and Astomatous Leaf Surfaces-Further Evidence for a Stomatal Pathway. *Physiol. Plant* **2008**, *132*, 491–502.

Eichert, T.; Kurtz, A.; Steiner, U.; Goldbach, H. E. Size Exclusion Limits and Lateral Heterogeneity of the Stomatal Foliar Uptake Pathway for Aqueous Solutes and Water Suspended Nanoparticles. *Physiol. Plant* **2008**, *134*, 151–160.

El-Temsah, Y. S.; Joner, E. J. Impact of Fe and Ag Nanoparticles on Seed Germination and Differences in Bioavailability during Exposure in Aqueous Suspension and Soil. *Environ. Toxicol.* **2012**, *27*, 42–49.

Etxeberria, E.; Gonzalez, P.; Baroja-Fernandez, E.; Romero, J. P. Fluid Phase Endocytic Uptake of Artificial Nanospheres and Fluorescent Quantum Dots by Sycamore Cultured Cells. *Plant Signal. Behav.* **2006,** *1*, 196–200.

Gao, X.; Avellan, A.; Laughton, S.; Vaidya, R.; Rodrigues, S. M.; Casman, E. A.; Lowry, G. V. CuO Nanoparticle Dissolution and Toxicity to Wheat (*Triticum aestivum*) in Rhizosphere Soil. *Environ. Sci. Technol.* **2018,** *52*, 2888–2897.

Gardea-Torresdey, J. L.; Gomez, E.; Peralta-Videa, J. R.; Parsons, J. G.; Troiani, H.; Jose-Yacaman, M. Alfalfa Sprouts: A Natural Source for the Synthesis of Silver Nanoparticles. *Langmuir.* **2003,** *19*, 1357–1361.

Gardea-Torresdey, J. L.; Parsons, J. G.; Gomez, E.; Peralta- Videa, J.; Troiani, H. E.; Santiago, P.; Yacaman, M. J. Formation and Growth of Au Nanoparticles Inside Live Alfalfa Plants. *Nano Lett.* **2002,** *2*, 397–401.

Gardea-Torresdey, J. L.; Rico, C. M.; White, J. C. Trophic Transfer, Transformation, and Impact of Engineered Environmental Science: Nano Critical Review Published on 31 October 2018. Downloaded on 1/20/2019 7:29:54 PM. View Article Online 54|*Environ. Sci. Nano* **2019,** *6*, 41–59. This journal is © The Royal Society of Chemistry 2019 Nanomaterials in Terrestrial Environments. *Environ. Sci. Technol.* **2014,** *48*, 2526–2540.

Garner, K. L.; Keller, A. A. Emerging Patterns for Engineered Nanomaterials in the Environment: A Review of Fate and Toxicity Studies. *J. Nanopart. Res.* **2014,** *16*, 2503.

Geisler-Lee, J.; Wang, Q.; Yao, Y.; Zhang, W.; Geisler, M.; Li, K.; Huang, Y.; Chen, Y.; Kolmakov, A.; Ma, X. Phytotoxicity, Accumulation and Transport of Silver Nanoparticles by *Arabidopsis thaliana. Nanotoxicology* **2013,** *7*, 323–337.

Ghosh, M.; Bandyopadhyay, M.; Mukherjee, A. Genotoxicity of Titanium Dioxide (TiO_2) Nanoparticles at Two Trophic Levels Plant and Human Lymphocytes. *Chemosphere* **2010,** *81*, 1253–1262.

Gilbert, B.; Fakra, S. C.; Xia, T.; Pokhrel, S.; Madler, L.; Nel, A. E. The Fate of ZnO Nanoparticles Administered to Human Bronchial Epithelial Cells. *ACS Nano.* **2012,** *6*, 4921–4930.

Gimbert, L. J.; Hamon, R. E.; Casey, P. S.; Worsfold, P. J. Partitioning and Stability of Engineered ZnO Nanoparticles in Soil Suspensions Using Flow Field-Flow Fractionation. *Environ. Chem.* **2007,** *4*, 8–10.

Glenn, J. B.; White, S. A.; Klaine, S. J. Interactions of Gold Nanoparticles with Freshwater Aquatic Macrophytes are Size and Species Dependent. *Environ. Toxicol. Chem.* **2012,** *31*, 194–201.

Gomez-Rivera, F.; Field, J. A.; Brown, D.; Sierra-Alvarez, R. Fate of Cerium Dioxide (CeO_2) Nanoparticles in Municipal Wastewater during Activated Sludge Treatment. *Bioresour. Technol.* **2012,** *108*, 300–304.

Gottschalk, F.; Nowack, B. The Release of Engineered Nanomaterials to the Environment. *J. Environ. Monit.* **2011,** *13*, 1145–1155.

Gottschalk, F.; Sonderer, T.; Scholz, R. W.; Nowack, B. Modeled Environmental Concentrations of Engineered Nanomaterials (TiO_2, ZnO, Ag, CNT, Fullerenes) for Different Regions. *Environ. Sci. Technol.* **2009,** *43*, 9216–9222.

Guzman, K. A. D.; Taylor, M. R.; Banfield, J. F. Environmental Risks of Nanotechnology: National Nanotechnology Initiative Funding 2000–2004. *Environ. Sci. Technol.* **2006,** *40*, 1401–1407.

Hernandez-Viezcas, J. A.; Castillo-Michel, H.; Andrews, J. C.; Cotte, M.; Rico, C.; Peralta-Videa, J. R.; Ge, Y.; Priester, J. H.;. Holden, P. A.; Gardea-Torresdey, J. L. In Situ Synchrotron X-ray Fluorescence Mapping and Speciation of CeO_2 and ZnO Nanoparticles in Soil Cultivated Soybean (*Glycine max*). *ACS Nano.* **2013,** *7*, 1415–1423.

Hernandez-Viezcas, J. A.; Castillo-Michel, H.; Servin, A. D.; Peralta-Videa, J. R.; Gardea-Torresdey, J. L. Spectroscopic Verification of Zinc Absorption and Distribution in the Desert Plant Prosopis juliflora-velutina (Velvet Mesquite) Treated with ZnO Nanoparticles. *Chem. Eng. J.* **2011**, *170*, 346–352.

Hochella, M. F.; Spencer, M. G.; Jones, K. L. Nanotechnology: Nature's Gift or Scientists' Brainchild? *Environ. Sci.: Nano.* **2015**, *2*, 114–119.

Hong, J.; Peralta-Videa, J. R.; Rico, C.; Sahi, S.; Viveros, M. N.; Bartonjo, J.; Zhao, L.; Gardea-Torresdey, J. L. Evidence of Translocation and Physiological Impacts of Foliar Applied CeO_2 Nanoparticles on Cucumber (*Cucumis sativus*) Plants. *Environ. Sci. Technol.* **2014**, *48*, 4376–4385.

Honour, S. L.; Bell, J. N.; Ashenden, T. W.; Cape, J. N.; Power, S. A. Responses of Herbaceous Plants to Urban Air Pollution: Effects on Growth, Phenology and Leaf Surface Characteristics. *Environ. Pollut.* **2009**, *157*, 1279–1286.

Huang, Y.; Zhao, L.; Keller, A. A. Interactions, Transformations, and Bioavailability of Nano-Copper Exposed to Root Exudates. *Environ. Sci. Technol.* **2017**, *51*, 9774–9783.

Huth, F.; Govyadinov, A.; Amarie, S.; Nuansing, W.; Keilmann, F.; Hilenbrand, R. Nano-FTIR Absorption Spectroscopy of Molecular Fingerprints at 20 nm Spatial Resolution. *Nano Lett.* **2012**, *12*, 3973–3978.

Jilling, A.; Keiluweit, M.; Contosta, A. R.; Frey, S.; Schimel, J.; Schnecker, J.; Smith, R. G.; Tiemann, L.; Grandy, A. S. Minerals in the Rhizosphere: Overlooked Mediators of Soil Nitrogen Availability to Plants and Microbes. *Biogeochemistry* **2018**, *139*, 103–122.

Jimenez-Lamana, J.; Wojcieszek, J.; Jakubiak, M.; Asztemborska, M.; Szpunar, J. Single Particle ICP-MS Characterization of Platinum Nanoparticles Uptake and Bioaccumulation by *Lepidium sativum* and *Sinapis alba* Plants. *J. Anal. At. Spectrom.* **2016**, *31*, 2321–2329.

Judy, J. D.; Unrine, J. M.; Rao, W.; Wirick, S.; Bertsch, P. M. Bioavailability of Gold Nanomaterials to Plants: Importance of Particle Size and Surface Coating. *Environ. Sci. Technol.* **2012**, *46*, 8467–8474.

Kaegi, R.; Voegelin, A.; Ort, C.; Sinnet, B.; Thalmann, B.; Krismer, J.; Hagendorfer, H.; Elumelu, M.; Mueller, E. Fate and Transformation of Silver Nanoparticles in Urban Wastewater Systems. *Water Res.* **2013**, *47*, 3866–3877.

Kah, M. Nanopesticides and Nanofertilizers: Emerging Contaminants or Opportunities for Risk Mitigation? *Front. Chem.* **2015**, *3*, 64.

Kah, M.; Hofmann, T. Nanopesticide Research: Current Trends and Future Priorities. *Environ. Int.* **2014**, *63*, 224–235.

Kah, M.; Kookana, R. S.; Gogos, A.; Bucheli, T. D. A Critical Evaluation of Nanopesticides and Nanofertilizers against Their Conventional Analogues. *Nat. Nanotechnol.* **2018**, *13*, 677–684.

Kang, J. W.; Nguyen, F. T.; Lue, N.; Dasari, R. R.; Heller, D. A. Measuring Uptake Dynamics of Multiple Identifiable Carbon Nanotube Species via High-Speed Confocal Raman Imaging of Live Cells. *Nano Lett.* **2012**, *12*, 6170–6174.

Keller, A. A. Adeleye, A. S.; Conway, J. R.; Garner, K. L.; Zhao, L. J.; Cherr, G. N.; Hong, J.; Gardea-Torresdey, J. L.; Godwin, H. A.; Hanna, S. S. et al. Comparative Environmental Fate and Toxicity of Copper Nanomaterials. *Nanoimpact* **2017**, *7*, 28–40.

Keller, A. A.; Huang, Y. X.; Nelson, J. Detection of Nanoparticles in edible Plant Tissues Exposed to Nanocopper Using Single-Particle ICP-MS. *J. Nanopart. Res.* **2018**, *20*, 101.

Khan, F. R.; Laycock, A.; Dybowska, A.; Larner, F.; Smith, B. D.; Rainbow, P. S.; Luoma, S. N.; Rehkamper, M.; Valsami-Jones, E. Stable Isotope Tracer to Determine Uptake and

318 Diverse Applications of Nanotechnology in the Biological Sciences

Efflux Dynamics of ZnO Nano- and Bulk Particles and Dissolved Zn to an Estuarine Snail. *Environ. Sci. Technol.* **2013,** *47,* 8532–8539.

Kim, J. H.; Oh, Y.; Yoon, H.; Hwang, I.; Chang, Y. S. Iron Nanoparticle-Induced Activation of Plasma Membrane HIJ+)- ATPase Promotes Stomatal Opening in *Arabidopsis thaliana. Environ. Sci. Technol.* **2015,** *49,* 1113–1119.

Koelmel, J.; Leland, T.; Wang, H. H.; Amarasiriwardena, D.; Xing, B. S. Investigation of Gold Nanoparticles Uptake and Their Tissue Level Distribution in Rice Plants by Laser Ablation-Inductively Coupled-Mass Spectrometry. *Environ. Pollut.* **2013,** *174,* 222–228.

Kurepa, J.; Paunesku, T.; Vogt, S.; Arora, H.; Rabatic, B. M.; Lu, J. J.; Wanzer, M. B.; Woloschak, G. E.; Smalle, J. A. Uptake and Distribution of Ultrasmall Anatase TiO_2 Alizarin Red S Nanoconjugates in *Arabidopsis thaliana. Nano Lett.* **2010,** *10,* 2296–2302.

Larue, C.; Castillo-Michel, H.; Sobanska, S.; Cecillon, L.; Bureau, S.; Barthes, V.; Ouerdane, L.; Carriere, M.; Sarret, G. Foliar Exposure of the Crop *Lactuca sativa* to Silver Nanoparticles: Evidence for Internalization and Changes in Ag Speciation. *J. Hazard. Mater.* **2014,** *264,* 98–106.

Larue, C.; Khodja, H.; Herlin-Boime, N.; Brisset, F.; Flank, A. M.; Fayard, B.; Chaillou, S.; Carrière, M. Investigation of Titanium Dioxide Nanoparticles Toxicity and Uptake by Plants. *J. Phys.: Conf. Ser.* **2011,** *304,* 012057.

Larue, C.; Laurette, J.; Herlin-Boime, N.; Khodja, H.; Fayard, B.; Flank, A. M.; Brisset, F.; Carriere, M. Accumulation, Translocation and Impact of TiO_2 Nanoparticles in Wheat (*Triticum aestivum* spp.): Influence of Diameter and Crystal Phase. *Sci. Total. Environ.* **2012,** *431,* 197–208.

Laycock, A.; Romero-Freire, A.; Najorka, J.; Svendsen, C.; van Gestel, C. A. M.; Rehkamper, M. Novel Multi-Isotope Tracer Approach to Test ZnO Nanoparticle and Soluble Zn Bioavailability in Joint Soil Exposures. *Environ. Sci. Technol.* **2017,** *51,* 12756–12763.

Layet, C.; Auffan, M.; Santaella, C.; Chevassus-Rosset, C.; Montes, M.; Ortet, P.; Barakat, M.; Collin, B.; Legros, S.; Bravin, M. N.; Angeletti, B.; Kieffer, I.; Proux, O.; Hazemann, J. L.; Doelsch, E. Evidence That Soil Properties and Organic Coating Drive the Phytoavailability of Cerium Oxide Nanoparticles. *Environ. Sci. Technol.* **2017,** *51,* 9756–9764.

Levard, C.; Hotze, E. M.; Lowry, G. V.; Brown, Jr., G. E. Environmental Transformations of Silver Nanoparticles: Critical Review Environmental Science: Nano Published on 31 October 2018. Downloaded on 1/20/2019 7:29:54 PM.

Li, H.; Ye, X.; Guo, X.; Geng, Z.; Wang, G. Effects of Surface Ligands on the Uptake and Transport of Gold Nanoparticles in Rice and Tomato. *J. Hazard. Mater.* **2016,** *314,* 188–196.

Lin, D. H.; Tian, X. L.; Wu, F. C.; Xing, B. S. Fate and Transport of Engineered Nanomaterials in the Environment. *J. Environ. Qual.* **2010,** *39,* 1896–1908.

Lin, D. H.; Xing, B. S. Root Uptake and Phytotoxicity of ZnO Nanoparticles. *Environ. Sci. Technol.* **2008,** *42,* 5580–5585.

Lindow, S. E.; Brandl, M. T. Microbiology of the Phyllosphere. *Appl. Environ. Microbiol.* **2003,** *69,* 1875–1883.

Liu, L.; He, B.; Liu, Q.; Yun, Z.; Yan, X.; Long, Y.; Jiang, G. Identification and Accurate Size Characterization of Nanoparticles in Complex Media. *Angew. Chem., Int. Ed.* **2014,** *53,* 14476–14479.

Lombi, E.; Donner, E.; Tavakkoli, E.; Turney, T. W.; Naidu, R.; Miller, B. W.; Scheckel, K. G. Fate of Zinc Oxide Nanoparticles during Anaerobic Digestion of Wastewater and Post-Treatment Processing of Sewage Sludge. *Environ. Sci. Technol.* **2012,** *46,* 9089–9096.

Toxicity of Nanoparticles in Plants 319

Lopez-Moreno, M. L.; de la Rosa, G.; Hernandez-Viezcas, J. A.; Castillo-Michel, H.; Botez, C. E.; Peralta-Videa, J. R.; Gardea-Torresdey, J. L. Evidence of the Differential Biotransformation and Genotoxicity of ZnO and CeO$_2$ Nanoparticles on Soybean (*Glycine max*) Plants. *Environ. Sci. Technol.* **2010,** *44,* 7315–7320.

Lopez-Moreno, M. L.; de la Rosa, G.; Hernandez-Viezcas, J. A.; Peralta-Videa, J. R.; Gardea-Torresdey, J. L. X-ray Absorption Spectroscopy (XAS) Corroboration of the Uptake and Storage of CeO$_2$ Nanoparticles and Assessment of Their Differential Toxicity in Four Edible Plant Species. *J. Agric. Food Chem.* **2010,** *58,* 3689–3693.

Lough, T. J.; Lucas, W. J. Integrative plant biology: Role of Phloem Long-Distance Macromolecular Trafficking. *Annu. Rev. Plant. Biol.* **2006,** *57,* 203–232.

Lowry, G. V.; Espinasse, B. P.; Badireddy, A. R.; Richardson, C. J.; Reinsch, B. C.; Bryant, L. D.; Bone, A. J.; Deonarine, A.; Chae, S.; Therezien, M.; Colman, B. P.; Hsu- Kim, H.; Bernhardt, E. S.; Matson, C. W.; Wiesner, M. R. Long-Term Transformation and Fate of Manufactured Ag Nanoparticles in a Simulated Large Scale Freshwater Emergent Wetland. *Environ. Sci. Technol.* **2012,** *46,* 7027–7036.

Lowry, G. V.; Gregory, K. B.; Apte, S. C.; Lead, J. R. Transformations of Nanomaterials in the Environment. *Environ. Sci. Technol.* **2012,** *46,* 6893–6899.

Lu, D.; Liu, Q.; Zhang, T.; Cai, Y.; Yin, Y.; Jiang, G. Stable Silver Isotope Fractionation in the Natural Transformation Process of Silver Nanoparticles. *Nat. Nanotechnol.* **2016,** *11,* 682–686.

Lucas, W. J. Plasmodesmata: Intercellular Channels for Macromolecular Transport in Plants. *Curr. Opin. Cell Biol.* **1995,** *7,* 673–680.

Lucas, W. J.; Ham, L. K.; Kim, J. Y. Plasmodesmata Bridging the Gap between Neighboring Plant Cells. *Trends Cell Biol.* **2009,** *19,* 495–503.

Lucas, W. J.; Lee, J. Y. Plant cell biology-Plasmodesmata as a Supracellular Control Network in Plants. *Nat. Rev. Mol. Cell Biol.* **2004,** *5,* 712–726.

Luu, D. T.; Maurel, C. Aquaporins in a Challenging Environment: Molecular Gears for Adjusting Plant Water Status. *Plant Cell Environ.* **2005,** *28,* 85–96.

Lv, J. T.; Zhang, S. Z.; Luo, L.; Han, W.; Zhang, J.; Yang, K.; Christie, P. Dissolution and Microstructural Transformation of Zno Nanoparticles under the Influence of Phosphate. *Environ. Sci. Technol.* **2012,** *46,* 7215–7221.

Lv, J. T.; Zhang, S. Z.; Luo, L.; Zhang, J.; Yang, K.; Christie, P. Accumulation, Speciation and Uptake Pathway of ZnO Nanoparticles in Maize. *Environ. Sci. Nano.* **2015,** *2,* 68–77.

Lv, J. T.; Zhang, S. Z.; Wang, S. S.; Luo, L.; Huang, H. L.; Zhang, J. Chemical Transformation of Zinc Oxide Nanoparticles as a Result of Interaction with Hydroxyapatite. *Colloids Surf. A.* **2014,** *461,* 126–132.

M. Baalousha, B. Stolpe and J. R. Lead, Flow Field-Flow Fractionation for the Analysis and Characterization of Natural Colloids and Manufactured Nanoparticles in Environmental Systems: A Critical Review. *J. Chromatogr. A* **2011,** *1218,* 4078–4103.

Ma, X. M.; Geiser-Lee, J.; Deng, Y.; Kolmakov, A. Interactions between Engineered Nanoparticles (ENPs) and Plants: Phytotoxicity, Uptake and Accumulation. *Sci. Total. Environ.* **2010,** *408,* 3053–3061.

Ma, Y. H.; He, X.; Zhang, P.; Zhang, Z. Y.; Guo, Z.; Tai, R. Z.; Xu, Z. J.; Zhang, L. J.; Ding, Y. Y.; Zhao, Y. L.; Chai, Z. F. Phytotoxicity and Biotransformation of La$_2$O$_3$ Nanoparticles in a Terrestrial Plant Cucumber (*Cucumis sativus*). *Nanotoxicology* **2011,** *5,* 743–753.

Ma, Y. H.; Zhang, P.; Zhang, Z. Y.; He, X.; Zhang, J. Z.; Ding, Y. Y.; Zhang, J.; Zheng, L. R.; Guo, Z.; Zhang, L. J.; Chai, Z. F.; Zhao, Y. L. Where does the Transformation of

Precipitated Ceria Nanoparticles in Hydroponic Plants Take Place? *Environ. Sci. Technol.* **2015,** *49*, 10667–10674.

Ma, Y.; He, X.; Zhang, P.; Zhang, Z.; Ding, Y.; Zhang, J.; Wang, G.; Xie, C.; Luo, W.; Zhang, J.; Zheng, L.; Chai, Z.; Yang, K. Xylem and Phloem Based Transport of CeO_2 Nanoparticles in Hydroponic Cucumber Plants. *Environ. Sci. Technol.* **2017,** *51*, 5215–5221.

Martin, M. C.; Schade, U.; Lerch, P.; Dumas, P. Recent Applications and Current Trends in Analytical Chemistry Using Synchrotron-Based Fourier-Transform Infrared Microspectroscopy. *TrAC, Trends Anal. Chem.* **2010,** *29*, 453–463.

Martinez-Criado, G.; Villanova, J.; Tucoulou, R.; Salomon, D.; Suuronen, J. P.; Laboure, S.; Guilloud, C.; Valls, V.; Barrett, R.; Gagliardini, E.; Dabin, Y.; Baker, R.; Bohic, S.; Cohen, C.; Morse, J. ID16B: A Hard X-ray Nanoprobe Beamline at the ESRF for Nano-Analysis. *J. Synchrotron Radiat.* **2016,** *23*, 344–352.

Matczuk, M.; Anecka, K.; Scaletti, F.; Messori, L.; Keppler, B. K.; Timerbaev, A. R.; Jarosz, M. Speciation of Metal-Based Nanomaterials in Human Serum Characterized by Capillary Electrophoresis Coupled to ICP-MS: A Case Study of Gold Nanoparticles. *Metallomics* **2015,** *7*, 1364–1370.

Maynard, A. D.; Kuempel, E. D. Airborne Nanostructured Particles and Occupational Health. *J. Nanopart. Res.* **2005,** *7*, 587–614.

McCully, M. How Do Real Roots Work-Some New Views of Root Structure. *Plant. Physiol.* **1995,** *109*, 1–6.

Miralles, P.; Church, T. L.; Harris, A. T.; Toxicity, Uptake, and Translocation of Engineered Nanomaterials in Vascular Plants. *Environ. Sci. Technol.* **2012,** *46*, 9224–9239.

Mitrano, D. M.; Motellier, S.; Clavaguera, S.; Nowack, B. Review of Nanomaterial Aging and Transformations through the Life Cycle of Nano-Enhanced Products. *Environ. Int.* **2015,** *77*, 132–147.

Monico, L.; Janssens, K.; Alfeld, M.; Cotte, M.; Vanmeert, F.; Ryan, C. G.; Falkenberg, G.; Howard, D. L.; Brunetti, B. G.; Miliani, C. Full Spectral XANES Imaging Using the Maia Detector Array as a New Tool for the Study of the Alteration Process of Chrome Yellow Pigments in Paintings by Vincent van Gogh. *J. Anal. At. Spectrom.* **2015,** *30*, 613–626.

Moore, K. L.; Lombi, E.; Zhao, F. J.; Grovenor, C. R. M. Elemental Imaging at the Nanoscale: Nano SIMS and Complementary Techniques for Element Localisation in Plants. *Anal. Bioanal. Chem.* **2012,** *402*, 3263–3273.

Mortimer, M.; Gogos, A.; Bartolome, N.; Kahru, A.; Bucheli, T. D.; Slaveykova, V. I. Potential of Hyperspectral Imaging Microscopy for Semi-Quantitative Analysis of Nanoparticle Uptake by Protozoa. *Environ. Sci. Technol.* **2014,** *48*, 8760–8767.

N. Geldner, D.; Roppolo, B.; De Rybel, V. D.; Tendon, A.; Pfister, J.; Alassimone, J. E. M.; Vermeer, M.; Yamazaki, Y.; Stierhof, D.; Beeckman, T. A Novel Protein Family Mediates Environmental Science: Nano Critical Review Published on 31 October 2018. Downloaded on 1/20/2019 7:29:54 PM. View Article Online 56|*Environ. Sci. Nano* **2019,** *6*, 41–59. This journal is © The Royal Society of Chemistry 2019 Casparian Strip Formation in the Endodermis. *Nature* **2011,** *473*, 380–384.

Nair, R.; Varghese, S. H.; Nair, B. G.; Maekawa, T.; Yoshida, Y.; Kumar D. S. Nanoparticulate Material Delivery to Plants. *Plant Sci.* **2010,** *179*, 154–163.

Navarro, D. A.; Bisson, M. A.; Aga, D. S. Investigating Uptake of Water-Dispersible CdSe/ZnS Quantum Dot Nanoparticles by *Arabidopsis thaliana* Plants. *J. Hazard. Mater.* **2012,** *211*, 427–435.

Nguyen, T.; Yu, X. X.; Zhang, Z. M.; Liu, M. M.; Liu, X. H. Relationship between Types of Urban Forest and PM2.5 Capture at Three Growth Stages of Leaves. *J. Environ. Sci.* **2015**, *27*, 33–41.

Noori, A.; White, J. C.; Newman, L. A. Mycorrhizal Fungi Influence on Silver Uptake and Membrane Protein Gene Expression Following Silver Nanoparticle Exposure. *J. Nanopart. Res.* **2017**, *19*, 66.

Nowack, B.; Bucheli, T. D. Occurrence, Behavior and Effects of Nanoparticles in the Environment. *Environ. Pollut.* **2007**, *150*, 5–22.

Onelli, E.; Prescianotto-Baschong, C.; Caccianiga, M.; Moscatelli, A. Clathrin-Dependent and Independent Endocytic Pathways in Tobacco Protoplasts Revealed by Labelling with Charged Nanogold. *J. Exp. Bot.* **2008**, *59*, 3051–3068.

Oparka, K. J.; Turgeon, R. Sieve elements and Companion Cells-Traffic Control Centers of the Phloem. *Plant Cell.* **1999**, *11*, 739–750.

Pan, B.; Xing, B. S. Applications and Implications of Manufactured Nanoparticles in Soils: A Review. *Eur. J. Soil. Sci.* **2012**, *63*, 437–456.

Park, H. J.; Kim, S. H.; Kim, H. J.; Choi, S. H. A New Composition of Nanosized Silica-Silver for Control of Various Plant Diseases. *Plant. Pathol. J.* **2006**, *22*, 295–302.

Peng, C.; Duan, D.; Xu, C.; Chen, Y.; Sun, L.; Zhang, H.; Yuan, X.; Zheng, L.; Yang, Y.; Yang, J.; Zhen, X.; Chen, Y.; Shi, J. Translocation and biotransformation of CuO nanoparticles in rice (Oryza sativa L.) plants. *Environ. Pollut.* **2015**, *197*, 99–107.

Peng, C.; Xu, C.; Liu, Q.; Sun, L.; Luo, Y.; Shi, J. Fate and Transformation of CuO Nanoparticles in the Soil-Rice System Environmental Science: Nano Critical review Published on 31 October 2018. Downloaded on 1/20/2019 7:29:54 PM. View Article Online 58|*Environ. Sci. Nano* **2019**, *6*, 41–59 This journal is © The Royal Society of Chemistry 2019 during the Life Cycle of Rice Plants. *Environ. Sci. Technol.* **2017**, *51*, 4907–4917.

Pollard, M.; Beisson, F.; Li, Y. H.; Ohlrogge, J. B. Building Lipid Barriers: Biosynthesis of Cutin and Suberin. *Trends Plant Sci.* **2008**, *13*, 236–246.

Popp, C.; Burghardt, M.; Friedmann, A.; Riederer, M. Characterization of Hydrophilic and Lipophilic Pathways of Hedera helix L. Cuticular Membranes: Permeation of Water and Uncharged Organic Compounds. *J. Exp. Bot.* **2005**, *56*, 2797–2806.

Pozebon, D.; Scheffler, G. L.; Dressler, V. L. Recent Applications of Laser Ablation Inductively Coupled Plasma Mass Spectrometry (LA-ICP-MS) for Biological Sample Analysis: A Follow-up Review. *J. Anal. At. Spectrom.* **2017**, *32*, 890–919.

Pradas del Real, A. E.; Castillo-Michel, H.; Kaegi, R.; Sinnet, B.; Magnin, V.; Findling, N.; Villanova, J.; Carriere, M.; Santaella, C.; Fernandez-Martinez, A.; Levard, C.; Sarret, G. Fate of Ag-NPs in Sewage Sludge after Application on Agricultural Soils. *Environ. Sci. Technol.* **2016**, *50*, 1759–1768.

Reinsch, B. C.; Forsberg, B.; Penn, R. L.; Kim, C. S.; Lowry, G. V. Chemical Transformations during Aging of Zerovalent Iron Nanoparticles in the Presence of Common Groundwater Dissolved Constituents. *Environ. Sci. Technol.* **2010**, *44*, 3455–3461.

Rico, C. M.; Johnson, M. G.; Marcus, M. A. Cerium Oxide Nanoparticles Transformation at the Root-Soil Interface of Barley (*Hordeum vulgare* L.). *Environ. Sci. Nano* **2018**, *5*, 1807–1812.

Roth, G. A.; Tahiliani, S.; Neu-Baker, N. M.; Brenner, S. A. Hyperspectral Microscopy as an Analytical Tool for Nanomaterials. *Wires Nanomed. Nanobi.* **2015**, *7*, 565–579.

S. Pakrashi, N. Jain, S. Dalai, J. Jayakumar, P. T. Chandrasekaran, A. M. Raichur, N. Chandrasekaran and A. Mukherjee. In Vivo Genotoxicity Assessment of Titanium Dioxide

322 Diverse Applications of Nanotechnology in the Biological Sciences

Nanoparticles by Allium Cepa Root Tip Assay at High Exposure Concentrations. *PLoS One* **2014**, *9*, e87789.

Sabo-Attwood, T.; Unrine, J. M.; Stone, J. W.; Murphy, C. J.; Ghoshroy, S.; Blom, D.; Bertsch, P. M.; Newman, L. A. Uptake, Distribution and Toxicity of Gold Nanoparticles in Tobacco (*Nicotiana xanthi*) Seedlings. *Nanotoxicology* **2012**, *6*, 353–360.

Sankaran, S.; Mishra, A.; Ehsani, R.; Davis, C. A Review of Advanced Techniques for Detecting Plant Diseases. *Comput. Electron. Agric.* **2010**, *72*, 1–13.

Scheringer. M. Nanoecotoxicology-Environmental Risks of Nanomaterials. *Nat. Nanotechnol.* **2008**, *3*, 322–323.

Schwab, F.; Zhai, G. S.; Kern, M.; Turner, A.; Schnoor, J. L.; Wiesner, M. R. Barriers, Pathways and Processes for Uptake, Translocation and Accumulation of Nanomaterials in Plants-Critical Review. *Nanotoxicology* **2016**, *10*, 257–278.

Schymura, S.; Fricke, T.; Hildebrand, H.; Franke, K. Elucidating the Role of Dissolution in CeO_2 Nanoparticle Plant Uptake by Smart Radiolabeling. *Angew. Chem., Int. Ed.* **2017**, *56*, 7411–7414.

Seaton, A.; Tran, L.; Aitken R.; Donaldson K. Nanoparticles, Human Health Hazard and Regulation. *J. R. Soc. Interface* **2010**, *7* (Suppl 1), S119–S129.

Serag, M. S.; Kaji, M. F. N.; Gaillard, C.; Okamoto, Y.; Terasaka, K.; Jabasini, M.; Tokeshi, M.; Mizukami, H.; Bianco, A.; Baba, Y. Trafficking and Subcellular Localization of Multiwalled Carbon Nanotubes in Plant Cells. *ACS Nano* **2011**, *5*, 493–499.

Servin, A. D.; Castillo-Michel, H.; Hernandez-Viezcas, J. A.; Diaz, B. C.; Peralta-Videa J. R.; Gardea-Torresdey, J. L. Synchrotron Micro-XRE and Micro-XANES Confirmation of the Uptake and Translocation of TiO_2 nanoparticles in cucumber (*Cucumis sativus*) Plants. *Environ. Sci. Technol.* **2012**, *46*, 7637–7643.

Servin, A.; Elmer, W.; Mukherjee, A.; De la Torre-Roche, R.; Hamdi, H.; White, J. C.; Bindraban, P.; Dimkpa, C. A Review of the Use of Engineered Nanomaterials to Suppress Plant Disease and Enhance Crop Yield. *J. Nanopart. Res.* **2015**, *17*, 92.

Seshadri, B.; Bolan, N. S.; Naidu, R. Rhizosphere-Induced Heavy Metalloid Transformation in Relation to Bioavailability and Remediation. *J. Soil Sci. Plant Nutr.* **2015**, *15*, 524–548.

Shapiro, D. A.; Yu, Y. S.; Tyliszczak, T.; Cabana, J.; Celestre, R.; Chao, W. L.; Kaznatcheev, K.; Kilcoyne, A. L. D.; Maia, F.; Marchesini, S.; Meng, Y. S.; Warwick, T.; Yang, L. L.; Padmore, H. A. Chemical Composition Mapping with Nanometer Resolution by Soft X-ray Microscopy. *Nat. Photonics.* **2014**, *8*, 765–769.

Sharma, V. K.; Filip, J.; Zboril, R.; Varma, R. S. Natural Inorganic Nanoparticles-Formation, Fate, and Toxicity in the Environment. *Chem. Soc. Rev.* **2015**, *44*, 8410–8423.

Shi, J.; Peng, C.; Yang, Y.; Yang, J.; Zhang, H.; Yuan, X.; Chen, Y.; Hu, T. Phytotoxicity and Accumulation of Copper Oxide Nanoparticles to the Cu-Tolerant Plant Elsholtzia Splendens. *Nanotoxicology* **2014**, *8*, 179–188.

Slomberg, D. L.; Schoenfisch, M. H. Silica Nanoparticle Phytotoxicity to *Arabidopsis thaliana*. *Environ. Sci. Technol.* **2012**, *46*, 10247–10254.

Spielman-Sun, E.; Lombi, E.; Donner, E.; Avellan, A.; Etschmann, B.; Howard, D.; Lowry, G. V. Temporal Evolution of Copper Distribution and Speciation in Roots of *Triticum aestivum* Exposed to CuO, $CuOH_2$, and CuS nanoparticles. *Environ. Sci. Technol.* **2018**, *52*, 9777–9784.

Spielman-Sun, E.; Lombi, E.; Donner, E.; Howard, D.; Unrine, J. M.; Lowry, G. V. Impact of Surface Charge on Cerium Oxide Nanoparticle Uptake and Translocation by Wheat (*Triticum aestivum*). *Environ. Sci. Technol.* **2017**, *51*, 7361–7368.

Toxicity of Nanoparticles in Plants

Stegemeier, J. P.; Schwab, F.; Colman, B. P.; Webb, S. M.; Newville, M.; Lanzirotti, A.; Winkler, C.; Wiesner, M. R.; Lowry, G. V. Speciation Matters: Bioavailability of Silver and Silver Sulfide Nanoparticles to Alfalfa (*Medicago sativa*). *Environ. Sci. Technol.* **2015**, *49*, 8451–8460.

Steudle, E.; Peterson, C. A. How Does Water Get through Roots? *J. Exp. Bot.* **1998**, *49*, 775–788.

Tan, Z. Q.; Liu, J. F.; Guo, X. R.; Yin, Y. G.; Byeon, S. K.; Moon, M. H.; Jiang, G. B. Toward full Spectrum Speciation of Silver Nanoparticles and Ionic Silver by On-line Coupling of Hollow Fiber Flow Field-Flow Fractionation and Minicolumn Concentration with Multiple Detectors. *Anal. Chem.* **2015**, *87*, 8441–8447.

Taylor, A. F.; Rylott, E. L.; Anderson, C. W.; Bruce, N. C. Investigating the Toxicity, Uptake, Nanoparticle Formation and Genetic Response of Plants to Gold. *PLoS One* **2014**, *9*, e93793.

Thygesen, L. G.; Lokke, M. M.; Micklander, E.; Engelsen, S. B. Vibrational Microspectroscopy of Food. Raman vs. FT-IR. *Trends Food Sci. Technol.* **2003**, *14*, 50–57.

Tiede, K.; Boxall, A. B. A.; Tiede, D.; Tear, S. P.; David, H.; Lewis, J. A Robust Size-Characterisation Methodology for Studying Nanoparticle Behaviour in 'Real' Environmental Samples, Using Hydrodynamic Chromatography Coupled to ICP-MS. *J. Anal. At. Spectrom.* **2009**, *24*, 964–972.

Torney, F.; Trewyn, B. G.; Lin, V. S. Y.; Wang, K. Mesoporous Silica Nanoparticles Deliver DNA and Chemicals into Plants. *Nat. Nanotechnol.* **2007**, *2*, 295–300.

Tripathi, D. K.; Tripathi, A.; Singh Shweta, S.; Singh, Y.; Vishwakarma, K.; Yadav, G.; Sharma, S.; Singh, V. K.; Mishra, R. K.; Upadhyay, R. G.; Dubey, N. K.; Lee, Y.; Chauhan, D. K. Uptake, Accumulation and Toxicity of Silver Nanoparticle in Autotrophic Plants, and Heterotrophic Microbes: A Concentric Review. *Front. Microbiol.* **2017**, *8*, 07.

Uzu, G.; Sobanska, S.; Sarret, G.; Munoz, M.; Dumat, C. Foliar Lead Uptake by Lettuce Exposed to Atmospheric Fallouts. *Environ. Sci. Technol.* **2010**, *44*, 1036–1042.

Vorholt, J. A. Microbial Life in the Phyllosphere. *Nat. Rev. Microbiol.* **2012**, *10*, 828–840.

Wagner, S.; Gondikas, A.; Neubauer, E.; Hofmann, T.; von der Kammer, F. Spot the Difference: Engineered and Natural Nanoparticles in the Environment-Release, Behavior, and Fate. *Angew. Chem., Int. Ed.* **2014**, *53*, 12398–12419.

Wang, F.; Liu, X.; Shi, Z.; Tong, R.; Adams, C. A.; Shi, X. Arbuscular Mycorrhizae Alleviate Negative Effects of Zinc Oxide Nanoparticle and Zinc Accumulation in Maize Plants—a Soil Microcosm Experiment. *Chemosphere* **2016**, *147*, 88–97.

Wang, P.; Lombi, E.; Sun, S. K.; Scheckel, K. G.; Malysheva, A.; McKenna, B. A.; Menzies, N. W.; Zhao, F. J.; Kopittke, P. M. Characterizing the Uptake, Accumulation and Toxicity of Silver Sulfide Nanoparticles in Plants. *Environ. Sci. Nano.* **2017**, *4*, 448–460.

Wang, P.; Lombi, E.; Zhao, F. J.; Kopittke, P. M. Nanotechnology: A New Opportunity in Plant Sciences. *Trends Plant Sci.* **2016**, *21*, 699–712.

Wang, P.; Menzies, N. W.; Lombi, E.; McKenna, B. A.; Johannessen, B.; Glover, C. J.; Kappen, P.; Kopittke, P. M. Fate of ZnO Nanoparticles in Soils and Cowpea (*Vigna unguiculata*). *Environ. Sci. Technol.* **2013**, *47*, 13822–13830.

Wang, S.; Kurepa J.; Smalle, J. A. Ultra-small TiO_2 Nanoparticles Disrupt Microtubular Networks in Arabidopsis Thaliana. *Plant Cell Environ.* **2011**, *34*, 811–820.

Wang, S.; Lv, J.; Ma, J.; Zhang, S. Cellular Internalization and Intracellular Biotransformation of Silver Nanoparticles in *Chlamydomonas reinhardtii*. *Nanotoxicology* **2016**, *10*, 1129–1135.

324 Diverse Applications of Nanotechnology in the Biological Sciences

Wang, W. N.; Tarafdar, J. C.; Biswas, P. Nanoparticle Synthesis and Delivery by an Aerosol Route for Watermelon Plant Foliar Uptake. *J. Nanopart. Res.* **2013**, *15*, 1417.

Wang, Z.; Xie, X.; Zhao, J.; Liu, X.; Feng, W.; White, J. C.; Xing, B. Xylem- and Phloem-Based Transport of CuO Nanoparticles in Maize (*Zea mays* L.). *Environ. Sci. Technol.* **2012**, *46*, 4434–4441.

Wei, G. T.; Liu, F. K.; Wang, C. R. Shape Separation of Nanometer Gold Particles by Size-Exclusion Chromatography. *Anal. Chem.* **1999**, *71*, 2085–2091.

Wiesner, M. R.; Lowry, G. V.; Jones, K. L.; Hochella, M. F.; Di Giulio, R. T.; Casman, E.; Bernhardt, E. S. Decreasing Uncertainties in Assessing Environmental Exposure, Risk, and Ecological Implications of Nanomaterials. *Environ. Sci. Technol.* **2009**, *43*, 6458–6462.

Wild, E.; Jones, K. C. Novel Method for the Direct Visualization of in Vivo Nanomaterials and Chemical Interactions in Plants. *Environ. Sci. Technol.* **2009**, *43*, 5290–5294.

Yin, L. Y.; Cheng, Y. W.; Espinasse, B.; Colman, B. P.; Auffan, M.; Wiesner, M.; Rose, J.; Liu, J.; Bernhardt, E. S. More Than the Ions: The Effects of Silver Nanoparticles on *Lolium multiflorum*. *Environ. Sci. Technol.* **2011**, *45*, 2360–2367.

Yin, Y.; Tan, Z.; Hu, L.; Yu, S.; Liu, J.; Jiang, G. Isotope Tracers to Study the Environmental Fate and Bioaccumulation of Metal-Containing Engineered Nanoparticles: Techniques and Applications. *Chem. Rev.* **2017**, *117*, 4462–4487.

Yu, S. J.; Yin, Y. G.; Zhou, X. X.; Dong, L. J.; Liu, J. F. Transformation Kinetics of Silver Nanoparticles and Silver Ions in Aquatic Environments Revealed by Double Stable Isotope Labeling. *Environ. Sci. Nano.* **2016**, *3*, 883–893.

Zambryski, P.; Crawford, K. Plasmodesmata: Gatekeepers for Cell-to-Cell Transport of Developmental Signals in Plants. *Annu. Rev. Cell Dev. Biol.* **2000**, *16*, 393–421.

Zhai, G. S.; Walters, K. S.; Peate, D. W.; Alvarez, P. J. J.; Schnoor, J. L. Transport of Gold Nanoparticles through Plasmodesmata and Precipitation of Gold Ions in Woody Poplar. *Environ. Sci. Technol. Lett.* **2014**, *1*, 146–151.

Zhalnina, K.; Louie, K. B.; Hao, Z.; Mansoori, N.; da Rocha, U. N.; Shi, S.; Cho, H.; Karaoz, U.; Loque, D.; Bowen, B. P.; Firestone, M. K.; Northen, T. R.; Brodie, E. L. Dynamic Root Exudate Chemistry and Microbial Substrate Preferences Drive Patterns in Rhizosphere Microbial Community Assembly. *Nat. Microbiol.* **2018**, *3*, 470–480.

Zhang, P.; Ma, Y. H.; Zhang, Z. Y.; He, X.; Guo, Z.; Tai, R. Z.; Ding, Y. Y.; Zhao, Y. L.; Chai, Z. F. Comparative Toxicity of Nanoparticulate/Bulk Yb_2O_3 and $YbCl_3$ to Cucumber (*Cucumis sativus*). *Environ. Sci. Technol.* **2012**, *46*, 1834–1841.

Zhang, P.; Ma, Y. H.; Zhang, Z. Y.; He, X.; Zhang, J.; Guo, Z.; Tai, R. Z.; Zhao, Y. L.; Chai, Z. F. Biotransformation of Ceria Nanoparticles in Cucumber Plants. *ACS Nano* **2012**, *6*, 9943–9950.

Zhang, P.; Ma, Y.; Zhang, Z.; He, X.; Guo, Z.; Tai, R.; Ding, Y.; Zhao, Y.; Chai, Z. Comparative Toxicity of Nanoparticulate/Bulk Yb_2O_3 and $YbCl_3$ to Cucumber (*Cucumis sativus*). *Environ. Sci. Technol.* **2012**, *46*, 1834–1841.

Zhao, L. J.; Hu, Q. R.; Huang, Y. X.; Fulton, A. N.; Hannah- Bick, C.; Adeleye, A. S.; Keller, A. A. Activation of Antioxidant and Detoxification Gene Expression in Cucumber Plants Exposed to a $CuOH_2$ Nanopesticide. *Environ. Sci. Nano* **2017**, *4*, 1750–1760.

Zhao, L. J.; Huang, Y. X.; Zhou, H. J.; Adeleye, A. S.; Wang, H. T.; Ortiz, C.; Mazer, S. J.; Keller, A. A. GC-TOF-MS Based Metabolomics and ICP-MS Based Metallomics of Cucumber (*Cucumis sativus*) Fruits Reveal Alteration of Metabolites Profile and Biological Pathway Disruption Induced by Nano Copper. *Environ. Sci. Nano* **2016a**, *3*, 1114–1123.

Zhao, L. J.; Huang, Y. X.; Hu, J.; Zhou, H. J.; Adeleye, A. S.; Keller, A. A. H-1 NMR and GC-MS Based Metabolomics Reveal Defense and Detoxification Mechanism of Cucumber Plant under Nano-Cu Stress. *Environ. Sci. Technol.* **2016b**, *50*, 2000–2010.

Zhao, L.; Ortiz, C.; Adeleye, A. S.; Hu, Q.; Zhou, H.; Huang, Y.; Keller, A. A. Metabolomics to Detect Response of Lettuce (*Lactuca sativa*) to CuOH2 Nanopesticides: Oxidative Stress Response and Detoxification Mechanisms. *Environ. Sci. Technol.* **2016c**, *50*, 9697–9707.

Zhao, L. J.; Peng, B.; Hernandez-Viezcas, J. A.; Rico, C.; Sun, Y. P.; Peralta-Videa, J. R.; Tang, X. L.; Niu, G. H.; Jin, L. X.; Varela-Ramirez, A.; Zhang, J. Y.; Gardea-Torresdey, J. L. Stress Response and Tolerance of Zea Mays to CeO_2 Nanoparticles: Cross Talk among H_2O_2, Heat Shock Protein, and Lipid Peroxidation. *ACS Nano* **2012**, *6*, 9615–9622.

Zhao, L. J.; Peralta-Videa, J. R.; Varela-Ramirez, A.; Castillo-Michel, H.; Li, C. Q.; Zhang, J. Y.; Aguilera, R. J.; Keller, A. A.; Gardea-Torresdey, J. L. Effect of Surface Coating and Organic Matter on the Uptake of CeO_2 NPs by Corn Plants Grown in Soil: Insight into the Uptake Mechanism. *J. Hazard. Mater.* **2012**, *225*, 131–138.

Zhou, D. M.; Jin, S. Y.; Li, L. Z.; Wang, Y.; Weng, N. Y. Quantifying the Adsorption and Uptake of CuO Nanoparticles by Wheat Root Based on Chemical Extractions. *J. Environ. Sci.* **2011**, *23*, 1852–1857.

Zhou, X. X.; Liu, J. F.; Geng, F. L. Determination of Metal Oxide Nanoparticles and Their Ionic Counterparts in Environmental Waters by Size Exclusion Chromatography Coupled to ICP-MS. *NanoImpact.* **2016**, *1*, 13–20.

Zhu, Y.; Zhang, J. C.; Li, A. G.; Zhang, Y. Q.; Fan, C. H. Synchrotron-Based X-Ray Microscopy for Sub-100 nm Resolution Cell Imaging. *Curr. Opin. Chem. Biol.* **2017**, *39*, 11–16.

CHAPTER 12

PLANT-BASED NANOPARTICLES AND THEIR APPLICATIONS

ISHANI CHAKRABARTTY

Department of Science, P. A. First Grade College (affiliated to Mangalore University, Mangalore), Nadupadav, Mangalore, Karnataka, India

ABSTRACT

Nanotechnology and biotechnology have made remarkable advances in the recent times; the two fields have been made since time immemorial to reap huge benefits in numerous sectors like medicine, industry, human health, and welfare. A fast, sustainable and nontoxic process for the synthesis of metallic nanoparticles (NPs) is of vital importance in the field of nanobiotechnology. Synthesis of metallic NPs, using biological systems, particularly plants, is an emerging field. For the production of metallic NPs, extracts from plants, tissues, extrudes, and other plant parts have been used widely. The benefit of using plant-based NPs can be exponentially scaled up if such particles are produced extracellularly and their size, shape, and dispersion be controlled. "Green" synthesis of NPs is highly very cost-effective and therefore, it can be proposed to be a suitable prospect, if scaled up suitably, for the production of NPs on an industrial scale. In every field of science—from agriculture, to health as well as to solving the global energy crisis—plant-based NPs have defined a niche to prove their applicability. In this view, the current chapter attempts to highlight the numerous uses of "green" nanomaterials in various sectors.

12.1 INTRODUCTION

Nanobiotechnology is a wide branch of science; principles and techniques are applied in this branch, in the nanoscale range, in order to understand the creation and functioning of life-forms and nonliving systems, and then recreate its efficient its modified version. It uses basic principles that govern biological systems and utilizes tools to generate a completely new device, with increased efficiency (Mashwani et al., 2016). Nanotechnology has been growing exponentially to keep pace with the advancement of technology. "Green" and eco-friendly techniques in chemistry are growing progressively and are highly demanding worldwide due to the associated environmental and ecological concerns (Devatha and Thalla, 2018). A very significant aspect of current nanotechnology research is the development of suitable processes (that can be relied upon for a long duration of time) for synthesis of metallic nanoparticles (NPs). Ideally, metal cannot function as a catalyst in the bulk form; however, due to their high surface-area-to-volume ratio and their interface-dominated properties, nanosized particles behave as excellent catalysts, which no doubt differ remarkably from the bulk-sized material (Shen et al., 2013; Usman et al., 2013). Noble NPs are very significant as they have gained sufficient importance over the past few years because they have been and continue to be used in fields and areas that were previously unimaginable like medicine, chemistry, biology, in addition to physics and material sciences (Conde et al., 2012; Klebowski et al., 2018). Due to the high surface area: volume ratio and enhanced surface activity, nanomaterials are predominantly harmful. They can very easily porate/perforate and move into biological systems through the cell walls and membranes and can remain lodged inside long enough to perform their desired activities. They can delay or prolong the toxicity effects and can cause many such effects that are normally hard to predict; even neurotoxicity caused by nanomaterials is not unheard of (Sun et al., 2007; Navarro et al., 2008; Misra et al., 2012; Schwirn et al., 2014; Gao and Jiang, 2017). Also, the synthesis of metallic nanomaterials by physical and chemical methods that use extremely reactive and toxic reducing agents is highly hazardous both to human health and the surroundings in which he/she dwells; these methods are also very expensive (Varma, 2012; Rajasekhar and Kanchi, 2018). Apparently, due to the extensive focus on the application and assets of nanomaterials, investigations on their undesirable properties or associated health hazards are often delayed (Pallas et al., 2018). Thus, compliance with basic principles of chemistry and engineering, incorporated with "greeness," is essential to ensure effective applications of nanomaterials with minimal or no impact on the surroundings (Eckelman et al., 2008; Salame et al., 2018).

Ideally, nanomaterials are synthesized by two different methods: (1) top-down in which nanomaterials are synthesized by a diverse range of techniques like milling, sputtering, etching, etc. and (2) bottom-up in which nanomaterials are "grown" from simple forms using different methods

Plant-Based Nanoparticles and Their Applications

like condensation, precipitation, pyrolysis, etc. "Green" synthesis of nanomaterials belongs to the latter category. In this method, nanomaterials are prepared through control, cleanup regulation so as to promote more environmental applicability and overcome the existing limitations, associated with toxic nanomaterial synthesis (Singh et al., 2018). "Green" synthesis is highly significant because it helps in the reduction of unwanted or harmful by-products by using suitable organic solvent systems.

12.2 PLANT-BASED NANOMATERIALS

Nature, which is endowed with a huge diversity of plants and microorganisms, is what scientists have turned to for "green" synthesis of nanomaterials. After all, the long-term association of man with nature always promises low toxicity. Plants and plant parts are bestowed with a wealth of chemicals, known as secondary metabolites, such as flavonoids, alkaloids, saponins, and tannins (Ghosh and Rangan, 2013; Singh et al., 2016; Basak et al., 2017; Chakrabartty et al., 2019); these secondary metabolites are first synthesized and then produced by the plants to adapt themselves to the ever-changing environment or as a mechanism or defense against insects, pests, hungry animals, etc. (Dewick, 2002; Colegate and Molyneux, 2008). Green synthesis of nanomaterials, also called biogenic nanomaterials by using plants or plant extracts, is considered to be a rather simple and easy process relative to microorganism-mediated synthesis; effective phytochemicals, like ketones, phenols, aldehydes, flavones, carboxylic acids, ascorbic acids, etc., are present in all parts of plant biodiversity, predominantly leaves, and these chemicals act as reducing agents to convert metallic salts into metallic NPs (Doble et al., 2007). Not only that, primary metabolites like glucose, fructose, sucrose, etc. that are present in plant extracts are also responsible for the formation of NPs—fructose-mediated synthesis of NPs often leads to the formation of monodispersed NPs (Panigrahi et al., 2004; Shan et al., 2008). This synthesis, like all other methods, is also based on a number of physical parameters like pH, temperature, pressure, solvent used, etc. Green synthesis is a one-pot, single stop, synthesis method using bioreduction that is 100% eco-friendly and requires relatively low amount of energy to initiate the process; this method is also highly cost-efficient (Dahoumane et al., 2016; Wadhwani et al., 2016; Singh et al., 2018). Solvent system is one of the most important parameters that need to be considered in the synthesis process of any material; particularly biological. According to Shanker et al. (2016), the best method is always using "no solvent" but if the use of a solvent is indispensible, water is always the best choice—this is in accordance with the synthesis of ZnO NPs by a solvent-free and nontoxic method (Saravanan et al., 2012).

Though toxicity of "green" NPs is many-fold lower than the chemically synthesized ones, stability can be something of an issue—this is mainly because these NPs tend to form aggregations or dissolute in the solution. Stability of NPs is often affected by surface complexation processes—these can be regulated by modifying the particle size, using appropriate capping agents and proper techniques for functionalization (Tejamaya et al., 2012; Sharma et al., 2014). The use of biocompatible stabilizing agents like biodegradable polymers, etc. has opened newer and "greener" avenues for nanoparticle surface engineering—together with the addition of suitable functional groups; these agents not only increase the stability but also lower the toxicity of bio-based NPs (Banerjee et al., 2014).

12.3 APPLICATIONS

12.3.1 ANTIMICROBIAL APPLICATIONS

Certain precious metals like gold (Au), silver (Ag), platinum (Pt), palladium (Pd), etc. are known to have very strong antimicrobial properties; however, high amount of these metals can exert toxic effects in the human body or any other host organism (Mubarakali et al., 2011; Tahir et al., 2017; Rajendran et al., 2020). As discussed above, metal NPs have managed to attain special focus due to their distinctive properties, like high conductivity, amazing chemical stability, strong catalytic property, and potent antibacterial activity. Silver has a long history of being used in every household for wound washing/cleaning as well as sanitization; this metal is known to bind with S=S bonds of the glycoprotein/protein backbone the membrane of microorganisms and other pathogens. Both AgNPs and Ag^{2+} can alter and modify the 3-D structure of these membrane proteins; since the S=S bonds are interfered with at the membrane level, further downstream cell signaling is prevented and, thus, metabolic functions in the single-celled/minuscule microorganisms remain blocked/impaired (Sadeghi and Gholamhoseinpoor, 2015; Ahmed et al., 2016). A lot of reports are available on the antimicrobial effects of biosynthesized or rather plant synthesized metallic NPs—these particles are effective in controlling oxidative stress, toxicity

Plant-Based Nanoparticles and Their Applications 331

at the gene level, and changes related to apoptosis and necrosis (Kim et al., 2007; Arab et al., 2014). A large number of plants and their parts are used for the synthesis of NPs (usually by using expensive metals) to be used as antimicrobial agents; most of these plants are known to have some pharmacological use among ethnic tribes as traditional knowledge viz. *Parthenium hysterophorus, Alpinia nigra, Acalypha indica, Avena sativa, Boerhaavia diffusa, Vitis vinifera, Psidium guajava, Cinnamomum camphora, Medicago sativa, Pelargonium graveolens, Hovenia dulcis, Azadirachta indica, Tamarindus indica, Emblica officinalis, Aloe vera, Coriandrum sativum, Carica papaya, Triticum vulgare, Acanthella elongata, Sesuvium portulacastrum,* and AuNPs are also synthesized by biological products like honey. Out of all the plant parts used, leaves are the mostly used (Krishnaraj et al., 2010; Al Akeel et al., 2014; Basavegowda et al., 2014; Gupta et al., 2014; Kumar et al., 2014).

Plant-based Ag and AuNPs are the most widely used and synthesized NPs against a large number of microbes like harmful bacteria, fungi, protozoans, and other parasites. These NPs are found to exhibit stronger antibacterial potential against Gram-negative bacteria than Gram-positive bacteria, for example, *Escherichia coli* and *Klebsiella pneumonia* would be more susceptible to plant-based NPs unlike *Staphylococcus aureus*; this may be attributed to the fact that the thin lipoprotein layer in the cell wall of Gram-negative bacteria—metallic NPs can break this thin cell wall, protrude through the cell membrane, and then interact with the genetic machinery in the DNA or the functional machinery of the proteins. In total contrast, the cell wall of Gram-positive bacteria contains a thick peptidoglycan layer, made up of polysaccharide chains, arranged in a linear fashion and crosslinked by peptides made up of very few amino acids; this kind of arrangement presents a very robust and rigid shell or structure that makes it difficult for NPs to break and penetrate through it (Krishnaraj et al., 2010; Prabhu and Poulose, 2012; Khan et al., 2019). Not only that are these plant-based NPs used as antibacterial agents, but also they are also used as antifungal agents against a number of pathogenic fungi like *Candida albicans* and other dermatophytes. Like bacteria, the membrane potential of the cell wall of the fungi are disrupted by the metallic NPs—it may be hypothesized that NPs perturb the membrane lipid bilayers, that lead to the leakage of ions and other materials through the pores formed on the fungal membrane by the NPs and thus, distorting the electrical potential of the membrane. Due to the action of these NPs, the cell cycle, growth as well as budding mechanism of the fungus also get disrupted (Kim et al., 2008, 2009). The phenolic compounds extracted from spice plant, *Rosemary officinalis*, is known to inhibit HSV-1 infection in monkeys; AgNPs produced with the extract of this plant exhibited anitiviral potential against herpes simplex virus-1 and 2 and HPIV-3 viruses. These small-sized plant-based NPs attach to the virus, preventing its attachment with the host cells, and thus preventing viral replication (Elechiguerra et al., 2005; Gaikwad et al., 2013).

12.3.2 AGRICULTURAL APPLICATIONS

As already stated, NPs find a wide range of applicability and use in every area possible. Currently, researchers are employing plant-based metallic NPs, particularly Au and AgNPs to enhance agricultural output and enrich nutrient uptake by plants. An increase in the germination potential and root: shoot ratio was observed in plants like *Eruca sativa, Silybum marianum,* etc., when treated with nonmetallic NPs and NPs prepared from bimetallic alloys (Gomes et al., 2012; Zaka et al., 2016). It was also hypothesized that silver NPs can cause pore in the roots of plants that may be responsible for the influx of essential nutrients from the soil, resulting in rapid and healthy growth of plants (Hussain et al., 2017). However, it was also reported that while lower doses of plant-based NPs can extensively promote seed germination and growth of plants like *Arabidopsis thaliana* and *Brassica* sp., higher doses of the same had and inhibitory effect (Mishra et al., 2019). In 2009, Tang et al. had reported the production of sensor-based "green" NPs (mixed with plant extracts) that can efficiently detect mycotoxins in fruits, vegetables, and other food products; thus, plant-based NPs are the most suitable candidate for preventing the damage and destruction of food products because of strong antimicrobial properties (Tang et al., 2009). Silver NPs synthesized by using the extracts of *Fatsia japonica* are used for the protection and promotion of longer shelf life of *Citrus* fruits (Zhang et al., 2017).

12.3.3 PHARMACOLOGICAL APPLICATIONS

NPs, particularly noble metal NPs, are considered as the most versatile agents because they have a wide range of biomedical applications; they are proposed and known to be the nontoxic carriers and delivery agents of drug and genes (Garcia-Bennett et al., 2011; Bhattacharyya et al., 2012). Because of the "nontoxic" nature of plant-based NPs, they are used in a large number of pharmaceutical applications like gene therapy, drug delivery, imaging, radiotherapy, tumor cell targeting assays, gene silencing, etc. in the case of cancer treatment. However, the biocompatibility of these NPs needs to be tested well in advance in order to eliminate any damage to healthy tissue, guard against alliterations that may occur in the genetic make-up and function of the proteins while killing those specific cells that are either abnormal or infected (Bellucci, 2009; Conde et al., 2012). An interesting point to be kept in mind is that the presence of certain phenolic compounds in the plant extracts is extremely beneficial and necessary for the human body; hence, metal NPs synthesized for pharmaceutical and pharmacological applications (using plants) can be very useful not only from the treatment point of view but also for immunology (Yadi et al., 2018).

Plants have been an essential source for the synthesis and development of antimalarial drugs. Metallic NPs, synthesized by using extracts from such plants, have been very useful to suppress the malarial infections. Au and AgNPs that were synthesized by using the plant, *Sphaeranthus amaranthoides*, were able to lower a-amylase and other sugars in animal models, having induced diabetes; later, it was found that certain components which inhibit a-amylase are present in organic extract of *S. amaranthoides*; higher amount of these compounds are obtained by using ethanol (Barathmanikanth et al., 2010; Kuppusamy et al., 2016). Metallic NPs are also known to have very strong antioxidant properties; in addition to scavenging free radicals, these NPs are able to "devour" certain radicals that are produced in the cells (like NO, ABTS) as a part of cellular processes (Kumar and Yadav, 2009). Well-known antioxidants like catalase, superoxide dismutase (SOD) and glutathione peroxidase (GSHPx), malondialdehyde (MDA) have the ability to protect cells from damage and injuries. Biosynthesized AuNPs using *Azolla microphylla* have been reported to lower the levels of free radicals present, scavenge them, and increase the inherent basal levels of antioxidants mentioned above in fishes; it may be assumed that these antioxidants tend to stabilize the cell membrane, thus, repairing and further preventing any damage caused to the hepatocytes of fishes. Thus, plant-based NPs have hepatoprotective activity; however, no report is available till date on the effect of plant-based metallic NPs on human hepatocytes (Kunjiappan et al., 2015; Khan et al., 2019).

12.3.4 INDUSTRIAL AND COMMERCIAL APPLICATIONS

It was previously discussed that plant-based metallic NPs have a very strong free radical scavenging and antioxidant activity. Due to this such NPs are considered to be of much use, with adequate success, in industries where they can be used to formulate various sunscreen and anti-aging creams (Royer et al., 2013; Khan et al., 2019). Researchers are currently working on the development of plant-based nanomaterials as biosensors or other kinds of "sensing materials." Nanoproducts have recently been reported to be incorporated in the indispensible items of one's every-day life. Various nontoxic products, impregnated with plant-based NPs, are available in commercial market that are known to function with adequate seamlessness such as dental fillings, water coolers (armed with purifiers), creams, bone cements, and homemade products. For instance, Ag, Si, and Pt NPs have various applications in personal care and cosmetics; they are used as the "core" ingredient in number of commercial products such as sunscreens, anti-aging creams, toothpastes, mouthwash, hair care products, and perfumes (Kumar and Yadav, 2009; Sanzari et al., 2019). Apart from the use of Si nanomaterials as commercial products, these nanomaterials are modified and used as excellent insecticide, pest control agents as well as in a variety of nonagricultural applications.

12.3.5 ENERGY APPLICATIONS

The quality of human life largely depends on the generation, utilization, and expenditure of energy; at present, this is in jeopardy unless renewable sources of energy are considerably developed in the near future. The increasing demand for a clean source of energy (that is sustainable) is a matter of huge concern globally; the use of conventional fuels comes with a baggage of disturbances to the environment. Hence, it is imperative to develop suitable technologies that are concerned with efficient energy conversion techniques like thermoelectrics. Thermoelectric devices have the ability to scavenge a lot of waste heat generated from industrial applications and vehicle exhaust and recover it in an environment-friendly and cost-effective manner (Huang et al., 2016). Though the energy consumption of the world is globally dominated by fossil sources, the share of renewable sources has drastically increased while coal, gas, and oil have steadily decreased (Locatelli, 2009). At present, a lot of emphasis is laid on the use and utilization of solar and hydro energy sources as a lot of awareness and consciousness is being created for a clean, "green" and sustainable environment (Gratzel, 2005). In the search for "green" and "clean" energy, an important role is played by nanomaterials, synthesized by "green" or plant-based techniques. Being nontoxic, these materials are highly advantageous and can possibly provide a solution to the ever-increasing global energy crisis.

Researchers agree that the safest possible future, using advanced nanotechnology in a sustainable world, can be targeted only with the help of biological chemistry. Chemistry, using biology, comprises formulating chemical processes and generating products in a way that reduces or eliminates hazardous substances from the very beginning to the very end of the generated product's life cycle. Catalysis is one among the basic 12 principles that govern the concept of green chemistry; nanocatalysis has become an emerging field in science due to its high activity, selectivity, and productivity (Singh and Tandon, 2014). In the past decade, nanomaterials, armed with high catalytic properties, have been designed and constructed as highly efficient light-harvesting devices for solar energy conversion in three ways (Kamat, 2007):

- Recreating the process photosynthesis by using clusters and assemblies of donor–acceptor molecules
- Photocatalysis assisted by semiconductor in order to generate clean fuels like H_2
- Producing solar cells that are based on nanostructured semiconductor

Nanotechnologies provide a much-needed improvement for the development of both nonrenewable sources of energy (e.g., fossil fuels and nuclear fuels) and renewable energy sources like geothermal energy, sun, wind, water, tides, or biomass. The overall yield of energy from the conversion of chemical energy by using fuel cells can be increased to a great extent by nanosized electrodes, micro-membranes as well as catalysts; application possibilities of these nanostructured energy devices in various electronics, vehicles and in the construction purposes can have tremendous economic implications. Perovskite nanomaterials, synthesized by "green" approach using genetically modified viruses, exhibited very high photocatalytic and photovoltaic performance; these were also able to extend their application to solar energy conversion (Nuraje et al., 2012). In addition to solar cells, plant-based nanotechnology has a tremendous impact on conventional fuel cells—being able to convert chemical energy, efficiently, into electricity (Rajasekhar and Kanchi, 2018). Carbon nanomaterials and nanofibers are much more energy intensive than alumina; nanostructured carbon materials play a significant role in the innovative rise of sustainable and clean alternatives for energy technologies; they are used as electrodes in supercapacitors, lithium-ion batteries, etc. and for hydrogen storage (Candelaria et al., 2012; Zhang et al., 2013). Also, TiO_2-RGO nanocomposites that were synthesized in a "green" way, by using plants, without the use of toxic reducing agents—these composites, that exhibited enhanced photocatalytical properties, are anticipated to be used in lithium-ion batteries (Shah et al., 2012). To achieve energy supply that is sustainable and is optimized to be at-par with the energy sources that are currently in use, it is necessary to promote efficient use of energy and to avoid its unnecessary consumption—such measures needs to be applicable to all households, private, and public sectors, including different industries. Interestingly, nanotechnologies provide a myriad of useful approaches to reduce consumption of energy and, hence, lead to energy saving (Kamat, 2007).

However, nanomanufacturing is not devoid of drawbacks—these methods are often found to have environmental and human health impacts—when nanoproducts are assessed and evaluated for their "greenness," the perspective of its associated impacts needs to be kept in mind. Making well-thought and informed choices/decisions during the designing stage of the product by incorporating the concept of "life cycle", in combination with life cycle analysis and risk assessment, utilizing manufacturing processes that are sustainable, and using suitable alternatives with extensive green chemistry approach can be viewed as possible solutions (Dhingra et al., 2010). Regardless of the path of synthesis take, photovoltaics with "green" nanomaterials have shown a path and assured a possible role in meeting the huge energy demand, currently in existence on Earth.

KEYWORDS

- **agriculture**
- **antimicrobial applications**
- **commercial use**
- **energy crisis**
- **plant-based nanoparticles**
- **pharmacological aspect**

REFERENCES

Ahmed, S.; Ahmad, M.; Swami, B. L.; Ikram, S. A Review on Plants Extract Mediated Synthesis of Silver Nanoparticles for Antimicrobial Applications : A Green Expertise. *J. Adv. Res.* **2016,** *7*, 17–28. doi: 10.1016/j.jare.2015.02.007

Al Akeel, R.; Al-sheikh, Y.; Mateen, A. et al. Evaluation of Antibacterial Activity of Crude Protein Extracts from Seeds of Six Different Medical Plants against Standard Bacterial Strains. *Saudi J. Biol. Sci.* **2014,** *21*, 147–151. doi: 10.1016/j.sjbs.2013.09.003

Arab, M. M.; Yadollahi, A.; Hosseini-Mazinani, M.; Bagheri, S. Effects of Antimicrobial Activity of Silver Nanoparticles on in Vitro Establishment of G N15 (Hybrid of Almond Peach) Rootstock. *J. Genet. Eng. Biotechnol.* **2014,** *12,* 103–110. doi: 10.1016/j.jgeb.2014.10.002

Banerjee, P.; Satapathy, M.; Mukhopahayay, A.; Das, P. Leaf Extract Mediated Green Synthesis of Silver Nanoparticles from Widely Available Indian Plants: Synthesis, Characterization, Antimicrobial Property and Toxicity Analysis. *Bioresour. Bioprocessinh* **2014,** *1,* 1–10. doi: 10.1186/s40643-014-0003-y

Barathmanikanth, S.; Kalishwaralal, K.; Sriram, M. et al. Anti-Oxidant Effect of Gold Nanoparticles Restrains Hyperglycemic Conditions in Diabetic Mice. *J. Nanobiotechnol.* **2010,** *8,* 1–15.

Basak, S.; Kesari, V.; Ramesh, A. M. et al. Assessment of Genetic Variation among Nineteen Turmeric Cultivars of Northeast India: Nuclear DNA Content and Molecular Marker Approach. *Acta Physiol. Plant* **2017,** *39,* 45. doi: 10.1007/s11738-016-2341-1

Basavegowda, N.; Idhayadhulla, A.; Lee, Y. R. Phyto-Synthesis of Gold Nanoparticles Using Fruit Extract of *Hovenia dulcis* and Their Biological Activities. *Ind. Crop Prod.* **2014,** *52,* 745–751. doi: 10.1016/j.indcrop.2013.12.006

Bellucci, S. *Nanoparticles and Nanodevices in Biological Applications*; Springer: Berlin Heidelberg, 2009.

Bhattacharyya, S.; Kudgus, R. A.; Bhattacharya, R.; Mukherjee, P. Inorganic Nanoparticles in Cancer Therapy. *Pharm. Res.* **2012,** *28,* 237–259. doi: 10.1007/s11095-010-0318-0.Inorganic

Chakrabartty, I.; Vijayasekhar, A.; Rangan, L. Therapeutic Potential of Labdane Diterpene Isolated from *Alpinia nigra*: Detailed Hemato- Compatibility and Antimicrobial Studies. *Nat Prod Res.* **2019**. doi: 10.1080/14786419.2019.1610756

Colegate, S. M.; Molyneux, R. J. *Bioactive Natural Products. Detection, Isolation, and Structural Determination*, 2nd ed; CRC Press: Boca Raton, FL, 2008.

Conde, J.; Doria, G.; Baptista, P. Noble Metal Nanoparticles Applications in Cancer. *J. Drug Deliv.* **2012,** *2012,* 1–12. doi: 10.1155/2012/751075

Dahoumane, S. A.; Yéprémian, C.; Djédiat, C.; et al. Improvement of Kinetics, Yield, and Colloidal Stability of Biogenic Gold Nanoparticles Using Living Cells of *Euglena gracilis* Microalga. *J. Nanoparticle Res.* **2016,** *18,* 79–99.

Devatha, C. P.; Thalla, A. K. Green Synthesis of Nanomaterials. In *Synthesis of Inorganic Nanomaterials*; Elsevier Ltd., 2018; pp 169–184.

Dewick, P. M. *Medicinal Natural Products: A Biosynthetic Approach*, 2nd ed.; John Wiley & Sons Ltd: England, UK, 2002.

Doble, M.; Rollins, K.; Kumar, A. *Green Chemistry and Engineering*, 1st ed.; Academic Press, Elsevier: Cambridge, UK, 2007.

Eckelman, M. J.; Zimmerman, J. B.; Anastas, P. T. Toward Green Nano: E-factor Analysis of Several Nanomaterial Syntheses. *J. Ind. Ecol.* 2008, *12,* 316–328. doi: 10.1111/j.1530-9290.2008.00043.x

Elechiguerra, J. L.; Burt, J. L.; Morones, J. R. et al. Interaction of Silver Nanoparticles with HIV-1. *J. Nanobiotechnol.* **2005,** *3,* 1–10. doi: 10.1186/1477-3155-3-6

Gaikwad S, Ingle A, Gade A, et al. Antiviral Activity of Mycosynthesized Silver Nanoparticles against Herpes Simplex Virus and Human Parainfluenza Virus Type 3. *Int. J. Nanomed.* **2013,** *8,* 4303–4314.

Gao, H.; Jiang, X. *Neurotoxicity of Nanomaterials and Nanomedicine*; Academic Press, Elsevier: Cambridge, 2017.

Garcia-Bennett, A.; Nees, M.; Fadeel, B. In Search of the Holy Grail : Folate-targeted Nanoparticles for Cancer Therapy. *Biochem. Pharmacol.* **2011**, *81*, 976–984. doi: 10.1016/j.bcp.2011.01.023

Ghosh, S.; Rangan, L. *Alpinia*: The Gold Mine of Future Therapeutics. *3 Biotech* **2013**, *3*, 173–185. doi: 10.1007/s13205-012-0089-x

Gomes, S. I. L.; Novais, S.; Gravato, C. Effect of Cu-Nanoparticles Versus One Cu-Salt : Analysis of Stress Biomarkers Response in *Enchytraeus albidus* (Oligochaeta). *Nanotoxicology* **2012**, *6*, 134–143. doi: 10.3109/17435390.2011.562327

Gupta, K.; Hazarika, S. N.; Saikia, D. et al. One Step Green Synthesis and Anti-Microbial and Anti-Biofilm Properties of *Psidium guajava* L. Leaf Extract-Mediated Silver Nanoparticles. *Mater. Lett.* **2014**, *125*, 67–70. doi: 10.1016/j.matlet.2014.03.134

Hussain, M.; Iqbal, N.; Muhammad, R.; Sumaira, I. Applications of Plant Flavonoids in the Green Synthesis of Colloidal Silver Nanoparticles and Impacts on Human Health. *Iran J. Sci. Technol. Trans. A: Sci.* **2017**, *43*, 1381–1392. doi: 10.1007/s40995-017-0431-6

Khan, T.; Ullah, N.; Ali, M. et al. Plant-Based Gold Nanoparticles: A Comprehensive Review of the Decade-Long Research on Synthesis, Mechanistic Aspects and Diverse Applications. *Adv. Colloid Interface Sci.* **2019**, *272*, 2–8. doi: 10.1016/j.cis.2019.102017

Kim, J. S.; Kuk, E.; Yu, N. et al. Antimicrobial Effects of Silver Nanoparticles. *Nanomedicine* **2007**, *3*, 95–101. doi: 10.1016/j.nano.2006.12.001

Kim, K-J.; Sung, W. S.; Moon, S.; et al. Antifungal Effect of Silver Nanoparticles on Dermatophytes. *J. Microbiol. Biotechnol.* **2008**, *18*, 1482–1484.

Kim, K-J.; Sung, W. S.; Suh, B. K.; et al. Antifungal Activity and Mode of Action of Silver Nano-Particles on *Candida albicans*. *Biometals* **2009**, *22*, 235–242. doi: 10.1007/s10534-008-9159-2

Klebowski, B.; Depciuch, J.; Parlinska-Wojtan, M.; Baran, J. Applications of Noble Metal-Based Nanoparticles in Medicine. *Int. J. Mol. Sci.* **2018**. doi: 10.3390/ijms19124031

Krishnaraj, C.; Jagan, E. G.; Rajasekar, S.; et al. Synthesis of Silver Nanoparticles Using *Acalypha indica* Leaf Extracts and Its Antibacterial Activity against Water Borne Pathogens. *Colloids Surf. B Biointerf.* **2010**, *76*, 50–56. doi: 10.1016/j.colsurfb.2009.10.008

Kumar, P. P. N. V.; Pammi, S. V. N.; Kollu, P.; et al. Green Synthesis and Characterization of Silver Nanoparticles Using *Boerhaavia diffusa* Plant Extract and Their Anti Bacterial Activity. *Ind. Crop Prod.* **2014**, *52*, 562–566. doi: 10.1016/j.indcrop.2013.10.050

Kumar, V.; Yadav, S. K. Plant-Mediated Synthesis of Silver and Gold Nanoparticles and Their Applications. *J. Chem. Technol. Biotechnol.* **2009**, *176061*, 151–157. doi: 10.1002/jctb.2023

Kunjiappan, S.; Bhattacharjee, C.; Chowdhury, R. In vitro Antioxidant and Hepatoprotective Potential of *Azolla microphylla* Phytochemically Synthesized Gold Nanoparticles on Acetaminophen—Induced Hepatocyte Damage in *Cyprinus carpio* L. *Vitr. Cell Dev. Biol—Anim.* **2015**, *51*, 630–643. doi: 10.1007/s11626-014-9841-3

Kuppusamy, P.; Yusoff, M. M.; Maniam, G. P.; Govindan, N. Biosynthesis of Metallic Nanoparticles Using Plant Derivatives and Their New Avenues in Pharmacological Applications—an Updated Report. *Saudi Pharm. J.* **2016**, *24*, 473–484. doi: 10.1016/j.jsps.2014.11.013

Mashwani, Z.; Khan, M. A.; Khan, T.; Nadhman, A. Applications of Plant Terpenoids in the Synthesis of Colloidal Silver Nanoparticles. *Adv. Colloid Interf. Sci.* **2016**, *234*, 132–141. doi: 10.1016/j.cis.2016.04.008

Mishra, V. K.; Husen, A.; Rahman, Q. I. et al. Plant-Based Fabrication of Silver Nanoparticles and Their Application. In *Nanomaterials and Plant Potential*, 2019; pp 135–175.

Misra, S. K.; Dybowska, A.; Berhanu, D. et al. The Complexity of Nanoparticle Dissolution and Its Importance in Nanotoxicological Studies. *Sci. Total Environ.* **2012,** *438,* 225–232. doi: 10.1016/j.scitotenv.2012.08.066

Mubarakali, D.; Thajuddin, N.; Jeganathan, K.; Gunasekaran, M. Plant Extract Mediated Synthesis of Silver and Gold Nanoparticles and Its Antibacterial Activity against Clinically Isolated Pathogens. *Colloids Surf. B Biointerf.* **2011,** *85,* 360–365. doi: 10.1016/j.colsurfb.2011.03.009

Navarro, E.; Baun, A.; Behra, R.; et al. Environmental Behavior and Ecotoxicity of Engineered Nanoparticles to Algae, Plants and Fungi. *Ecotoxicology* 2008, *17,* 372–386. doi: 10.1007/s10646-008-0214-0

Pallas, G.; Peijnenburg, W. J. G. M.; Heijungs, R.; Vijver, M. G. Green and Clean : Reviewing the Justification of Claims for Nanomaterials from a Sustainability Point of View. *Sustainibility* **2018,** *10,* 689. doi: 10.3390/su10030689

Panigrahi, S.; Kundu, S.; Ghosh, S. K.; et al. General Method of Synthesis for Metal Nanoparticles. *J. Nanoparticle Res.* **2004,** *6,* 411–414.

Prabhu, S.; Poulose, E. K. Silver Nanoparticles: Mechanism of Antimicrobial Action, Synthesis, Medical Applications, and Toxicity Effects. *Int. Nano Lett.* **2012,** *2,* 32. doi: 10.1186/2228-5326-2-32

Rajasekhar, C.; Kanchi, S. Green Nanomaterials for Clean Environment. In *Handbook of Economics*; Martínez, L., Kharissova, O., Kharisov, B., Eds.; Springer: Cham, 2018; pp 63–79.

Rajendran, S.; Prabha, S. S.; Rathish, R. J. et al. Antibacterial Activity of Platinum Nanoparticles. In *Nanotoxicity: Prevention and Antibacterial Applications of Nanomaterials,* 2020; pp 275–281.

Royer, M.; Prado, M.; Garcia-Perez, M. E. et al. Study of Nutraceutical, Nutricosmetics and Cosmeceutical Potentials of Polyphenolic Bark Extracts from Canadian Forest Species. *Pharma Nutr* **2013,** *1,* 158–167.

Sadeghi, B.; Gholamhoseinpoor, F. A Study on the Stability and Green Synthesis of Silver Nanoparticles Using *Ziziphora tenuior* (Zt) Extract at Room Temperature. *Spectrochim. Acta Part A: Mol. Biomol. Spectrosc.* **2015,** *134,* 310–315. doi: 10.1016/j.saa.2014.06.046

Salame, P. H.; Pawde, V. B.; Bhanvase, B. A. Characterization Tools and Techniques for Nanomaterials. In *Nanomaterials for Green Energy,* 2018; pp 83–111.

Sanzari, I.; Leone, A.; Ambrosone, A. Nanotechnology in Plant Science : To Make a Long Story Short. *Front. Bioeng. Biotechnol.* **2019,** *7,* 1–12. doi: 10.3389/fbioe.2019.00120

Saravanan, M.; Dhivakar, S.; Jayanthi, S. S. An Eco Friendly and Solvent Free Method for the Synthesis of Zinc Oxide Nano Particles Using Glycerol as Organic Dispersant. *Mater. Lett.* **2012,** *67,* 128–130. doi: 10.1016/j.matlet.2011.09.009

Schwirn, K.; Tietjen, L.; Beer, I. Why Are Nanomaterials Different and How Can They Be Appropriately Regulated Under REACH ? **2014,** 1–9.

Shan, G.; Andrews, M. P.; Gonzalez, T.; Djeghelian, H. A Simple, Low-Temperature Route to Synthesize Aqueous-Dispersible Yb 3 + – Er 3 + Co-Doped Hexagonal LaF 3 Nano-Crystals for Up-conversion Fluorescence. *Mater. Lett.* **2008,** *62,* 4187–4190. doi: 10.1016/j.matlet.2008.06.024

Shanker, U.; Jassal, V.; Rani, M. et al. Towards Green Synthesis of Nanoparticles : From Bio-Assisted Sources to Benign Solvents . A Review. *Int. J. Environ. Anal. Chem.* **2016,** *96,* 801–835. doi: 10.1080/03067319.2016.1209663

Sharma, V. K.; Siskova, K. M.; Zboril, R.; Gardea-torresdey, J. L. Organic-Coated Silver Nanoparticles in Biological and Environmental Conditions: Fate, Stability and Toxicity. *Adv. Colloid Interf. Sci.* **2014,** *204,* 15–34. doi: 10.1016/j.cis.2013.12.002

Shen, W.; Zhang, X.; Huang, Q. et al. Preparation of Solid Silver Nanoparticles for Inkjet Printed Flexible Electronics with High Conductivity. *Nanoscale* **2013**. doi: 10.1039/c3nr05479a

Singh, A.; Jahan, I.; Sharma, M. et al. Structural Characterization, *in Silico* Studies and in Vitro Antibacterial Evaluation of a Furanoflavonoid from Karanj. *Planta Medica Lett.* **2016**, *3*, e91–e95. doi: 10.1055/s-0042-105159

Singh, J.; Dutta, T.; Kim, K. H. et al. 'Green' Synthesis of Metals and Their Oxide Nanoparticles : Applications for Environmental Remediation. *J. Nanobiotechnol.* **2018**, *16*, 1–24. doi: 10.1186/s12951-018-0408-4

Sun, H.; Zhang, X.; Niu, Q. et al. Enhanced Accumulation of Arsenate in Carp in the Presence of Titanium Dioxide Nanoparticles. *Water, Air Soil Pollut.* **2007**, *178*, 245–254. doi: 10.1007/s11270-006-9194-y

Tahir, K.; Nazir, S.; Ahmad, A. et al. Facile and Green Synthesis of Phytochemicals Capped Platinum Nanoparticles and in Vitro Their Superior Antibacterial Activity. *J. Photochem. Photobiol. B: Biol.* **2017**, *166*, 246–251. doi: 10.1016/j.jphotobiol.2016.12.016

Tang, D.; Sauceda, J. C.; Lin, Z. et al. Magnetic Nanogold Microspheres-Based Lateral-Flow Immune Dipstick for Rapid Detection of Aflatoxin B2 in Food. *Biosens. Bioelectron.* **2009**, *25*, 514–518.

Tejamaya, M.; Romer, I.; Merrifield, R. C.; Lead, J. R. Stability of Citrate, PVP, and PEG Coated Silver Nanoparticles in Ecotoxicology Media. *Environ. Sci. Technol.* **2012**, *46*, 7011–7017.

Usman, M. S.; Zowalaty, M. E.; El Shameli, K. et al. Synthesis, Characterization, and Antimicrobial Properties of Copper Nanoparticles. *Int. J. Nanomed.* **2013**, *3*, 4467–4479.

Varma, R. S. Greener Approach to Nanomaterials and Their Sustainable. *Curr. Opin. Chem. Eng.* **2012**, *1*, 123–128. doi: 10.1016/j.coche.2011.12.002

Wadhwani, S. A.; Shedbalkar, U. U.; Singh, R.; Chopade, B. A. Biogenic Selenium Nanoparticles: Current Status and Future Prospects. *Appl. Microbiol. Biotechnol.* **2016**, *100*, 2555–2566. doi: 10.1007/s00253-016-7300-7

Yadi, M.; Mostafavi, E.; Saleh, B. et al. Current Developments in Green Synthesis of Metallic Nanoparticles Using Plant Extracts: A Review. *Artif. Cells, Nanomed., Biotechnol.* **2018**. doi: 10.1080/21691401.2018.1492931

Zaka, M.; Abbasi, B. H.; Rahman, L. et al. Synthesis and Characterisation of Metal Nanoparticles and Their Effects on Seed Germination and Seedling Growth in Commercially Important *Eruca sativa. IET Nanobiotechnol.* **2016**, *10*, 1–7. doi: 10.1049/iet-nbt.2015.0039

Zhang, J.; Si, G.; Zou, J. et al. Antimicrobial Effects of Silver Nanoparticles Synthesized by *Fatsia japonica* Leaf Extracts for Preservation of *Citrus* Fruits. *J. Food Sci.* **2017**, *00*, 1–6. doi: 10.1111/1750-3841.13811

INDEX

A

Agents in imaging, 114
  coating types, 117–118
  interaction and clearance routes, 119–120
  nanoparticle selection, 115–116
  size and synthetic strategies, 116–117
  targeting strategies, 118–119
Agricultural applications, 332
Agroecosystem, 181–184
Algae, 221–222
Aluminum nanoparticles (AL_2O_3NPS), 179
Antibacterial coatings, 143
Anti-inflammatory in NPs
  alimentary tract distribution of, 91
Antimicrobial properties, 330–332
Antitumor and anticancer agents, 249–250
Applications of nanoscience in
  biotechnology, 248
  agriculture, 255–256
  antimicrobial activity, 252–253
  antitumor and anticancer agents, 249–250
  cosmetics, 253–254
  in diagnostics, 251
  in drug-delivery systems, 250–251
  food industry, 254
  food packaging and labeling, 255
  food process, 254–255
  in nano fabrics, 253
  in nanobioremediation, 257–258
  nanobiosensors, 251–252
  nanofertilizers, 256
  in nanomedicine, 249
  nanopesticides, 256–257
Artificial intelligence (AI), 25–26
Atomic force microscope (AFM), 4

B

Bacterial species, 222
Barcode amplification, 149–150
Biocompatible and biodegradable
  nanocarrier, 44

Biocompatible magnetic micro and
  nanodevices
  cancer immunotherapy, 16–17
  in cancer therapy, 14–16
  cell imaging, 18
  gas nanomedicine, 19–20
  gene editing, 17–18
  Glancing Angle Deposition (GLAD)
    method, 13
  iron platinum (FePt) alloy, 13
  localized treatment, 14
  magnetic nanoparticles (MNPs), 13
  in vivo imaging, 19
Biomolecular templates, 224–225
Blood–brain barrier (BBB), 27
Bottom-up method, 244
Bowel imaging, 91–92

C

Cancer immunotherapy, 16–17
Carbon nanotubes (CNTs), 54–55
Cell imaging, 18
Challenges in nanomedicine
  blood–brain barrier (BBB), 27
  clinical translation of, 29–30
  crossing biological barriers, 27
  nanoformulations designed, 29
  regulatory and ethical issues, 30
  specifying target, 28
Chemotherapeutics
  types and roles in, 284
    dendrimers, 286–287
    gold nanoparticles, 288–289
    liposomes, 285–286
    micelles, 287–288
    nanostructured lipid carriers (NLC),
      291–292
    silver NPs (AgNPs), 289–290
    solid lipid nanoparticles/carrier (SLN/
      SLC), 290–291
Computed tomography scanning, 121
  fluorescence imaging, 124–125
  nanoparticles for CT, 122–123

342   Index

positron emission tomography (PET),
123–124
Computer axial tomography and
nanoparticles
CT/CAT photo-slides, 77
Contrast design for imaging, 86
Core–shell NPs, 82–83
Crossing biological barriers, 27

D

Damascus, 3
Drug delivery
nanocarriers for drug delivery
biocompatible and biodegradable
nanocarrier, 44
nanocapsules, 45–46
nanomicelles, 46
nanospheres, 44
polymersomes, 47
nanomedicine in clinics, 56
US FDA approved nanomedicines, 57–62
nanomedicine, potential advantages of,
42–43
nanosystems for medical use, 41
novel nanotechnology-based drug-
delivery systems
carbon nanotubes (CNTs), 54–55
dendrimers, 52–53
exosomes, 51
graphene, 55
injectable nanoparticle generator
(iNG), 51–52
liposomes, 53–54
nanobots, 48–49
nanobubbles, 51
nanoclew or nanococoon, 50
nanoclusters, 50
nanoghost, 49
nano-terminators, 52
niosomes, 54

F

Functionalization of nanoparticles (NPS)
controlled release, 275–276
improved bioavailability, 274–275
surface conjugation, 274
targeting, 273–274
tunability, 276

G

Gas nanomedicine, 19–20
Gene editing, 17–18
Glancing Angle Deposition (GLAD)
method, 13
Gold nanoparticle
MRI/CT modality, 79–81
and ultrasound, 78–79
and X-ray imaging equations II, 78
Graphene, 6, 55

I

Injectable nanoparticle generator (iNG),
51–52
Iron platinum (FePt) alloy, 13

L

Lipid–polymer hybrid nanoparticles (LPH),
7

M

Macromolecules, 7
Magnetic nanoparticles (MNPs), 13
Magnetic resonance imaging (MRI), 120
contrast agents for, 120–121
Malignancies, plant product-based
nanomedicine
chemotherapeutics, types and roles in, 284
dendrimers, 286–287
gold nanoparticles, 288–289
liposomes, 285–286
micelles, 287–288
nanostructured lipid carriers (NLC),
291–292
silver NPs (AgNPs), 289–290
solid lipid nanoparticles/carrier (SLN/
SLC), 290–291
functionalization of nanoparticles (NPS)
controlled release, 275–276
improved bioavailability, 274–275
surface conjugation, 274
targeting, 273–274
tunability, 276
nanoformulation-based phytochemicals
in, 278, 284
nanomedicine, 272
nanomedicine, 270–272

Index

343

nanotechnology and phytochemicals, 277–278

phytochemicals, sources, and their applications in, 279–280

potential advantages and applications oral drug delivery, 276–277

Metal salt concentration, 227–228

N

Nanobiotechnology in plants
agroecosystem, use of, 181–184
application of
aluminum nanoparticles (AL$_2$O$_3$NPS), 179
factories and potential use of, 185–189
quantification of agrochemical, 176–177
silicon nanoparticles (SINPS), 178 179
silver nanomaterials, 178
titanium nanoparticles (TIO$_2$NPS), 180
zinc nanoparticles (ZNONPS), 179–180
bioactive molecules, 175
green-NP assembler, 180, 184, 190
metal nanoparticle modification, mechanisms of, 187
refined silver nanoparticles, 190

Nanofertilizers and nanopesticides
cultivation, restrictions in, 205–206
and nutrient use efficiency, 206–207
pesticide in agriculture, 202
nanofertilizer technology, 203–204
nanopesticides, 204–205

Nanomedicine
biocompatible magnetic micro and nanodevices
cancer immunotherapy, 16–17
in cancer therapy, 14–16
cell imaging, 18
gas nanomedicine, 19–20
gene editing, 17–18
Glancing Angle Deposition (GLAD) method, 13
iron platinum (FePt) alloy, 13
localized treatment, 14
magnetic nanoparticles (MNPs), 13
in vivo imaging, 19
challenges
blood–brain barrier (BBB), 27
clinical translation of, 29–30

crossing biological barriers, 27
nanoformulations designed, 29
regulatory and ethical issues, 30
specifying target, 28

nanomaterials
development of, 8, 9–10
graphene, 6
lipid–polymer hybrid nanoparticles (LPH), 7
liposomes, 6–7
macromolecules, 7
metal nanostructures, 5–6
nanobiotechnology, 7–8
organic and inorganic nanocarriers, 6
polyfunctionalized carbon materials, 5
polymeric micelles, 7
quantum dots (QD), 6
silica nanoparticles, 6
solid lipid nanoparticles (SLNs), 7
superparamagnetic iron oxide nanoparticles (SPIONS), 6
synthesis, approaches of, 4–5
three-dimensional nanostructures, 9

nanotechnology applications
atomic force microscope (AFM), 4
Damascus, 3
scanning probe microscopes (SPM), 4
scanning tunneling microscope (STM), 4
U.S. National Nanotechnology Initiative (NNI), 2
use in 1974, 3

nanotechnology in obesity control, 20
artificial intelligence (AI), 25–26
bioelectronic medicines, 24–25
combination therapy, 25
inflammatory diseases, 22–23
microbial-nanohybrids, 21–22
nanoinformatics, 24
in neurodegenerative disease, 23–24
regenerative medicine, 21

promises and prospects, 10
nanobots in, 12–13
targeted drug delivery, 11

Nanoparticle synthesis, greener methods
advantages of, 228
applications, 229
bottom-up approach, 216–217
factors affecting

metal salt concentration, 227–228
pH reaction medium, 226–227
reaction time, 226
temperature, 225–226
volume/concentration, 227
green synthesis, 220
actinomycetes, 221
algae, 221–222
bacterial species, 222
biomolecular templates, 224–225
fungi, 222–223
plants, 223–224
process of nanoparticle synthesis, 225
yeasts, 224
limitations, 228–229
methods of
biological methods, 219–220
chemical methods, 219
physical methods, 218–219
nanoparticles (NPs), 215–216
top-down approach, 216
Nanoparticles in medical imaging, 110
agents in imaging, 114
coating types, 117–118
interaction and clearance routes, 119–120
nanoparticle selection, 115–116
size and synthetic strategies, 116–117
targeting strategies, 118–119
computed tomography scanning, 121
fluorescence imaging, 124–125
nanoparticles for CT, 122–123
positron emission tomography (PET), 123–124
magnetic resonance imaging (MRI), 120
contrast agents for, 120–121
nonradiolabeled in biological imaging, 111–112
radiotracer techniques, 113–114
Nanoparticles (NPs), 74
anti-inflammatory
alimentary tract distribution of, 91
and bowel imaging, 91–92
clearance routes, physiology of, 93
coatings, types, 88
computer axial tomography and nanoparticles
CT/CAT photo-slides, 77
contrast design for imaging, 86

contrasts X-rays, 77–78
core–shell NPs, 82–83
FE oxide nanoproteins, 84–85
gold nanoparticle
MRI/CT modality, 79–81
and ultrasound, 78–79
and X-ray imaging equations II, 78
limitations in, 93
macromolecular dendrimers, 83
magnetic nanoparticle, 78
micelles, 82
in MRI operations, 83–84
nanobiopsy, 85
nanogels, 74–75
nanopyramids, 75
nanoshells for, 85
nanowires and its use in diagnostics
single-walled CNTs (SWCNTs), 76
SiNW bioinformatic sensors, 75
STAGE I manifestation, 77
superparamagnetic iron oxide (SPIO) NP (SPION), 76
and nuclear imaging, 89–90
photoacoustic imaging, 79
potency of, 87–88
quantum dots, 86–87
synthetic and targeting strategies, 88–89
theragnostic dendrimers, 75
toxicity, effects, 93–94
Nanoscience in biotechnology
applications of, 248
agriculture, 255–256
antimicrobial activity, 252–253
antitumor and anticancer agents, 249–250
cosmetics, 253–254
in diagnostics, 251
in drug-delivery systems, 250–251
food industry, 254
food packaging and labeling, 255
food process, 254–255
in nano fabrics, 253
in nanobioremediation, 257–258
nanobiosensors, 251–252
nanofertilizers, 256
in nanomedicine, 249
nanopesticides, 256–257
nanoparticles (NPs)
biological synthesis, 246–247
bottom-up method, 244

Index

chemical synthesis, 245–246
extracellular method, 247
intracellular method, 247
one-dimensional nanoparticles, 243
physical synthesis, 246
three-dimensional nanoparticles, 244
top-down method, 244
two-dimensional nanoparticles, 244
Nanostructured lipid carriers (NLC), 291–292
Nanotechnology in healthcare management, 141
nanomedical technical tools/devices
dentifrobots, 158
nanocomputer, 159
nanorobots, 158
nanotweezers, 159
respirocyte, 158
role
antibacterial coatings, 143
barcode amplification, 149–150
imaging technology, 146–148
implantable nanosensors, 148–149
nanobioassay, 144–145
nanochips, 145
nanofilters, 142–143
nanomask, 142
nano/microfluidic technologies, 146
nanosensors, 143–144
pregnancy test, 150
toxicity and safety, 159
treatment, role in, 151
cancers, 152–153
diabetes, 153–154
neurodegenerative disorders, 156–158
respiratory system, 154–156

O

Oral drug delivery, 276–277

P

Plant-based nanoparticles
applications
agricultural applications, 332
antimicrobial properties, 330–332
pharmacological applications, 332–333

industrial and commercial, 333
energy applications, 334–336
Positron emission tomography (PET), 123–124

Q

Quantification of agrochemical, 176–177
Quantum dots (QD), 6

S

Scanning probe microscopes (SPM), 4
Scanning tunneling microscope (STM), 4
Silica nanoparticles, 6
Silver NPs (AgNPs), 289–290
Single-walled CNTs (SWCNTs), 76
Solid lipid nanoparticles (SLNs), 7
Superparamagnetic iron oxide nanoparticles (SPIONS), 6
Superparamagnetic iron oxide (SPIO) NP (SPION), 76

T

Three-dimensional nanostructures, 9
Toxicity of nanoparticles in plants
phytotoxicity, mechanism of, 305–306
translocation and uptake routes
components impacting root take-up, 311–312
coordinated root take-up and, 312–313
foliar take-up and translocation of, 307–309
phyllo-sphere factors, 310
plant leaves as captors, 310–311

U

US FDA approved nanomedicines, 57–62
U.S. National Nanotechnology Initiative (NNI), 2

Z

Zinc nanoparticles (ZNONPS), 179–180